The Art of Cyber Defense

The Art of Cyber Defense: From Risk Assessment to Threat Intelligence offers a comprehensive exploration of cybersecurity principles, strategies, and technologies essential for safeguarding digital assets and mitigating evolving cyber threats. This book provides invaluable insights into the intricacies of cyber defense, guiding readers through a journey from understanding risk assessment methodologies to leveraging threat intelligence for proactive defense measures.

Delving into the nuances of modern cyber threats, this book equips readers with the knowledge and tools necessary to navigate the complex landscape of cybersecurity. Through a multidisciplinary approach, it addresses the pressing challenges organizations face in securing their digital infrastructure and sensitive data from cyber-attacks.

This book offers comprehensive coverage of the most essential topics, including:

- Advanced malware detection and prevention strategies leveraging artificial intelligence (AI)
- Hybrid deep learning techniques for malware classification
- Machine learning solutions and research perspectives on Internet of Services (IoT) security
- Comprehensive analysis of blockchain techniques for enhancing IoT security and privacy
- Practical approaches to integrating security analysis modules for proactive threat intelligence

This book is an essential reference for students, researchers, cybersecurity professionals, and anyone interested in understanding and addressing contemporary cyber defense and risk assessment challenges. It provides a valuable resource for enhancing cybersecurity awareness, knowledge, and practical skills.

The Art of Cyber Defense

From Risk Assessment to Threat Intelligence

Edited By

Youssef Baddi, Mohammed Amin Almaiah, Omar Almomani, and Yassine Maleh

CRC Press
Taylor & Francis Group
Boca Raton London New York

CRC Press is an imprint of the
Taylor & Francis Group, an **informa** business

Cover illustration: Shutterstock – image number 1188038749

First edition published 2025
by CRC Press
2385 Executive Center Drive, Suite 320, Boca Raton, FL 33431

and by CRC Press
4 Park Square, Milton Park, Abingdon, Oxon, OX14 4RN

CRC Press is an imprint of Taylor & Francis Group, LLC

ISBN: 9781032714783 (hbk)
ISBN: 9781032714790 (pbk)
ISBN: 9781032714806 (ebk)

DOI: 10.1201/9781032714806

Typeset in Sabon
by codeMantra

Contents

Preface

This book on cybersecurity risk assessment and threat intelligence will address significant issues in the context of the rapidly advancing cybersecurity landscape. For its value to be practically and scientifically realized, this book's goal is to provide chapters on recent advances and ideas on the progress of research and the practicality of risk analysis, threat assessments, and management of cybersecurity risks to address the main issues and challenges that face people, organizations, and countries, which are now called "Cyberwar".

The Art of Cyber Defense: From Risk Assessment to Threat Intelligence embodies a collection of cutting-edge insights and practical strategies curated by cybersecurity experts and practitioners. Designed to serve as a comprehensive reference for a diverse audience, including students, researchers, engineers, and cybersecurity professionals, this book delves into the intricate intersections of technology and security, offering invaluable perspectives on mitigating cyber threats and enhancing digital resilience.

This book consists of 16 chapters, each exploring key topics for understanding and addressing modern cybersecurity challenges. From advanced malware detection techniques to the intricacies of IoT security, every chapter offers a unique perspective on the evolving landscape of cyber defense.

Moreover, the books proposed in this context will not delve into the various applications of cybersecurity risk analysis across different domains. Through this book, we aim to address this gap by compiling chapters authored by professionals and experts. We aim to offer readers a comprehensive theoretical and practical foundation, enabling them to understand this subject matter deeply.

By laying out a blueprint for big data infrastructure and emphasizing the paramount importance of security and privacy in the digital age, this book seeks to empower readers with the knowledge and tools necessary to navigate the complexities of cyber defense effectively. Each chapter is vital in building a resilient cyber future, offering actionable insights and practical guidance for fortifying digital infrastructure against emerging cyber threats.

Through its contributors' collective expertise and dedication, *The Art of Cyber Defense: From Risk Assessment to Threat Intelligence* aspires to contribute significantly to the ongoing cybersecurity discourse, serving as a beacon of knowledge and insight in the ever-evolving cyber defense landscape.

We want to take this opportunity to thank the contributors of this volume and the editorial board for their tremendous efforts in reviewing and providing interesting feedback. The editors would like to thank Gabriella Williams and Gupta Radhika from Taylor and Francis for their editorial assistance and support in producing this important scientific work. Without this collective effort, completing this book would not have been possible.

Prof. Youssef Baddi
El Jadida, Morocco

Prof. Mohammed Amin Almaiah
Aqaba, Jordan

Prof. Omar Almomani
Amman, Jordan

Prof. Yassine Maleh
Khouribga, Morocco

Introduction

In the contemporary age of digitalization, characterized by ubiquitous technological integration across various domains, the imperative for resilient cybersecurity protocols emerges as an unequivocal necessity. As industries embrace the paradigm of digital transformation and interconnected operational frameworks, they concurrently encounter a multifaceted spectrum of cyber threats. These threats, from data breaches to operational disruptions, pose formidable challenges to organizational integrity, data security, and stakeholder trust. Consequently, in this era marked by escalating cyber vulnerabilities, acquiring proficiency in cyber defense assumes paramount significance. Mastery of cybersecurity principles and practices becomes instrumental in fortifying critical assets, preempting potential breaches, and cultivating organizational resilience amid the dynamic landscape of evolving threats.

The Art of Cyber Defense: From Risk Assessment to Threat Intelligence embarks on a comprehensive journey through the intricacies of cybersecurity, offering insights, strategies, and practical approaches to fortify defenses and mitigate risks. With a multidisciplinary approach encompassing Artificial Intelligence (AI), machine learning, cryptography, and blockchain, this book delves deep into the foundational pillars of cyber defense while charting a course toward proactive threat intelligence.

The book is structured into three cohesive parts, each addressing a distinct facet of cyber defense and threat intelligence within critical industries.

SECTION I: FOUNDATIONS OF CYBER DEFENSE AND RISK ASSESSMENT

Section I of this book lays the groundwork by delving into the essential foundations of cyber defense, equipping readers with the knowledge and tools necessary to navigate the complex challenges malicious actors pose. From leveraging AI-powered strategies for advanced malware detection to exploring innovative techniques for malware classification, this section offers a comprehensive overview of the cutting-edge approaches shaping the future of cyber defense.

Chapter 1: AI-Powered Strategies for Advanced Malware Detection and Prevention elucidates the pivotal role of AI in fortifying information systems against malware attacks. This chapter showcases how AI-driven approaches can enhance traditional security measures and enable real-time threat detection by harnessing the capabilities of machine learning algorithms, supervised and unsupervised.

Chapter 2: Advancing Malware Classification with Hybrid Deep Learning presents a groundbreaking study on integrating DenseNet and LSTM for malware classification.

This hybrid deep learning model demonstrates exceptional accuracy and precision through meticulous training on diverse malware datasets, offering a potent tool for combating evolving threats.

Chapter 3: A Comprehensive Overview of AI-Driven Behavioral Analysis for Security in Internet of Things provides an in-depth examination of AI techniques for behavioral analysis, encompassing machine learning algorithms, deep learning models, and federated learning frameworks.

Chapter 4: A Deep Dive into IoT Security: Machine Learning Solutions and Research Perspectives explores the expanding field of IoT security, highlighting the vulnerabilities present in interconnected systems. With a focus on machine learning solutions, this chapter explores proactive defense mechanisms to strengthen IoT security against cyber threats.

Chapter 5: Exploring Blockchain Techniques for Enhancing IoT Security and Privacy comprehensively analyses blockchain's potential in bolstering IoT security. This chapter paves the way for innovative approaches to safeguarding IoT ecosystems by examining security risks, privacy concerns, and blockchain-based remedies.

SECTION II: ANALYZING AND RESPONDING TO EMERGING THREATS

This section explores the dynamic field of cybersecurity, providing insights into the latest strategies and technologies for combating evolving cyber threats. From proactive threat intelligence to comprehensive security audits, these chapters provide invaluable guidance for organizations seeking to bolster their defenses and protect against emerging cyber threats.

Chapter 6: Integrating Security Analysis Module for Proactive Threat Intelligence proposes an integrated safety analysis module to enhance organizational cybersecurity posture. By amalgamating cutting-edge technologies such as Docker, MISP, Wazuh, and OpenSearch, this chapter lays the groundwork for preemptive threat management and incident response.

Chapter 7: Security Study of Web Applications through a White Box Audit Approach: A Case Study and Chapter 8: Case Study Method: A Step-by-Step Black Box Audit for Security Study of Web Applications delve into the intricacies of web application security auditing. These chapters provide comprehensive insights into enhancing web application security by leveraging white-box and black-box audit approaches.

Chapter 9: Security in Cloud-Based IoT: A Survey comprehensively overviews cloud-based IoT security issues and challenges. This chapter equips readers with a deeper understanding of securing cloud-based IoT deployments by identifying key security issues and proposing preferred solutions.

Chapter 10: Exploring IoT Penetration Testing: From Fundamentals to Practical Setup provides a detailed exploration of IoT penetration testing fundamentals. By delving into firmware security, web application security, and hardware security, this chapter offers practical guidance on setting up IoT testing laboratories.

Chapter 11: A Fuzzy Logic-Based Trust System for Detecting Selfish Nodes and Encouraging Cooperation in Optimized Link State Routing Protocol. This chapter presents a trust system based on fuzzy logic to identify selfish nodes and foster collaboration within the Optimized Link State Routing protocol (OLSR). Evaluating global trust assists in selecting nodes to act as Multipoint Relays (MPRs). It is determined by a fuzzy multicriteria function incorporating direct, indirect, and past global trust, performance across throughput, end-to-end delay, overhead, and network lifetime. Furthermore, it demonstrates swift adaptability to an increasing number of selfish nodes and changes in network mobility.

Chapter 12: Collaborative Cloud-SDN Architecture for IoT Privacy-Preserving Based on Federated Learning introduces a novel architecture for enhancing IoT security through federated learning. This chapter establishes a decentralized, autonomous system capable of detecting and characterizing attacks within a collaborative framework by leveraging cloud computing and SDN.

Chapter 13: An Adaptive Cybersecurity Strategy Based on Game Theory to Manage Emerging Threats in the SDN Infrastructure presents an innovative approach to cybersecurity by integrating game-theoretic principles into an adaptive architecture. By leveraging the Software-Defined Network (SDN) concept, it introduces an innovative framework that proactively models threat scenarios, enabling networks to adapt their defense strategies dynamically.

SECTION III: HUMAN-CENTRIC RISK MITIGATION APPROACHES

This section underscores the critical role of human factors in cybersecurity and offers strategies for building a resilient defense.

Chapter 14: A Human-Centric Approach to Cyber Risk Mitigation emphasizes the importance of human awareness and training in cybersecurity defense. This chapter aims to foster a culture of cyber-safe practices within organizations by providing best practices for different categories of individuals.

Chapter 15: Human Factors in Cyber Defense explores the intricate relationship between human behavior and cybersecurity. By focusing on user actions, perception, and psychology, this chapter offers insights into designing user-centered security systems and mitigating insider threats.

Chapter 16: Security Operation Center: Towards a Maturity Model presents a practical maturity model for Security Operation Centers (SOCs). By offering a systematic framework for near-real-time threat detection and response, this chapter aims to maximize the effectiveness of SOC investments in safeguarding against cyber threats.

The Art of Cyber Defense: From Risk Assessment to Threat Intelligence is more than just a theoretical exploration; it is a practical guide for cybersecurity practitioners, researchers, and professionals seeking to navigate the complexities of cyber defense in today's digital landscape. This book aims to empower readers to build resilient defenses and adapt to the evolving threat landscape by integrating cutting-edge technologies, proactive strategies, and human-centric approaches.

Prof. Yassine Maleh
Khouribga, Morocco

Prof. Youssef Baddi
El Jadida, Morocco

About the Editor

Prof. Youssef Baddi is a full-time Associate Professor at Chouaib Doukkali University UCD EL Jadida, Morocco. Ph.D. thesis degree in Computer Science at the ENSIAS School, University Mohammed V Souissi of Rabat, Morocco, since 2016. He also holds a Research Master's degree in networking obtained in 2010 from the High National School for Computer Science and Systems Analysis—ENSIAS-Morocco-Rabat. He is a member of Information and Communication Sciences and Technologies Laboratory (STIC), since 2017. He is a guest member of Information Security Research Team (ISeRT) and Innovation on Digital and Enterprise Architectures Team, ENSIAS, Rabat, Morocco. Dr Baddi was awarded as the best PhD student in the University Mohammed V Souissi of Rabat in 2013. Dr Baddi has made contributions in the fields of group communications and protocols, information security and privacy, software-defined network, the Internet of Things, mobile and wireless networks security, and Mobile IPv6. His research interests include information security and privacy, Internet of Things, networks security, software-defined network, software-defined security, IPv6, and mobile IP. He has served and continues to serve on executive and technical program committees and as a reviewer of numerous international conferences and journals such as Elsevier *Pervasive and Mobile Computing* (*PMC*), *International Journal of Electronics and Communications* (*AEÜ*), and *Journal of King Saud University – Computer and Information Sciences*. He was the General Chair of the IWENC 2019 Workshop and the Secretary member of the ICACIN 2020 Conference.

Prof. Mohammed Amin Almaiah obtained his MSc in Computer Information System from Middle East University (MEU), Jordan, in 2011, and his PhD in Computer Science from University Malaysia Terengganu, Malaysia, in 2017. He is now working as an Associate Professor in the Department of Computer Networks and Communications at King Faisal Saudi Arabia. He has published over 100 research papers in highly reputed journals such as *Engineering and Science Technology*, *International Journal of Education and Information Technologies*, and *Journal of Educational Computing Research*. Most of his publications were indexed under the ISI Web of Science and Scopus. His current research interests include mobile learning, software quality, network security, and technology acceptance. He is a certified recognized reviewer in IEEE, Elsevier, and Springer.

Prof. Omar Almomani received his Bachelor's and Master's degrees in Telecommunication Technology from the Institute of Information Technology at the University of Sindh in 2002 and 2003, respectively. In 2010, he received his PhD in Computer Networking from the University Utara Malaysia (UUM). Currently, he is a Professor at the Information Technology Faculty of the World Islamic Sciences and Education University. His research interests include computer networks, the Internet of Things, and network security.

Prof. Yassine Maleh is Associate Professor of Cybersecurity and IT Governance at Sultan Moulay Slimane University, Morocco, since 2019. He is the founding chair of IEEE Consultant Network Morocco and the founding president of the African Research Center of Information Technology & Cybersecurity. He is a former CISO at the National Port Agency between 2012 and 2019. He is a senior member of IEEE and a member of the International Association of Engineers (IAENG) and the Machine Intelligence Research Labs. Dr Maleh has made contributions in the fields of information security and privacy, Internet of Things security, and wireless and constrained networks security. His research interests include information security and privacy, Internet of Things, networks security, information system, and IT governance. He has published over 200 papers (book chapters, international journals, and conferences/workshops), 40 edited books, and 5 authored books. He is the editor-in-chief of *International Journal of Information Security and Privacy* (IJISP, IF: 0.8) and *International Journal of Smart Security Technologies* (IJSST). He serves as an associate editor for IEEE Access since 2019 (Impact Factor 4.098), *International Journal of Digital Crime and Forensics* (IJDCF), and *International Journal of Information Security and Privacy* (IJISP). He is a series editor of *Advances in Cybersecurity Management*, by CRC Taylor & Francis. He was also a guest editor for many special issues with prestigious journals (*IEEE Transactions on Industrial Informatics*, *EEE Engineering Management Review*, *Sensors*, *Big Data Journal*). He has served and continues to serve on executive and technical program committees and as a reviewer of numerous international conferences and journals such as *Ad Hoc Networks*, *IEEE Network Magazine*, *IEEE Sensor Journal*, *ICT Express*, and *Cluster Computing*. He was the general chair and publication chair of many international conferences (BCCA 2019, MLBDACP 19, ICI2C'21, ICACNGC 2022, CCSET'22, IEEE ISC2 2022, ISGTA'24, etc.). He received Publons Top 1% Reviewer Award for the years 2018 and 2019.

Contributors

Qasem Abu Al-Haija
Department of Cybersecurity, Faculty of Computer & Information Technology
Jordan University of Science and Technology
Irbid, Jordan

Anas Anouar
ENSA Berrechid
Hassan 1st University
Settat, Morocco

Syed Immamul Ansarullah
Department of Computer Applications
Government Degree College Sumbal
Sumbal, India

Youssef Baddi
Department of Computer Sciences
Chouaib Doukkali University
El Jadida, Morocco

Hafssa Benaboud
IPSS, Faculty of Sciences
Mohammed V University in Rabat
Rabat, Morocco

Faycal Bensalah
Department of Computer Sciences
Chouaib Doukkali University
El Jadida, Morocco

Achraf Samir Chamkar
LaSTI Laboratory
Sultan Moulay Slimane University
Beni Mellal, Morocco

Imane Chlioui
ENSA Berrechid
Hassan 1st University
Settat, Morocco

Khalid Chougdali
ENSA Kenitra
Ibn Tofail University
Kenitra, Morocco

Muhammad Firdaus Darmawan
Kulliyyah of Information and Communication Technology (KICT),
International Islamic University Malaysia
Kuala Lumpur, Malaysia

Chaima Dhiba
ENSA Berrechid
Hassan 1st University
Settat, Morocco

Rabii El Hakouni
ENSA Kenitra
Ibn Tofail University
Kenitra, Morocco

Mohamed Salem Eleze
Faculty of Economics and Management of Sfax
University of Sfax
Sfax, Tunisia

Yousra Fadili
Department of Computer Sciences
Chouaib Doukkali University
El Jadida, Morocco

Nur Adila Ahmad Faizul
Electrical and Computer Engineering (ECE) Department
International Islamic University Malaysia
Kuala Lumpur, Malaysia

Noreddine Gherabi
LaSTI Laboratory
Sultan Moulay Slimane University
Beni Mellal, Morocco

Anas Harchi
Faculty of Science Ben M'sik
Hassan II University
Casablanca, Morocco

Abdelhalim Hnini
ENSA Berrechid
Hassan 1st University
Settat, Morocco

Shahmie Abd Jalil
Kulliyyah of Information and Communication Technology (KICT)
International Islamic University Malaysia
Kuala Lumpur, Malaysia

Zakariaa Jamal
Department of Computer Sciences
Chouaib Doukkali University
El Jadida, Morocco

Muhammad Hafizudin Jamhari
Kulliyyah of Information and Communication Technology (KICT)
International Islamic University Malaysia
Kuala Lumpur, Malaysia

Najib el Kamoun
Department of Computer Sciences
Chouaib Doukkali University
El Jadida, Morocco

Hiba Kandil
IPSS, Faculty of Sciences
Mohammed V University in Rabat
Rabat, Morocco

Ayoub Khadrani
ENSA Berrechid
Hassan 1st University
Settat, Morocco

Jihad Kilani
Department of Computer Sciences
Chouaib Doukkali University
El Jadida, Morocco

Ouidad Labouidya
Department of Computer Sciences
Chouaib Doukkali University
El Jadida, Morocco

Fatima Lakrami
Department of Computer Sciences
Chouaib Doukkali University
El Jadida, Morocco

Afef Jmal Maâlej
ReDCAD Laboratory, National School of Engineers of Sfax
University of Sfax
Sfax, Tunisia

Yassine Maleh
LaSTI Laboratory
Sultan Moulay Slimane University
Beni Mellal, Morocco

Salmaa Naffah
ENSA Berrechid
Hassan 1st University
Settat, Morocco

Irshad Rasheed
Department of Computer Applications
Government Degree College Sumbal
Sumbal, India

Nur Alya Aqilah Razak Ratne
Electrical and Computer Engineering (ECE) Department
International Islamic University Malaysia
Kuala Lumpur, Malaysia

Peer Zada Rayees
Department of Computer Applications
Government Degree College Sumbal
Sumbal, India

Aliah Maisarah Roslee
Electrical and Computer Engineering (ECE) Department
International Islamic University Malaysia
Kuala Lumpur, Malaysia

Nuramiratul Aisyah Ruzaidi
Electrical and Computer Engineering (ECE) Department
International Islamic University Malaysia
Kuala Lumpur, Malaysia

Abdelekbir Sahid
ENCG Settat
Hassan 1st University
Settat, Morocco

Nur Fatnin Izzati Sidik
Electrical and Computer Engineering (ECE) Department
International Islamic University Malaysia
Kuala Lumpur, Malaysia

Hind Sounni
Department of Computer Sciences
Chouaib Doukkali University
El Jadida, Morocco

Muhammad Syafiq
Kulliyyah of Information and Communication Technology (KICT)
International Islamic University Malaysia
Kuala Lumpur, Malaysia

Mohamed Talea
Faculty of Science Ben M'sik
Hassan II University
Casablanca, Morocco

Hicham Toumi
Department of Computer Sciences
Higher School of Technology – Sidi Bennour
Chouaib Doukkali University
El Jadida, Morocco

Ediomo Udofia
Jubaili Bros Engineering Limited
Ikeja, Nigeria

Abdul Wahid Wali
Department of Computer Applications
Government Degree College
Sumbal, India

Muhammad Ziyad Fathullah Mohd Yazid
Electrical and Computer Engineering (ECE) Department
International Islamic University Malaysia
Kuala Lumpur, Malaysia

Hicham Yzzogh
IPSS, Faculty of Sciences
Mohammed V University in Rabat
Rabat, Morocco

Ahmad Anwar Zainuddin
Electrical and Computer Engineering (ECE) Department
Kulliyyah of Information and Communication Technology (KICT)
International Islamic University Malaysia
Kuala Lumpur, Malaysia

Muhammad Aizzul Izzuddin Zulhazizi
Kulliyyah of Information and Communication Technology (KICT)
International Islamic University Malaysia
Kuala Lumpur, Malaysia

Part I

Foundations of cyber defense and risk assessment

Chapter 1

AI-powered strategies for advanced malware detection and prevention

Syed Immamul Ansarullah, Abdul Wahid Wali, Irshad Rasheed, and Peer Zada Rayees

1.1 INTRODUCTION

The growing threat of malware represents a digital epidemic in our interconnected world. Malicious software, or malware, is crafted with the intent to infiltrate, compromise, and exploit computer systems, posing risks to individuals, businesses, and even governments [1]. As technology evolves, so does the arsenal of cybercriminals, who continually develop more sophisticated and elusive forms of malware. From ransomware that locks up vital data to stealthy viruses that silently siphon sensitive information, the impact of malware is far-reaching, causing financial losses, privacy breaches, and operational disruptions [2]. In this era of constant connectivity, understanding and mitigating the escalating threat of malware have become paramount to safeguarding our digital lives and ensuring the resilience of our technological infrastructure.

As the digital landscape becomes increasingly complex and hostile, the adoption of advanced detection and prevention strategies is imperative for organizations to fortify their cybersecurity posture, protect sensitive information, and mitigate the evolving threats posed by malware and other cyber threats [3]. Artificial intelligence (AI) plays a pivotal role in strengthening cybersecurity by providing advanced threat detection, dynamic adaptability, automated response mechanisms, and improved insights into evolving cyber threats [4]. As cyber threats become more sophisticated, the integration of AI technologies becomes increasingly essential for maintaining robust and effective cybersecurity defenses.

The purpose of this chapter is to explore the role of AI in combating the growing threat of malware. By delving into advanced detection and prevention strategies powered by AI, this chapter seeks to provide insights into how organizations can enhance their cybersecurity posture to effectively mitigate the risks posed by malware and other cyber threats. Through the examination of machine learning (ML) algorithms, deep learning methodologies, and real-time monitoring techniques, this chapter seeks to showcase the transformative potential of AI in bolstering cybersecurity defenses and safeguarding digital assets against the ever-evolving landscape of cyber threats.

This chapter is organized as follows: we begin with an exposition of the research methodology (Section 1.2), detailing the approaches employed in data collection and analysis. Following this, we delve into a comprehensive examination of malware and its various types (Section 1.3), laying the foundation for understanding the evolving threat landscape. In Section 1.4, we explore the challenges inherent in traditional cybersecurity approaches, leading to a discussion on the relevance of AI in cybersecurity (Section 1.5). Section 1.6 focuses on ML techniques for malware detection, discussing their efficacy and limitations. Subsequently, Section 1.7 addresses the complexities of integrating AI into existing cybersecurity frameworks, while Section 1.8 delves into the ethical considerations surrounding

DOI: 10.1201/9781032714806-2

AI-driven malware detection, with a particular emphasis on privacy and data handling. Furthermore, Section 1.9 explores emerging technologies in AI for malware detection, highlighting innovative approaches to combating evolving cyber threats. Finally, we synthesize our findings and insights in Section 1.10, offering concluding remarks and outlining directions for future research in this dynamic and critical domain.

1.2 RESEARCH METHODOLOGY

This study employed a structured methodology to investigate how AI can effectively detect and prevent malware threats. The Preferred Reporting Items for Systematic Reviews and Meta-Analyses (PRISMA) methodology was employed, encompassing a systematic approach from paper selection, filtering, and screening to inclusion, as depicted in Figure 1.1.

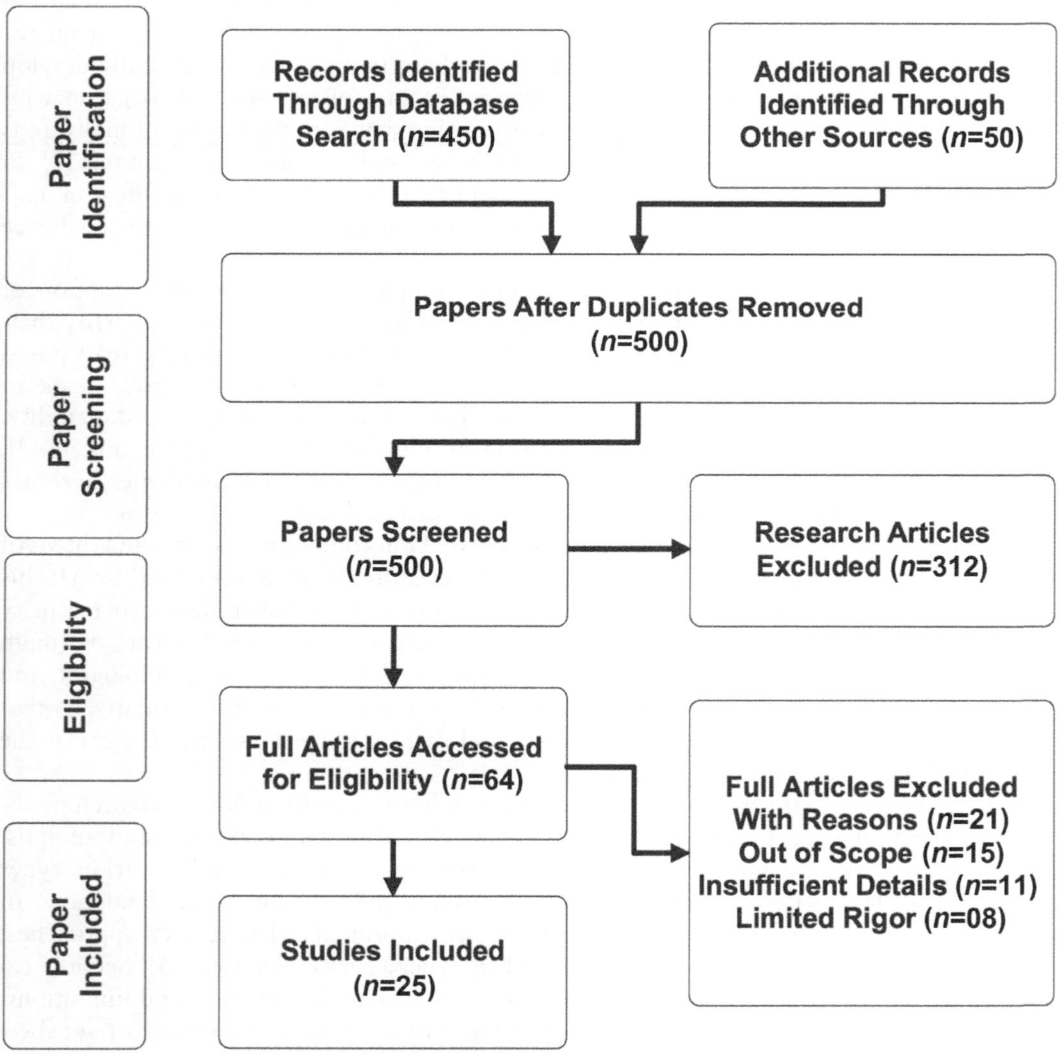

Figure 1.1 Schematic representation of the PRISMA methodology used in this study.

- **Research questions:** Clear research questions were formulated to guide the selection of relevant literature and ensure a comprehensive examination of the topic.
- **Search strategy:** A robust search strategy was developed to minimize bias and retrieve relevant literature from various sources such as databases, journals, and search engines.
- **Inclusion and exclusion criteria:** By carefully selecting search terms and defining inclusion and exclusion criteria, the study aimed to gather a comprehensive range of scholarly articles and reports.
- **Screening and selection:** Identified papers underwent rigorous screening based on predefined criteria to identify high-quality literature. Initially, titles and abstracts were screened to identify potentially relevant studies, followed by a full-text review to assess eligibility for inclusion.
- **Data extraction:** Pertinent information from selected papers was systematically extracted using a standardized template. This included study objectives, methodology, key findings, and conclusions. Standardizing the data extraction process helped ensure consistency and facilitated the synthesis of findings across studies.
- **Quality assessment:** Selected papers were evaluated for methodological rigor and bias using established tools or checklists. By rigorously assessing the quality of included studies, the study aimed to provide robust and credible insights into AI-driven malware detection and prevention.
- **Analysis:** Synthesized findings were analyzed to identify patterns, themes, and trends in the literature. This analysis provided valuable insights into the current state of knowledge on the effectiveness of AI in combating malware threats.
- **Transparency and reporting:** Results were transparently reported following established guidelines such as PRISMA.

1.3 UNDERSTANDING MALWARE AND ITS TYPES

Malware, short for "malicious software," refers to any software specifically designed to harm, exploit, or compromise the functionality of a computer system, network, or device. The intent behind malware is typically malicious, ranging from stealing sensitive information and financial data to disrupting system operations or gaining unauthorized access [5]. There are various types of malware (shown in Figure 1.2), each with distinct characteristics and purposes. Here are some common categories of malware:

- **Virus:** A computer virus is a type of malware that attaches itself to legitimate executable files and replicates when the infected file is executed. They often spread through infected files, email attachments, or removable media [6].
- **Worms:** Worms are standalone malicious programs that can replicate and spread independently across networks, often exploiting vulnerabilities in operating systems or software to infect other systems [6].
- **Trojans (Trojan horses):** Trojans disguise themselves as legitimate software to deceive users into installing them. Once inside the system, they can create backdoors, steal data, or provide remote access to attackers [7].
- **Ransomware:** Ransomware encrypts files on a victim's system, rendering them inaccessible. The attacker then demands a ransom, usually in cryptocurrency, in exchange for providing the decryption key [7].
- **Spyware:** Spyware is designed to gather information about a user's activities without their knowledge or consent. This may include keystrokes, browsing habits, login credentials, and other sensitive information [8].

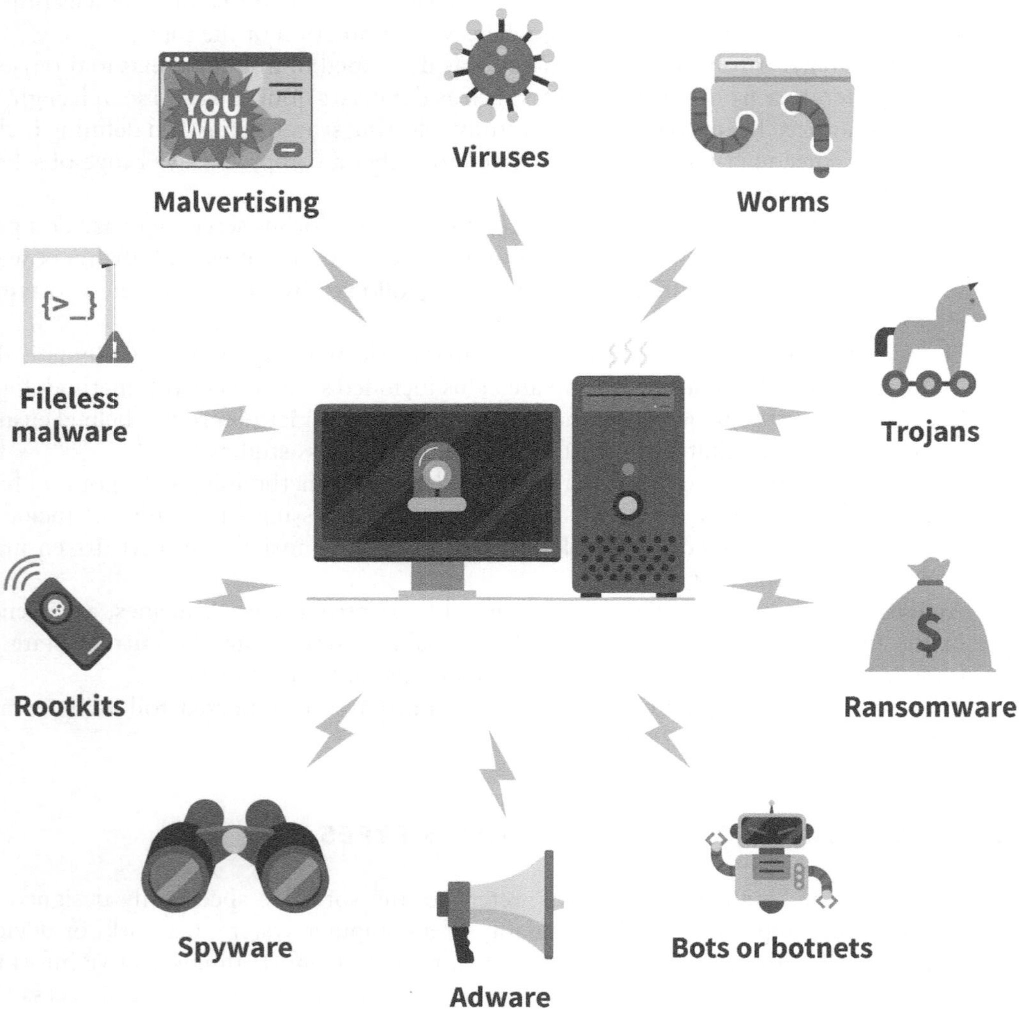

Figure 1.2 Malware and its types.

- **Adware:** Adware displays unwanted advertisements, often in the form of pop-ups or banners, to generate revenue for the malware creator. While not inherently malicious, it can be intrusive and negatively impact system performance [8].
- **Rootkits:** Rootkits are a type of malware that allows unauthorized access to a system while concealing its presence. They often exploit vulnerabilities to gain control over the operating system [8].
- **Botnets:** Botnets are networks of compromised computers (bots) controlled by a single entity, often used for various malicious activities such as launching distributed denial-of-service (DDoS) attacks or sending spam [8].

To protect against malware, individuals and organizations employ security measures such as antivirus software, firewalls, regular software updates, and user education on safe online practices. In addition, maintaining data backups and exercising caution when interacting with unknown or suspicious content are crucial for minimizing the impact of potential malware infections.

1.3.1 Common attack vectors

Common attack vectors refer to the various methods and pathways that cyber attackers use to exploit vulnerabilities and gain unauthorized access to computer systems, networks, or applications. These vectors serve as entry points for attacks, and attackers often choose the path of least resistance to achieve their objectives [9]. Understanding these common attack vectors is essential for developing effective cybersecurity strategies. Here are some examples of common attack vectors:

- **Phishing attacks:** Phishing is a form of cyber attack in which attackers use deceptive techniques to trick individuals into divulging sensitive information, such as login credentials, personal details, or financial information. Phishing attacks commonly involve email phishing, vishing (voice phishing), spear phishing, smishing (SMS phishing), pharming, social media phishing, etc. [10].
- **Malware infections:** Malware infections are instances where malicious software infiltrates and compromises a computer system, network, or device. Malware infections commonly involve drive-by downloads and malicious email attachments.
- **Social engineering:** Social engineering is a manipulative technique employed by cyber attackers to exploit human psychology and trick individuals into divulging sensitive information, providing unauthorized access, or taking actions that compromise security. Social engineering involves manipulation of trust and impersonation [11].
- **Brute force attacks:** Brute force attacks involve credential stuffing and dictionary attacks. In credential stuffing, attacker's use previously stolen username and password combinations to gain unauthorized access to user accounts on various platforms, whereas in dictionary attacks, attackers repeatedly try a list of common words or phrases to guess passwords.
- **Vulnerability exploitation:** Vulnerability exploitation may include software exploits or zero-day exploits. In software exploits, attackers target vulnerabilities in software or operating systems to gain unauthorized access or execute malicious code. In zero-day exploits, attackers exploit vulnerabilities that are unknown to the software vendor, giving attackers an advantage until a patch is released.
- **Man-in-the-middle (MitM) attacks:** MitM attacks include interception of communication and Wi-Fi eavesdropping. Earlier, attackers secretly intercept and possibly alter the communication between two parties without their knowledge. In Wi-Fi, eavesdropping, attackers monitor and capture wireless network traffic to gain unauthorized access to sensitive information [12].
- **Physical attacks:** Physical attacks include unauthorized access and hardware tampering. In unauthorized access, the intension is gaining physical access to devices, servers, or data centers to compromise security. In hardware tampering, intension of the hacker is modifying or manipulating hardware components to introduce vulnerabilities or gain unauthorized control.
- **Supply chain attacks:** Supply chain attacks consist of compromised software supply chains and third-party vendor risks. In an earlier form, attackers inject malicious code or compromise software updates during the development or distribution process. In third-party vendor risks, hackers exploit vulnerabilities in products or services provided by third-party vendors.

Understanding and addressing these common attack vectors is crucial for implementing effective cybersecurity measures, including network security, endpoint protection, user awareness training, and the regular application of security patches and updates.

1.3.2 Evolving landscape of malware threats

The evolving landscape of malware threats is marked by constant innovation and adaptation on the part of malicious actors. As technology advances, so do the techniques and strategies employed by cybercriminals to compromise systems, steal data, and disrupt operations. Understanding the evolving nature of malware threats is crucial for cybersecurity professionals and organizations to stay ahead of potential risks, adapt security measures, and implement proactive strategies to mitigate the impact of emerging cyber threats. Regular threat intelligence analysis and keeping systems updated with the latest security patches are critical components of a robust defense against the evolving landscape of malware [13]. Here are key aspects highlighting the dynamic nature of the malware threat landscape:

- **Increased sophistication:** Malware has become more sophisticated, with advanced evasion techniques, polymorphic code, and the ability to adapt to evolving cybersecurity measures.
- **Targeted attacks (advanced persistent threats – APTs):** Malicious actors now frequently engage in targeted attacks, tailoring malware to specific individuals, organizations, or industries. APTs often involve prolonged, covert campaigns with a specific goal, such as espionage or intellectual property theft [14].
- **Ransomware proliferation:** Ransomware attacks have surged with cybercriminals using increasingly sophisticated methods to encrypt data and demand ransoms for its release. Some ransomware operations operate as organized criminal enterprises.
- **Fileless malware:** Fileless malware operates in a system's memory, leaving minimal traces on disk. This makes detection challenging for traditional antivirus solutions as there are fewer conventional files to analyze.
- **Living off the land (LOL) attacks:** Cybercriminals leverage legitimate system tools and processes to carry out attacks, making it difficult for traditional security measures to distinguish between malicious and benign activities [15].
- **Internet of Things (IoTs) exploitation:** With the proliferation of connected devices, malware threats have expanded to target IoT devices. Weak security measures in these devices provide opportunities for attackers to compromise networks.
- **Zero-day exploits:** Malware often exploits vulnerabilities in software that are unknown to vendors or security experts. Zero-day exploits are particularly challenging to defend against as there are no available patches [16].
- **Phishing and social engineering evolution:** Phishing attacks have become more sophisticated, employing realistic-looking emails, messages, or websites. Social engineering techniques are continuously refined to exploit human vulnerabilities.
- **File encryption bypass:** Some malware strains have developed the capability to bypass or disable antivirus software and other security controls, allowing them to operate undetected for longer periods.
- **Multivector attacks:** Malware campaigns increasingly involve multiple attack vectors, combining tactics such as phishing, malware delivery, and exploitation of vulnerabilities for a more comprehensive and persistent assault.
- **Supply chain attacks:** Malicious actors target the software and hardware supply chain, compromising products during development or distribution to introduce malware at scale.
- **Blockchain exploitation:** Malware threats have expanded to exploit vulnerabilities in blockchain technology, targeting cryptocurrencies and blockchain-based systems [17].

- **AI- and ML-powered malware:** Malicious actors use AI and ML to develop more intelligent and adaptive malware, making detection and analysis more challenging.
- **Nation-state-sponsored malware:** State-sponsored entities deploy malware for espionage, cyber warfare, or political purposes, contributing to the complexity of the threat landscape.

1.4 CHALLENGES IN TRADITIONAL APPROACHES

Traditional approaches to cybersecurity face several challenges in effectively addressing the evolving threat landscape. Addressing these challenges requires a shift toward more proactive and adaptive cybersecurity strategies, including the incorporation of advanced technologies such as AI and behavioral analysis to complement traditional approaches. Here are some key challenges associated with conventional methods:

- **Signature-based detection:** It relies on known malware signatures, making it ineffective against zero-day threats or polymorphic malware that can change its code to evade detection. New and sophisticated malware variants may go undetected until a signature update is available.
- **Reactive nature:** Traditional methods often operate reactively, responding to known threats after they have been identified. There is a delay between the emergence of a new threat and the deployment of countermeasures, leaving systems vulnerable during this gap.
- **Limited behavioral analysis:** Traditional antivirus solutions may have limited capabilities for behavioral analysis, making it challenging to detect subtle or advanced threats that deviate from typical patterns. Malware employing sophisticated evasion techniques, such as fileless attacks, may go undetected.
- **Inability to handle polymorphism:** Polymorphic malware can change its code and appearance, rendering static signature-based detection ineffective. Signature-based antivirus solutions struggle to keep up with the dynamic and ever-changing nature of polymorphic malware.
- **False positives and negatives:** Traditional methods may generate false positives, flagging legitimate activities as malicious or false negatives, failing to detect actual threats. Security teams may waste time investigating nonthreats or miss critical alerts due to false negatives.
- **Overreliance on perimeter defense:** Traditional security models often focus on perimeter defense, assuming threats can be prevented from entering the network. As threats evolve, perimeter-focused defenses may not adequately protect against internal threats or sophisticated attacks that bypass traditional boundaries.
- **Single-layered defenses:** Relying solely on one layer of defense, such as antivirus software, can leave organizations vulnerable to multivector attacks that exploit various vulnerabilities. Cyber adversaries can exploit weaknesses in the security posture that a single-layered defense does not cover.
- **Inadequate response to insider threats:** Traditional approaches may struggle to detect and prevent insider threats, where authorized individuals misuse their access. Insiders with malicious intent may go undetected, leading to data breaches or unauthorized activities.

- **Legacy system compatibility:** Older security solutions may not be compatible with modern systems, applications, or cloud-based environments. Incompatibility hinders the integration of new security technologies, leaving legacy systems exposed.
- **Resource-intensive updates:** Regular updates and patches for traditional antivirus databases can be resource-intensive and may disrupt system operations. Delayed updates may leave systems vulnerable to emerging threats.
- **Lack of user awareness:** Users may lack awareness and training to recognize and respond appropriately to security threats. Successful social engineering attacks, such as phishing, may exploit user vulnerabilities.

1.5 OVERVIEW OF AI AND ITS RELEVANCE TO CYBERSECURITY

AI refers to the simulation of human intelligence in machines that are programmed to perform tasks that typically require human intelligence. These tasks include learning, reasoning, problem-solving, perception, language understanding, and even decision-making. The most important components of AI are ML, natural language processing (NLP), computer vision, and robotics. AI is of two fundamental types: narrow AI and general AI. Narrow or weak AI is designed and trained for a specific task, and its intelligence is limited to that particular domain [18]. Most of the AI applications currently in use, such as virtual assistants and recommendation systems, fall into this category. General or strong AI refers to a form of intelligence that can understand, learn, and apply knowledge across a broad range of tasks, similar to human intelligence. Achieving strong AI remains a long-term goal and is not yet realized [19].

The relevance of AI to cybersecurity lies in its ability to provide advanced, adaptive, and proactive defense mechanisms. By leveraging ML, behavioral analysis, and automation, AI empowers organizations to stay ahead of cyber threats, respond rapidly to incidents, and fortify their security posture in an increasingly complex and dynamic digital landscape.

- **Adaptive threat detection:** Traditional cybersecurity measures often struggle to keep pace with the evolving landscape of cyber threats. AI, particularly ML algorithms, excels in identifying patterns and anomalies within vast datasets, enabling more advanced and proactive threat detection.
- **Rapid incident response:** AI streamlines incident response by automating the analysis of security alerts and swiftly responding to potential threats. This reduces the time between detection and mitigation, which is crucial in preventing the escalation of cyberattacks.
- **Behavioral analysis and anomaly detection:** AI systems employ behavioral analysis to understand normal patterns of user and system behavior. Any deviations or anomalies that may indicate a potential security threat can be quickly identified and addressed.
- **Zero-day threat protection:** Traditional security solutions rely on known signatures of malware, making them ineffective against zero-day threats. AI, through its ability to detect deviations from normal behavior, provides a proactive defense against previously unknown and emerging threats.
- **Phishing and social engineering defense:** AI-powered tools can analyze communication patterns, detect anomalies, and identify phishing attempts or social engineering attacks. This enhances the ability to thwart attacks that exploit human vulnerabilities.
- **Predictive analysis for cyber threats:** AI can predict potential cyber threats by analyzing historical data and identifying trends. This enables organizations to proactively strengthen their defenses based on insights gained from predictive analytics.

- **Endpoint security enhancement:** AI plays a crucial role in endpoint security by continuously monitoring and analyzing activities on individual devices. This ensures swift detection and response to any malicious activities, reducing the risk of endpoint compromise.
- **Malware detection and classification:** AI, especially ML models, excels in the detection and classification of malware. By learning from large datasets of known malware samples, AI systems can identify new variants and even predict potential mutations.
- **Automated security operations:** AI-driven automation enhances security operations by handling routine tasks, allowing human analysts to focus on complex analysis and strategic decision-making. This improves overall efficiency and resource allocation.
- **Continuous learning and adaptation:** Cyber threats are dynamic and ever-changing. AI systems continuously learn from new data and adapt to evolving threat landscapes, ensuring that cybersecurity defenses remain effective against the latest and most sophisticated attacks.
- **Insider threat detection:** AI can identify unusual patterns of behavior among users, helping detect potential insider threats. This includes unauthorized access, data exfiltration, or other activities that deviate from normal behavior.
- **Network security optimization:** AI enhances network security by identifying vulnerabilities, analyzing network traffic patterns, and predicting potential weaknesses. This proactive approach aids in fortifying the overall security posture of an organization.

1.6 MACHINE LEARNING FOR MALWARE DETECTION

ML plays a pivotal role in malware detection, providing a proactive and adaptive approach to identifying malicious software. ML algorithms analyze patterns and features within datasets to learn from historical data and make predictions about the presence of malware. Here are some key aspects of using ML for malware detection.

1.6.1 Feature extraction

ML models rely on features extracted from data to make predictions. In the case of malware detection, these features might include file attributes, API calls, system calls, network behavior, and more. Metadata associated with files provides valuable information about their characteristics. Malicious files often exhibit certain patterns or anomalies in their attributes that can be detected through ML, such as file size, creation time, and file type [20].

One notable instance is the Stuxnet worm, discovered in 2010. Stuxnet was a sophisticated computer worm designed to target supervisory control and data acquisition (SCADA) systems, specifically those used in Iran's nuclear program [21]. Stuxnet used deceptive tactics to infiltrate systems and hide its presence. The malware creators were aware of the importance of file attributes in avoiding detection. The worm employed a technique known as "file system manipulation." It would manipulate file attributes, such as creation times, to make infected files appear unchanged and avoid raising suspicion. Despite its efforts to remain stealthy, Stuxnet was eventually discovered by researchers who analyzed file attributes and behavior through behavioral analysis. The worm exhibited specific patterns of interactions with SCADA systems that were inconsistent with normal behavior.

1.6.2 Supervised learning

Supervised learning is a type of ML where the algorithm is trained on a labeled dataset, which means that each input data point is paired with its corresponding output or target. The goal is for the algorithm to learn the mapping from inputs to outputs so that it can make accurate predictions or classifications on new, unseen data. Now, in the context of malware detection and prevention, various features can be used to characterize files and identify malicious behavior [22]. Examples of features include file size, file type (executable, script, etc.), creation/modification timestamps, digital signatures, and presence of certain keywords or strings. These features provide valuable information for training a supervised learning model to distinguish between normal and malicious files. The model learns patterns in the labeled data, allowing it to make predictions on new files based on their features.

Here is a simplified illustration of how supervised learning has been historically used by Kaspersky Lab for malware detection and prevention. Kaspersky Lab employs behavioral analysis to understand the patterns of behavior exhibited by both malicious and legitimate software [23]. This involves tracking actions such as file attributes, API calls, system calls, registry changes, and network activities. Supervised learning is employed during the training phase, utilizing labeled datasets comprising examples of known malware and benign software behaviors. Features extracted from these datasets serve as input variables for algorithms such as decision trees, random forests, support vector machines, and neural networks. The models learn to discern patterns and relationships between features, enabling them to make predictions about the nature of a file or activity. Once trained, these ML models are integrated into Kaspersky Lab's security solutions, providing real-time detection and prevention of malware by assessing features associated with files and activities. This approach complements traditional signature-based detection methods, adding an extra layer of protection based on behavior and features. The incorporation of a feedback loop from user-reported incidents contributes to refining and improving the ML models over time.

1.6.3 Unsupervised learning

Unsupervised learning methods have also been explored for their potential in detecting previously unknown or evolving malware threats. Unsupervised learning is particularly useful in scenarios where labeled datasets are scarce or when dealing with novel, zero-day attacks. Unsupervised learning is applied in cybersecurity, notably in detecting anomalies in network traffic for malware detection and prevention [24]. In this approach, unsupervised learning algorithms like clustering methods or autoencoders are utilized to analyze patterns within network traffic data. During the training phase, the model is exposed to a dataset that captures normal network behavior over an extended period in a secure environment. The unsupervised learning algorithm then endeavors to discern the intrinsic patterns and structures inherent in the benign network activity. Once the model is trained, it is deployed to analyze real-time network traffic. Deviations from the learned normal behavior are flagged as anomalies, and these anomalies may indicate potential malicious activity.

Unsupervised learning models are valued for their adaptability to new and previously unseen threats, as they don't rely on labeled datasets. This adaptability allows the model to identify novel attack patterns or behaviors that may not have been explicitly labeled during training. The deployment of unsupervised learning-based anomaly detection is commonly integrated into intrusion detection systems (IDSs) used in network environments. These systems continually monitor network activity and trigger alerts when they detect deviations that could signify malicious behavior.

It is essential to be aware that while unsupervised learning approaches offer adaptability to new threats, they may also be associated with higher false-positive rates compared with supervised methods, as they lack explicit knowledge of what constitutes malicious behavior.

The challenge lies in distinguishing between genuine anomalies indicative of a threat and false alarms that may be triggered by benign variations in network activity.

1.6.4 Behavioral analysis

Behavioral analysis for malware detection involves monitoring and analyzing the behavior of software or processes to identify malicious activities. Unlike traditional signature-based methods, which rely on known patterns of malware, behavioral analysis focuses on the actions and interactions of programs in real time [25].

WannaCry Ransomware serves as a notable real-world example of the effectiveness of behavioral analysis in malware detection. Its infection typically began through phishing emails containing malicious attachments or links. Once activated, WannaCry dynamically utilized the EternalBlue exploit, targeting vulnerabilities in Microsoft Windows SMB protocol for rapid propagation across vulnerable systems. Behavioral monitoring during execution revealed the ransomware's encryption of files and the display of a ransom note demanding Bitcoin payments for decryption keys. The malware demonstrated a propensity for lateral movement, scanning, and infecting other machines within the same network. Network activities included communication with command-and-control servers for instructions and key updates. WannaCry's self-propagation characteristics, exploiting Windows Management Instrumentation Command-line (WMIC) and PsExec tools, showcased its dual nature as both a worm and ransomware [26]. Anomaly detection through behavioral analysis could have identified the unusual patterns, such as mass file encryption, atypical lateral movement, and communication with known malicious servers, thereby playing a pivotal role in mitigating the impact of this widespread and rapidly evolving threat.

1.6.5 Deep learning

Deep learning, a subset of ML, involves neural networks with multiple layers. Convolutional neural networks (CNNs) and recurrent neural networks (RNNs) have been used for image and sequential data analyses, respectively, in malware detection. Deep learning models can automatically learn hierarchical representations of features, potentially improving detection accuracy [27]. Let's use an example to illustrate the application of long short-term memory networks (LSTMs) in malware detection and prevention. Consider a scenario where security researchers are dealing with a new strain of a banking Trojan that disguises itself within seemingly harmless executables. Traditional signature-based methods struggle to identify this polymorphic malware as it constantly evolves to avoid detection. Leveraging deep learning, specifically an LSTM model, proves to be effective in discerning the Trojan's behavior.

In this context, the LSTM is trained on a diverse dataset containing both normal software execution sequences and instances of the evolving banking Trojan. The LSTM learns the intricate temporal patterns inherent in the system call sequences, recognizing the nuances that distinguish benign software activities from the malicious behavior associated with the banking Trojan. During the detection phase, as users interact with various executables, the LSTM analyzes the real-time sequences of system calls. When confronted with an unknown executable, the LSTM identifies deviations from learned patterns, such as abnormal attempts to access sensitive files or establish unauthorized network connections. This prompts the system to raise an alert, signaling the potential presence of the banking Trojan.

The adaptability of the LSTM model ensures its effectiveness against new variants of the banking Trojan, even those with altered code or obfuscation techniques [28]. This example underscores how deep learning, particularly with LSTM networks, can provide a dynamic and robust defense against sophisticated and rapidly evolving malware threats in real-world cybersecurity scenarios.

1.6.6 Feature importance and explainability

Feature importance and explainability in the context of malware detection involve understanding which features or factors contribute most significantly to the identification of malicious behavior and providing explanations for the model's decisions. This transparency is crucial for building trust, aiding in the interpretation of results, and enhancing the overall efficacy of security systems [29]. Let's explore this concept with a real-world example of Random Forest for Malware Detection with Explainability.

The dataset consists of various features extracted from executables, such as file attributes, system call sequences, and network activities. The Random Forest model is trained on a labeled dataset comprising instances of both benign and malicious executables. It learns to discern patterns and associations between features that indicate the presence of malware. After training, the model can quantify the importance of each feature in making accurate predictions. Features that consistently contribute to the model's decision-making process are assigned higher importance scores. Techniques such as SHAP (SHapley Additive exPlanations) values or LIME (Local Interpretable Model-agnostic Explanations) can be employed to provide granular explanations for individual predictions.

Suppose the model identifies a file as potentially malicious due to its high frequency of suspicious system calls and unexpected network activity. The explainability aspect helps security analysts understand the rationale behind the classification, making it possible to investigate further and take appropriate actions. Insights from feature importance and explainability can guide improvements to the model or the feature extraction process. If certain features consistently prove critical, this knowledge can inform the refinement of detection strategies.

By combining feature importance analysis and explainability techniques, security practitioners can enhance their understanding of how malware detection models operate. This transparency fosters trust in the system and enables them to adapt their defense mechanisms effectively, improving the accuracy of their malware detection system in real-world scenarios.

1.7 LIMITATIONS AND CHALLENGES OF INTEGRATING AI INTO EXISTING CYBERSECURITY FRAMEWORKS

While AI offers significant advantages in enhancing malware detection and prevention, its implementation is not without challenges. Here are some key limitations and challenges associated with deploying AI for malware detection and prevention. Addressing these limitations and challenges requires a holistic approach, involving collaboration between cybersecurity experts, data scientists, ethicists, policymakers, and other stakeholders. By acknowledging and mitigating these challenges, organizations can leverage AI effectively to enhance security measures and defend against evolving cyber threats.

- **Data quality and quantity:** Cybersecurity data encompasses a wide range of sources, including network traffic logs, system event logs, threat intelligence feeds, and malware samples. However, this data is often noisy, imbalanced, and incomplete, making it challenging to extract meaningful insights. Moreover, labeling data for supervised learning tasks requires domain expertise and may be subjective, leading to inconsistencies and biases in the training dataset. Addressing data quality issues involves data preprocessing techniques such as normalization, outlier detection, and missing value imputation, as well as data augmentation strategies to increase dataset diversity and balance class distributions.

- **Adversarial attacks:** Adversarial attacks exploit vulnerabilities in AI models by perturbing input data in imperceptible ways to cause misclassification or incorrect predictions. These attacks can target various components of the AI pipeline, including feature extraction, model training, and inference. Defending against adversarial attacks requires a multilayered approach, including robust training strategies (e.g., adversarial training, defensive distillation), model verification techniques (e.g., adversarial robustness certification), and runtime detection mechanisms (e.g., adversarial example detection). In addition, ongoing research into adversarial robustness and adversarial attack techniques is essential for staying ahead of evolving threats.
- **Interpretability and explainability:** The lack of interpretability and explainability in AI models hinders their adoption and trustworthiness, particularly in safety-critical domains such as cybersecurity. Interpretability refers to the ability to understand and interpret the internal workings of AI models, while explainability refers to the ability to provide understandable explanations for model predictions. Techniques for enhancing interpretability and explainability include feature visualization, attention mechanisms, saliency maps, and counterfactual explanations. However, achieving both high performance and high interpretability remains a challenging trade-off in AI model design, requiring further research into interpretable model architectures and training algorithms.
- **Scalability and resource constraints:** Scalability is a critical consideration in AI-driven cybersecurity, as security operations generate vast amounts of data that must be processed and analyzed in real time. Scaling AI models to handle large datasets and high-dimensional feature spaces requires efficient algorithms and scalable infrastructure. Techniques such as model parallelism, data parallelism, and distributed computing enable AI models to be trained and deployed across multiple GPUs, CPUs, or cloud instances. Moreover, optimizing AI algorithms for memory and compute efficiency is essential for minimizing resource consumption and maximizing performance in resource-constrained environments.
- **Ethical and privacy concerns:** The deployment of AI in cybersecurity raises ethical and privacy concerns related to data privacy, fairness, accountability, and transparency. AI algorithms may inadvertently capture and analyze sensitive information, leading to privacy violations or breaches. Moreover, biased or discriminatory AI models can perpetuate unfair or discriminatory outcomes, leading to ethical dilemmas and reputational risks for organizations. Addressing these concerns requires adherence to ethical guidelines and regulatory requirements, transparency in AI decision-making processes, and mechanisms for ensuring fairness, accountability, and user consent. In addition, organizations must consider the potential impacts of AI integration on existing workflows, processes, and stakeholders to ensure a seamless transition and adoption of AI-driven security solutions.
- **Integration complexity:** Integrating AI technologies into existing cybersecurity frameworks is complex and challenging due to the heterogeneity of computing environments, legacy systems, and proprietary protocols. Legacy systems may use proprietary formats, protocols, and data structures that are incompatible with modern AI technologies. Ensuring compatibility and interoperability between AI systems and existing security infrastructure requires careful planning, testing, and validation to minimize disruptions and maximize effectiveness. In addition, organizations must consider the potential impacts of AI integration on existing workflows, processes, and stakeholders to ensure a seamless transition and adoption of AI-driven security solutions. Furthermore, ongoing maintenance and support are essential to address compatibility issues, security vulnerabilities, and performance optimizations in AI-driven cybersecurity deployments (Table 1.1).

Table 1.1 Key challenges, descriptions, and mitigation strategies in AI-driven cybersecurity

Challenges	*Description*	*Mitigation strategies*	*Key considerations*
Data quality and quantity	Cybersecurity data is often noisy, imbalanced, and incomplete, making it challenging to extract meaningful insights. Labeling data for supervised learning tasks requires domain expertise and may be subjective, leading to inconsistencies and biases.	• Preprocessing techniques: Normalize data, detect and handle outliers, impute missing values. - Data augmentation: Increase dataset diversity and balance class distributions. - Collaborative labeling: Involve domain experts to ensure accurate and consistent labeling.	• Ensure data quality: Regularly audit data sources and update preprocessing pipelines. - Data governance: Establish policies for data collection, storage, and labeling to maintain quality and consistency. - Continuous monitoring: Implement mechanisms to detect and address data quality issues in real time.
Adversarial attacks	Adversarial attacks exploit vulnerabilities in AI models by perturbing input data to cause misclassification or incorrect predictions. These attacks can target various components of the AI pipeline, including feature extraction, model training, and inference.	• Robust training strategies: Employ adversarial training, defensive distillation, and ensemble methods to enhance model resilience. - Model verification techniques: Certify model robustness using adversarial robustness certification and runtime verification. - Runtime detection mechanisms: Implement defenses such as adversarial example detection and anomaly detection to detect and mitigate adversarial attacks in real time.	• Stay updated on adversarial attack techniques and defense strategies through continuous research and collaboration with the cybersecurity community. - Test AI models against a variety of adversarial scenarios to evaluate their robustness and effectiveness. - Foster interdisciplinary collaboration between AI researchers, cybersecurity experts, and adversarial attack specialists to develop comprehensive defense mechanisms.
Interpretability and explainability	The lack of interpretability and explainability in AI models hinders their adoption and trustworthiness, particularly in safety-critical domains like cybersecurity. Enhancing interpretability and explainability involves techniques such as feature visualization and attention mechanisms.	• Feature visualization: Visualize learned representations and decision boundaries to interpret model behavior. - Attention mechanisms: Highlight important features or regions of input data to explain model predictions. - Saliency maps: Generate heatmaps to indicate the contribution of each input feature to the model output. - Counterfactual explanations: Provide alternative scenarios to explain how changing input features affects model predictions.	• Strive for a balance between model performance and interpretability based on the specific requirements of the cybersecurity application. - Incorporate user feedback and domain knowledge to improve the interpretability of AI models and ensure alignment with end-user needs. - Consider the trade-offs between different interpretability techniques in terms of computational complexity, accuracy, and ease of understanding.

(*Continued*)

Table 1.1 (Continued) Key challenges, descriptions, and mitigation strategies in AI-driven cybersecurity

Challenges	*Description*	*Mitigation strategies*	*Key considerations*
Scalability and resource constraints	Scalability is crucial in AI-driven cybersecurity, as security operations generate vast amounts of data that must be processed and analyzed in real time. Scaling AI models requires efficient algorithms and scalable infrastructure, along with techniques like model parallelism and distributed computing.	• Model parallelism: Distribute model parameters across multiple GPUs or CPUs to scale model training and inference. - Data parallelism: Parallelize data processing and model evaluation across distributed computing nodes to handle large datasets. - Distributed computing: Deploy AI models on cloud-based infrastructure to leverage elastic computing resources and scale on-demand. - Memory and compute optimization: Optimize AI algorithms for memory and compute efficiency to minimize resource consumption in resource-constrained environments.	• Conduct thorough performance testing and benchmarking to identify scalability bottlenecks and optimize resource utilization. - Consider the trade-offs between scalability and model complexity when designing AI-driven cybersecurity solutions. - Leverage cloud-based infrastructure and serverless computing to dynamically scale AI workloads in response to fluctuating demand and resource availability. - Implement resource monitoring and auto-scaling mechanisms to automatically adjust computing resources based on workload characteristics and system performance metrics.
Ethical and privacy concerns	Deploying AI in cybersecurity raises ethical and privacy concerns related to data privacy, fairness, accountability, and transparency. AI algorithms may inadvertently capture sensitive information, leading to privacy violations. Moreover, biased AI models can perpetuate unfair outcomes, raising ethical dilemmas and reputational risks.	• Adhere to ethical guidelines: Ensure compliance with legal and regulatory requirements governing data privacy and security in AI applications. - Transparency and accountability: Provide clear explanations of AI decision-making processes and establish mechanisms for accountability and oversight. - Fairness and bias mitigation: Evaluate AI models for fairness and bias using metrics and techniques such as demographic parity and equal opportunity. - User consent and control: Obtain informed consent from users and empower them with control over their data and privacy settings.	• Engage with stakeholders to understand their concerns and perspectives on ethical and privacy issues in AI-driven cybersecurity. - Conduct comprehensive privacy impact assessments and risk assessments to identify and mitigate potential ethical and privacy risks. - Establish clear policies and procedures for handling sensitive data and ensuring transparency and accountability in AI decision-making processes. - Foster a culture of ethical AI development and responsible innovation within organizations by promoting ethical awareness, education, and training among employees and stakeholders.

(Continued)

Table 1.1 (Continued) Key challenges, descriptions, and mitigation strategies in AI-driven cybersecurity

Challenges	*Description*	*Mitigation strategies*	*Key considerations*
Integration complexity	Integrating AI technologies into existing cybersecurity frameworks is complex due to heterogeneity, legacy systems, and proprietary protocols. Legacy systems may use incompatible formats and data structures. Ensuring compatibility and interoperability requires careful planning, testing, and validation.	• Compatibility testing: Assess the compatibility of AI systems with existing security infrastructure and protocols through comprehensive testing and validation. - Interoperability standards: Adopt industry standards and protocols to facilitate seamless integration and communication between AI systems and legacy security systems. - Modular architecture: Design AI systems with modular components that can be easily integrated and customized to meet specific cybersecurity requirements. - Legacy system support: Develop adapters and middleware to bridge the gap between AI systems and legacy security systems, ensuring compatibility and interoperability.	• Collaborate with IT and cybersecurity teams to understand existing infrastructure and identify integration challenges and requirements. - Develop a roadmap for AI integration that outlines specific objectives, milestones, and timelines for deployment and integration activities. - Implement robust testing and validation procedures to verify the compatibility, interoperability, and performance of AI systems in real-world cybersecurity environments. - Provide training and support to IT staff and security personnel to facilitate the seamless adoption and integration of AI-driven cybersecurity solutions into existing workflows and processes.

Table 1.1 provides a comprehensive comparative analysis of the key challenges associated with integrating AI into existing cybersecurity frameworks. Each challenge is accompanied by a detailed description of its implications, mitigation strategies, and key considerations for addressing it effectively. By understanding and addressing these challenges, organizations can successfully leverage AI to enhance their cybersecurity posture and defend against evolving threats in the digital landscape.

1.8 ETHICAL CHALLENGES OF AI IN MALWARE DETECTION: PRIVACY AND DATA HANDLING

The ethical implications of using AI in malware detection, particularly regarding privacy concerns and data handling, are multifaceted and require careful consideration to ensure responsible and ethical use of AI technologies in cybersecurity:

1. **Privacy concerns:**
    - Data collection: AI-driven malware detection systems often rely on vast amounts of data, including network traffic logs, system event logs, and file metadata, to train and improve their models. Collecting and storing such data may raise privacy concerns, as it could potentially include sensitive information about individuals or organizations.
    - Data sharing: Sharing cybersecurity data, particularly threat intelligence feeds and malware samples, among organizations and security researchers is essential for enhancing detection capabilities and responding to emerging threats. However, indiscriminate sharing of sensitive data may violate privacy regulations and expose individuals or organizations to unnecessary risks.
    - Data retention: Retaining cybersecurity data for extended periods raises concerns about data security, confidentiality, and potential misuse. Storing large volumes of sensitive data increases the risk of data breaches, unauthorized access, and exposure to malicious actors.
2. **Data handling:**
    - Data minimization: Minimizing the collection and retention of personally identifiable information (PII) and sensitive data is essential for reducing privacy risks and complying with data protection regulations such as General Data Protection Regulation (GDPR). AI-driven malware detection systems should only collect and use data that is necessary for their intended purpose and implement data anonymization and encryption techniques to protect sensitive information.
    - Data security: Ensuring the security and integrity of cybersecurity data is paramount to prevent unauthorized access, tampering, or exfiltration by malicious actors. Implementing robust encryption, access controls, and monitoring mechanisms can help mitigate the risk of data breaches and unauthorized data handling.
    - Data transparency: Providing transparency and accountability in data handling practices is essential for building trust and confidence in AI-driven malware detection systems. Organizations should clearly communicate their data collection and processing practices, obtain explicit consent from data subjects where applicable, and establish mechanisms for individuals to access, correct, or delete their data.
3. **Bias and fairness:**
    - Training data bias: AI models used in malware detection may exhibit biases if trained on datasets that are unrepresentative or biased toward certain demographics or characteristics. Biased models can lead to unfair or discriminatory outcomes, disproportionately affecting certain individuals or groups.

- Fairness measures: Implementing fairness-aware AI algorithms and evaluation metrics can help mitigate bias and ensure equitable outcomes in malware detection. Techniques such as fairness-aware training, bias detection, and model interpretability can help identify and address biases in AI models and decision-making processes.

4. **Accountability and oversight:**
   - Regulatory compliance: Adhering to relevant regulations and industry standards is essential for ensuring legal compliance and accountability in AI-driven malware detection. Organizations should be aware of applicable data protection laws, cybersecurity regulations, and ethical guidelines governing the use of AI technologies.
   - Ethical governance: Establishing ethical governance frameworks and oversight mechanisms can help mitigate risks and ensure responsible use of AI in malware detection. Ethical review boards, internal audits, and transparency reports can provide oversight and accountability for AI-driven cybersecurity initiatives.

In summary, addressing the ethical implications of using AI in malware detection requires a holistic approach that balances the benefits of AI technologies with privacy concerns, data handling practices, fairness considerations, and regulatory compliance requirements. By prioritizing transparency, accountability, and ethical governance, organizations can leverage AI responsibly to enhance cybersecurity while respecting individual privacy rights and ethical principles.

1.9 EMERGING TECHNOLOGIES IN AI FOR MALWARE DETECTION

As the threat landscape evolves, researchers and cybersecurity professionals are exploring innovative ways to leverage emerging technologies in AI for more effective malware detection. Here are some cutting-edge technologies that hold promise in enhancing the capabilities of AI for malware detection:

- **Explainable AI (XAI):** XAI focuses on developing AI models that provide clear and understandable explanations for their decision-making processes. XAI helps in making AI-driven malware detection more transparent, interpretable, and trustworthy. It enables security analysts to understand why a particular file or activity is flagged as malicious, enhancing overall confidence in the AI system [30].
- **Generative adversarial networks (GANs):** GANs consist of two neural networks, a generator and a discriminator, that work in tandem. They are commonly used for generating data that is similar to, but not the same as, the training data. GANs can be employed to generate realistic synthetic malware samples, aiding in the training of AI models to recognize new and evolving malware variants [30].
- **Homomorphic encryption:** Homomorphic encryption allows computations to be performed on encrypted data without the need for decryption, preserving the privacy of sensitive information. By using homomorphic encryption, AI models can analyze encrypted data, such as network traffic or file content, without exposing the raw data. This enhances privacy while allowing for effective malware detection.
- **Federated learning:** Federated learning is a decentralized ML approach where models are trained across multiple devices or servers without exchanging raw data. In the context of malware detection, federated learning enables collaborative model training on distributed endpoints. This approach enhances the ability to detect localized and novel threats without centralizing sensitive information.

- **Blockchain for threat intelligence sharing:** Blockchain, a decentralized and tamper-resistant ledger, is used for secure and transparent data sharing. Blockchain can facilitate secure sharing of threat intelligence among different entities in the cybersecurity ecosystem. This ensures that threat data is trustworthy, unaltered, and shared in a timely manner.
- **Differential privacy:** Differential privacy involves adding noise or randomness to individual data points to protect privacy while still enabling meaningful analysis. Differential privacy can be applied to share aggregated information about security incidents or malware prevalence across a network without revealing specific details about individual devices or users [30].
- **Zero Trust architecture:** Zero Trust assumes that threats may exist both outside and inside a network. It requires verification from everyone trying to access resources, regardless of their location. Zero Trust principles can enhance AI-based malware detection by continuously verifying the trustworthiness of users, devices, and applications, reducing the risk of unauthorized access and potential malware spread.
- **Quantum computing for cryptographic analysis:** Quantum computing has the potential to break certain cryptographic algorithms currently used in cybersecurity. While quantum computing poses a potential threat to existing cryptographic methods, it also opens opportunities for new, quantum-resistant encryption techniques that can enhance the security of communication and data storage in the context of malware detection.
- **Swarm intelligence:** Inspired by the collective behavior of social insects, swarm intelligence involves the collaboration of multiple agents to solve complex problems. Swarm intelligence algorithms can be used to create dynamic and adaptive defense mechanisms where AI agents collaborate to identify and neutralize malware threats in real time.
- **Edge AI for endpoint security:** Edge AI involves processing data locally on endpoint devices, reducing the reliance on centralized servers. Edge AI enables real-time analysis of data on individual devices, allowing for faster response to potential threats. It reduces the need for constant communication with centralized servers, improving efficiency and reducing latency in malware detection (Table 1.2).

Table 1.2 provides a comprehensive overview of the advantages, challenges, and implementation considerations associated with each technology, aiding in informed decision-making for leveraging these technologies in AI-driven malware detection and prevention strategies.

1.10 CONCLUSION

This research illustrates the critical role of AI in fortifying cybersecurity against the escalating threat of malware. By leveraging ML algorithms and deep learning methodologies, this study effectively demonstrates the adaptability and efficacy of AI in detecting known malware signatures and identifying emerging threats. The integration of these AI-driven strategies creates a dynamic defense framework that complements traditional cybersecurity measures. The ever-changing nature of cyber threats requires a continuous process of updating defensive strategies and leveraging cutting-edge technologies to ensure resilience against the latest malware tactics. The research highlights the importance of continuous innovation in cybersecurity to stay ahead of cybercriminals who deploy increasingly sophisticated and elusive forms of malware. The findings emphasize the necessity of proactive measures,

Table 1.2 Overview of emerging technologies in cybersecurity

Technology	*Advantages*	*Challenges*	*Implementation considerations*
Explainable AI (XAI)	• Provides clear and understandable explanations	• Complexity in implementing XAI models	• Ensure interpretability without sacrificing performance
Generative adversarial networks (GANs)	• Generates realistic synthetic malware samples	• Training instability and mode collapse	• Implement robust training techniques to mitigate adversarial attacks
Homomorphic encryption	• Allows computations on encrypted data	• High computational overhead	• Optimize encryption schemes for efficiency and scalability
Federated learning	• Preserves data privacy and security	• Synchronization and communication overhead	• Implement efficient communication protocols for federated learning
Blockchain for threat intelligence sharing	• Provides secure and transparent data sharing	• Scalability and performance limitations	• Explore scalability solutions like sharding and sidechains
Differential privacy	• Protects privacy while enabling meaningful analysis	• Balance between privacy protection and data utility	• Tune privacy parameters to achieve the desired trade-off
Zero Trust architecture	• Enhances security by continuous verification	• Implementation complexity and compatibility with existing infrastructure	• Adopt gradual deployment strategies to minimize disruption
Quantum computing for cryptographic analysis	• Potential for quantum-resistant encryption	• Current limitations in scalability and error rates of quantum computers	• Monitor advancements in quantum computing for practical applications
Swarm intelligence	• Dynamic and adaptive defense mechanisms	• Complexity in modeling and coordinating swarm behavior	• Design lightweight and scalable algorithms for swarm coordination
Edge AI for endpoint security	• Enables real-time analysis on endpoint devices	• Resource constraints on edge devices, potential security vulnerabilities	• Implement robust security measures for edge devices

regular threat intelligence analysis, and system updates to mitigate the impact of emerging cyber threats. The research concludes that continuous innovation and adaptation in cybersecurity are imperative, recognizing the ongoing nature of the fight against cyber threats, demanding vigilance and agility.

REFERENCES

1. Maleh, Y., Shojafar, M., Alazab, M., & Baddi, Y. (Eds.). (2021). *Machine Intelligence and Big Data Analytics for Cybersecurity Applications*. Cham, Switzerland: Springer.
2. Kok, S., Abdullah, A., Jhanjhi, N., & Supramaniam, M. (2019). Ransomware, threat and detection techniques: A review. *International Journal of Computer Science and Network Security*, 19(2), 136.

3. Maleh, Y., Sahid, A., & Belaissaoui, M. (2021). Optimized machine learning techniques for IoT 6LoWPAN cyber attacks detection. In *Proceedings of the 12th International Conference on Soft Computing and Pattern Recognition (SoCPaR 2020)* 12 (pp. 669–677). Springer International Publishing. DOI:10.1007/978-3-030-73689-7_64
4. Sedjelmaci, H., Guenab, F., Senouci, S. M., Moustafa, H., Liu, J., & Han, S. (2020). Cyber security based on artificial intelligence for cyber-physical systems. *IEEE Network*, 34(3), 6–7.
5. Ray, A., & Nath, A. (2016). Introduction to malware and malware analysis: A brief overview. *International Journal*, 4(10).
6. Pachhala, N., Jothilakshmi, S., & Battula, B. P. (2021, October). A comprehensive survey on identification of malware types and malware classification using machine learning techniques. In *2021 2nd International Conference on Smart Electronics and Communication (ICOSEC)* (pp. 1207–1214). IEEE. DOI:10.1109/ICOSEC51865.2021.9591763
7. Tahir, R. (2018). A study on malware and malware detection techniques. *International Journal of Education and Management Engineering*, 8(2), 20.
8. Khan, I. U. K. U., Ouaissa, M., Ouaissa, M., Abou El Houda, Z., & Ijaz, M. F. (Eds.). (2024). *Cyber Security for Next-Generation Computing Technologies*. Boca Raton, FL: CRC Press.
9. Tekouabou, S. C. K., Maleh, Y., & Nayyar, A. (2022). Towards to intelligent routing for DTN protocols using machine learning techniques. *Simulation Modelling Practice and Theory*, 117, 102475.
10. Basit, A., Zafar, M., Liu, X., Javed, A. R., Jalil, Z., & Kifayat, K. (2021). A comprehensive survey of AI-enabled phishing attacks detection techniques. *Telecommunication Systems*, 76, 139–154.
11. Salahdine, F., & Kaabouch, N. (2019). Social engineering attacks: A survey. *Future Internet*, 11(4), 89.
12. Conti, M., Dragoni, N., & Lesyk, V. (2016). A survey of man in the middle attacks. *IEEE Communications Surveys & Tutorials*, 18(3), 2027–2051.
13. Clarke, N., & Clarke, N. (2011). The evolving technological landscape. *Transparent User Authentication: Biometrics, RFID and Behavioural Profiling* (pp. 25–43). https://link.springer.com/book/10.1007/978-0-85729-805-8
14. Ghafir, I., & Prenosil, V. (2014). Advanced persistent threat attack detection: An overview. *International Journal of Computer Science and Network Security*, 4(4), 5054.
15. Ding, K., Zhang, S., Yu, F., & Liu, G. (2023, August). LOLWTC: A deep learning approach for detecting living off the land attacks. In *2023 IEEE 9th International Conference on Cloud Computing and Intelligent Systems (CCIS)* (pp. 176–181). IEEE. DOI:10.1109/CCIS59572.2023.10262997
16. Bilge, L., & Dumitraş, T. (2012, October). Before we knew it: An empirical study of zero-day attacks in the real world. In *Proceedings of the 2012 ACM Conference on Computer and Communications Security* (pp. 833–844). DOI:10.1145/2382196.2382284
17. Maleh, Y., Lakkineni, S., Tawalbeh, L. A., & AbdEl-Latif, A. A. (2022). Blockchain for cyber-physical systems: Challenges and applications. *Advances in Blockchain Technology for Cyber Physical Systems* (pp. 11–59). DOI:10.1007/978-3-030-93646-4
18. Tao, F., Akhtar, M. S., & Jiayuan, Z. (2021). The future of artificial intelligence in cybersecurity: A comprehensive survey. *EAI Endorsed Transactions on Creative Technologies*, 8(28), e3.
19. Jain, J. (2021). Artificial intelligence in the cyber security environment. *Artificial Intelligence and Data Mining Approaches in Security Frameworks* (pp. 101–117). DOI:10.1002/9781119760429.ch6
20. Ranveer, S., & Hiray, S. (2015). Comparative analysis of feature extraction methods of malware detection. *International Journal of Computer Applications*, 120(5), 1–7.
21. Teixeira, M. A., Salman, T., Zolanvari, M., Jain, R., Meskin, N., & Samaka, M. (2018). SCADA system testbed for cybersecurity research using machine learning approach. *Future Internet*, 10(8), 76.
22. Santos, I., Nieves, J., & Bringas, P. G. (2011). Semi-supervised learning for unknown malware detection. In *International Symposium on Distributed Computing and Artificial Intelligence* (pp. 415–422). Berlin, Heidelberg: Springer.

23. Gostev, A., Zaitsev, O., Golovanov, S., & Kamluk, V. (2011). *Kaspersky Security Bulletin. Malware Evolution* 2010. Kaspersky Lab (April 2009), p. 5.
24. Comar, P. M., Liu, L., Saha, S., Tan, P. N., & Nucci, A. (2013, April). Combining supervised and unsupervised learning for zero-day malware detection. In *2013 Proceedings IEEE INFOCOM*, Turin, Italy (pp. 2022–2030). IEEE.
25. Zolkipli, M. F., & Jantan, A. (2010, September). Malware behavior analysis: Learning and understanding current malware threats. In *2010 Second International Conference on Network Applications, Protocols and Services* (pp. 218–221). IEEE. DOI:10.1109/NETAPPS.2010.46
26. Kumar, M. S., Ben-Othman, J., & Srinivasagan, K. G. (2018, June). An investigation on wannacry ransomware and its detection. In *2018 IEEE Symposium on Computers and Communications (ISCC)* (pp. 1–6). IEEE. DOI:10.1109/ISCC.2018.8538354
27. Vinayakumar, R., Alazab, M., Soman, K. P., Poornachandran, P., & Venkatraman, S. (2019). Robust intelligent malware detection using deep learning. *IEEE Access*, 7, 46717–46738.
28. Diro, A., & Chilamkurti, N. (2018). Leveraging LSTM networks for attack detection in fog-to-things communications. *IEEE Communications Magazine*, 56(9), 124–130.
29. Liu, Y., Tantithamthavorn, C., Li, L., & Liu, Y. (2022, October). Explainable AI for android malware detection: Towards understanding why the models perform so well? In *2022 IEEE 33rd International Symposium on Software Reliability Engineering (ISSRE)* (pp. 169–180). IEEE. DOI:10.1109/ISSRE55969.2022.00026
30. Gaber, M. G., Ahmed, M., & Janicke, H. (2023). Malware detection with artificial intelligence: A systematic literature review. *ACM Computing Surveys, 56*, Article 148.

Chapter 2

Advancing malware classification with hybrid deep learning

A comprehensive analysis using DenseNet and LSTM

Chougdali Khalid and Rabii El Hakouni

2.1 INTRODUCTION

The persistent and escalating threat of malware attacks presents an ongoing challenge in the realm of cybersecurity. Despite continuous efforts by the cybersecurity industry to combat malware, malicious hackers relentlessly devise sophisticated evasion techniques. These techniques, including polymorphism, metamorphism, and code obfuscation, effectively enable malware to bypass conventional mitigation systems. Among the various malware types employed by attackers targeting businesses, backdoors, dialers, trojans, and password stealers (PWSs) emerge as the most prevalent threats. The development of effective malware detection systems constitutes an ongoing battle characterized by continuous advancements and countermeasures between attackers and security analysts.

Malware analysis can be broadly categorized into two primary types: static analysis and dynamic analysis. Each method offers distinct advantages and insights into understanding and detecting malware. Static analysis involves a thorough examination of the code and structure of a malware sample without its execution. This approach focuses on extracting features such as opcodes (assembly instructions), system calls, API calls, strings, and metadata from the binary or source code. It aids in identifying patterns, signatures, and characteristics of the malware, which are then employed for detection and classification purposes. Automated tools like disassemblers and decompilers assist in conducting static analysis. Conversely, dynamic analysis entails the execution of the malware sample within a controlled environment, commonly referred to as a sandbox, while observing its behavior. This method captures the runtime activities of the malware, including file system modifications, network communications, registry changes, process and thread interactions, and system-level operations. By monitoring the behavior of the malware, analysts can gain insights into its intentions, capabilities, and potential impact on the system. Dynamic analysis allows for the detection of evasive techniques employed by malware, such as anti-analysis and anti-debugging measures. Both static and dynamic analysis approaches have their respective strengths and weaknesses. Static analysis provides a rapid and preliminary understanding of the malware's characteristics without the need for execution, making it suitable for large-scale scanning and signature-based detection. However, it may encounter challenges when dealing with obfuscated or polymorphic malware that employs sophisticated code obfuscation techniques. In contrast, dynamic analysis offers a more comprehensive analysis by observing the actual behavior of the malware. It can capture complex and evasive techniques that may go unnoticed during static analysis. However, dynamic analysis can be resource-intensive, time-consuming, and may not always detect certain malware that remains dormant or requires specific conditions to activate. By combining static and dynamic analysis techniques, analysts can enhance their ability to detect and analyze

DOI: 10.1201/9781032714806-3

malware effectively. This hybrid approach allows for a more comprehensive understanding of the malware's functionality, behavior, and potential impact, leading to improved threat detection, incident response, and mitigation strategies. Visualizing malware as grayscale images presents a valuable strategy for harnessing the power of deep learning convolutional neural networks (CNNs). By representing malware binaries as images, we can leverage the capabilities of CNNs, which excel in extracting intricate patterns and features from visual data. Through grayscale representations, we can effectively capture the structural and textural characteristics inherent in malware samples. Notably, images from the same malware family often exhibit striking similarities in terms of layout and texture. This inherent visual similarity forms the basis of our proposed classification method. A key advantage of our approach is that it eliminates the need for disassembly or code execution during the classification process. Instead, we leverage standard image features extracted from the visual representation of malware. This simplifies the analysis process and reduces the complexity and potential risks associated with code-based approaches.

In the work by Igor et al. [1], a straightforward yet potent methodology is presented for the visualization and classification of malware, employing image processing methodologies. This approach entails the representation of malware binaries as grayscale images, considering the insightful observation that many malware families share a consistent layout and texture. This commonality gives rise to striking visual patterns within images belonging to the same malware family. Figure 2.1 displays the visual representations of six types of malware taken from the Malimg dataset.

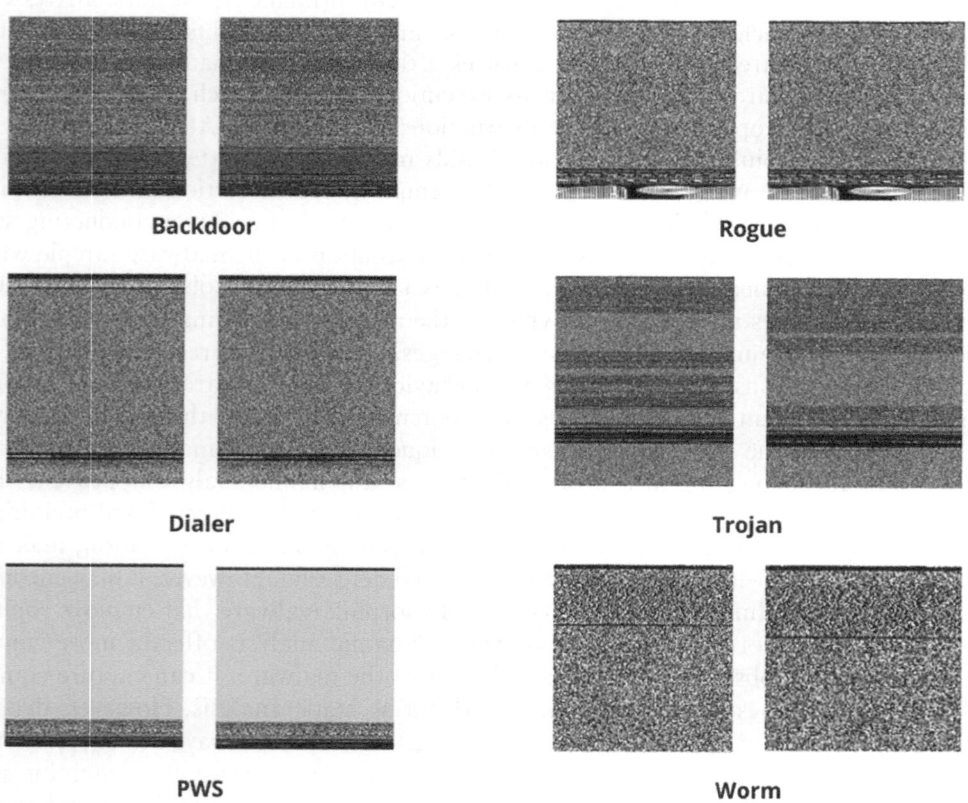

Figure 2.1 Malimg dataset.

2.2 BACKGROUND AND RELATED WORKS

2.2.1 Malware detection

2.2.1.1 Static malware analysis

The technique of static analysis represents a valuable method for the detection of malware by scrutinizing the observable characteristics of a binary file without its execution. This methodology encompasses various aspects, including the assessment of file entropy, the application of n-gram analysis, the inspection of imports and API calls, the analysis of embedded strings, and the examination of header information. It offers distinct advantages such as speed and minimal resource utilization, facilitating the swift identification of potentially malicious files without the necessity for execution.

An alternative static analysis approach in the realm of malware detection revolves around the search for a file's signature or hash. While highly effective when the hash is known and well-documented, this approach exhibits limitations in identifying modified or newly developed malware strains. Conversely, the examination of n-grams derived from byte sequences within the binary file has gained prominence as an effective static analysis technique. By scrutinizing the frequencies and patterns inherent in these sequences, even modified malware instances can be reliably detected. Consequently, n-gram analysis has become a widely embraced method in the field of malware detection.

In the research outlined in [1], the authors propose an innovative strategy for the identification of previously unknown malware families based on the frequency of opcode sequences. Their methodology entails the utilization of a support vector machine (SVM) classifier to ascertain the significance of individual opcodes and assess their frequency within malware samples. The SVM classifier undergoes training on a dataset comprising opcode sequences extracted from known malware families, enabling the classification of unknown malware executables. It is essential to recognize that this approach may not be as effective in detecting zero-day malware instances, which exploit vulnerabilities that were previously undisclosed. Furthermore, the authors of [2] implemented six distinct machine learning classifiers and leveraged two categories of features extracted from executable files: strings and portable executable header information. They conducted a comprehensive comparison of these classifiers with regard to detection accuracy and processing time.

2.2.1.2 Dynamic malware analysis

Dynamic analysis presents a distinctive and proactive approach when compared with static analysis, involving the active execution of binary files while closely monitoring their behavior. Typically conducted within a sandboxed environment, this technique ensures the isolation of the binary file, thereby mitigating potential threats to other system resources. Through the real-time observation of malware's execution behavior, researchers gather invaluable data that offers deeper insights into the intentions and attributes of suspicious binaries. Dynamic analysis facilitates a comprehensive comprehension of the binary's operations and aids in uncovering potential threats it may pose.

In reference to the research outlined in [3], dynamic analysis played a pivotal role in extracting the functionality of malware. This process involved the execution of malware within a controlled environment, with a vigilant focus on its behavior. The authors employed a behavior tracker and analyzer based on debugging techniques to extract dynamic features. In addition, IDA Pro, a decompiler and debugger, was utilized to capture instruction traces from executable files. The resulting dynamic feature set was subsequently employed to train Hidden Markov Models (HMMs) [4] for the purpose of malware detection. The authors noted that a fully dynamic approach often yielded the most optimal detection rates, underscoring the efficacy of dynamic analysis in the realm of malware detection.

Moreover, another work referenced in [5] introduces an innovative method for the detection and categorization of zero-day malware through data mining techniques that emphasize the frequency of Windows API calls. This proposed approach attains a notably high true positive (TP) rate and a low false positive (FP) rate, signifying a substantial advancement in zero-day malware detection.

In addition, the research presented in [6] puts forth a technique for detecting packed malware variants employing a sensitive system call-based principal component-initialized multilayer neural network. This approach demonstrates effectiveness in achieving both high accuracy and speed in classification. It proves to be on par with state-of-the-art methods, achieving a detection accuracy exceeding 95.6% with a minimal classification time cost of 0.048 seconds.

2.2.1.3 Deep learning

Deep learning techniques have undergone extensive exploration in various domains, including malware detection. They possess a remarkable ability to capture intricate and abstract data representations through multiple layers of abstraction. Within the realm of malware detection, neural networks have emerged as a promising departure from conventional machine learning methods [7]. Their advantages, such as incremental learning and the flexibility to train specific layers as needed, facilitate the development of automated and generalized models capable of proficiently identifying and classifying both known and unknown malware instances. The relentless advancements in deep learning algorithms have substantially improved the precision and efficiency of malware detection, thereby empowering the mitigation of ever-evolving digital threats.

In a related study [8], researchers investigate the application of deep learning algorithms, particularly the ResNet neural network, for detecting and categorizing malware. This research aims to leverage deep learning architectures to enhance the performance of malware detection systems.

Another research endeavor [9] introduces a hybrid approach that combines deep learning and visualization techniques to achieve effective malware detection. By utilizing image-based methods to identify suspicious system behaviors and incorporating these techniques into deep learning architectures, this approach enhances the accuracy of malware classification [10].

In addition, Wang et al., as cited in [11], introduce an innovative technique that significantly improves malware classification. They propose a reweighted class-balanced loss function within the final classification layer of the DenseNet model, effectively addressing the challenges posed by imbalanced data distributions.

In the context of the study [12], researchers present a novel method for detecting various variants of malicious code using deep learning. Their approach involves converting malicious code into grayscale images and utilizing a CNN for precise identification and categorization. To tackle the issue of imbalanced data distribution among various malware families, they incorporate a bat algorithm, effectively mitigating this challenge. Furthermore, the paper introduces a data equalization technique rooted in the bat algorithm and integrates data augmentation strategies. The authors provide comprehensive experimental evaluations of their proposed methodologies, demonstrating the efficacy of their approach in the detection and classification of malicious code variants.

2.3 THE PROPOSED METHODOLOGY

In this section, we present our improved malware detection method based on DenseNet+LSTM. Our approach involves mapping malware as grayscale images and utilizing the DenseNet+LSTM design for grayscale image detection. Figure 2.2 provides an overview of these two processes. First, we transform the binary files of malware into grayscale images. Then, we employ DenseNet+LSTM to identify and classify these images. Based on the results of image classification, we achieve automatic recognition and classification of malware.

2.3.1 Converting binary data into grayscale images

In general, there exists a variety of methods to transform binary code into visual images. The paper referenced as [13] explores a technique designed to visualize executable malware binary files by converting them into grayscale representations.

The procedure begins by segmenting the malware binary bit string into multiple substrings. In our research, we employ a specific method for converting binary data into grayscale images.

The conversion process entails breaking down the malware binary bit string into substrings, each composed of 8 bits. Each of these substrings can be equated to a pixel since the 8 bits can be expressed as an unsigned integer ranging from 0 to 255. For instance, consider the bit string 1011011100111010; its substrings are 10110111 and 00111010. When these

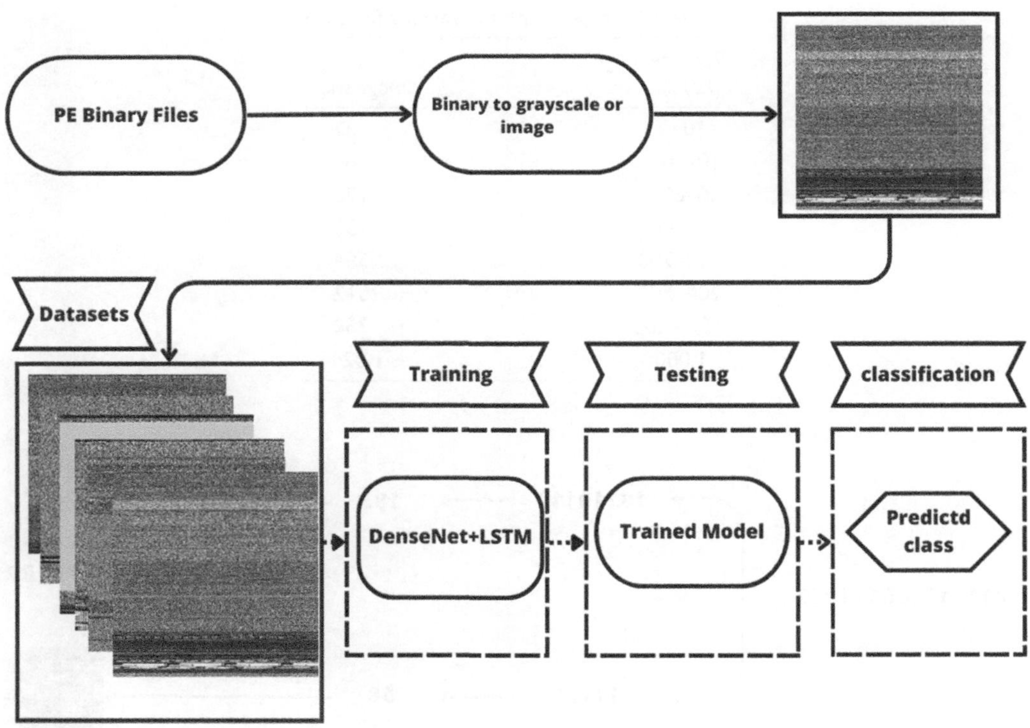

Figure 2.2 Structural diagram of the proposed model.

substrings are converted to unsigned integers, they become 183 and 58, respectively. These numerical values represent distinct colors. Figure 2.4 provides an illustrative overview of this conversion process.

Following the binary conversion, the malware's binary bit string can be further transformed into a one-dimensional array of decimal numbers, serving as a representation in grayscale. This resulting one-dimensional array can then be interpreted as a two-dimensional matrix with a specific width. Consequently, the malicious code matrix is perceived as a grayscale image. The width of this image varies depending on the file size, as outlined in Table 2.1, sourced from [14], which provides recommended image widths for different file sizes based on empirical observations.

The height can be calculated using the formula in Equation 2.1:

$$\text{height}=\text{int}(\text{size}/\text{width})+1 \tag{2.1}$$

where size is the file size and width is the corresponding image width. By dividing the total size of the binary file by the width, we obtain the number of complete rows in the image. Adding 1 accounts for the possibility of an incomplete row at the end of the image.

The images presented in Figure 2.3 (see Figure 2.4) illustrate converted representations of various types of malware. It is clear that images belonging to the same malware type share visual similarities while also being noticeably different from images associated with other types of malware. Leveraging this visual consistency among malware images, we have employed image recognition techniques for the purposes of classifying and detecting

Table 2.1 Image width for various file sizes

File size range (kB)	*Image width*
<10	32
10–30	64
30–60	128
60–100	256
100–200	384
200–500	512
500–1,000	768
>1,000	1,024

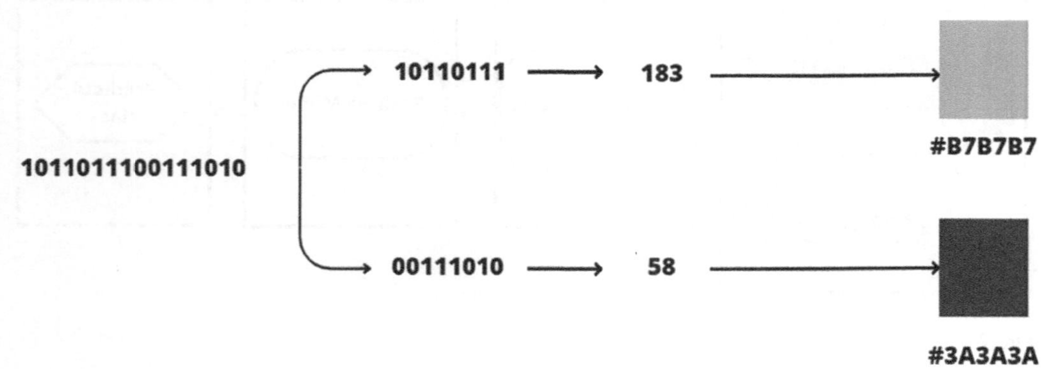

Figure 2.3 Binary to grayscale conversion.

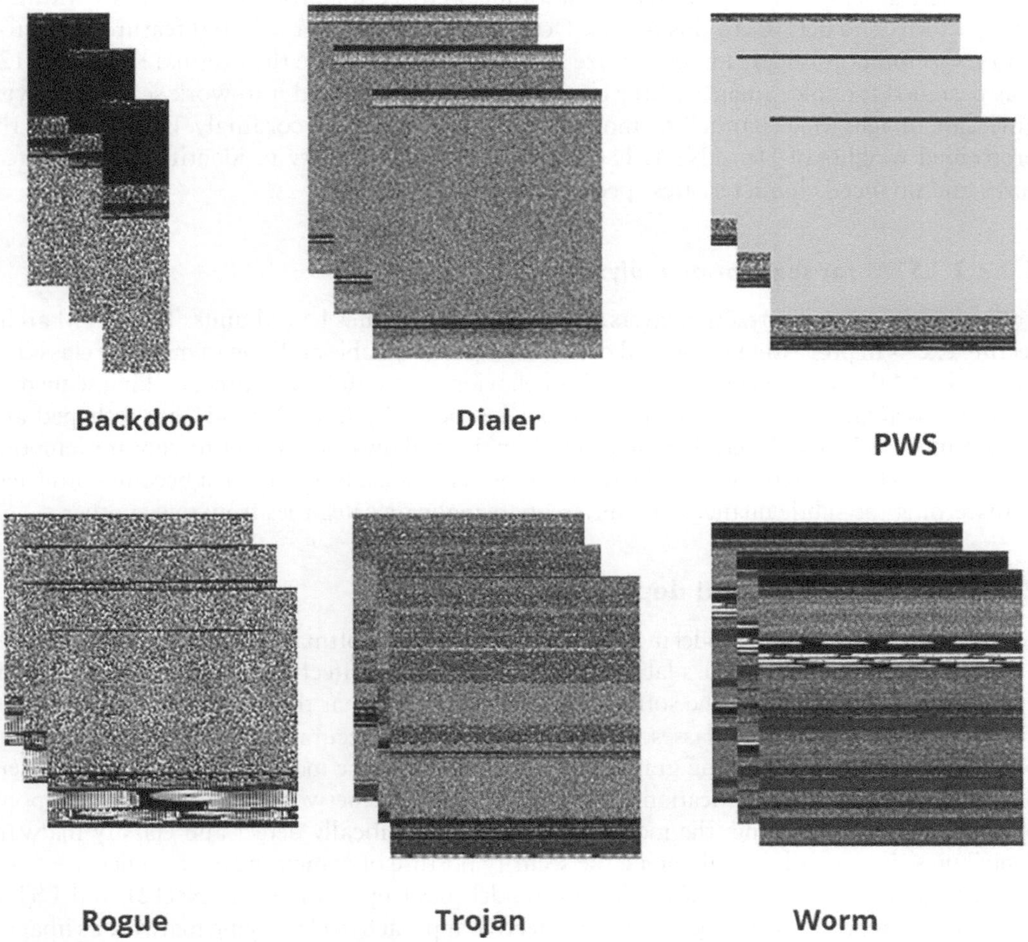

Figure 2.4 Illustration of malware as grayscale images.

malware. In our present research, we have utilized the sophisticated DenseNet+LSTM model as our classifier. This approach helps us prevent plagiarism by rephrasing the original text while retaining its essential meaning.

2.3.2 Malware detection model using DenseNet 121 and LSTM

This section introduces a robust malware detection model that leverages the power of deep learning, combining the DenseNet121 CNN architecture with long short-term memory (LSTM) layers [15,16]. This model is designed to analyze and classify malware samples into different categories, enhancing the overall security posture of computing environments.

Our malware classification model is a sophisticated neural network architecture that leverages the strengths of two fundamental components.

2.3.2.1 DenseNet121 for feature extraction

The foundation of our model lies in the DenseNet121 architecture, a powerful CNN pre-trained on the extensive ImageNet dataset. DenseNet121 has demonstrated remarkable feature extraction capabilities, making it an ideal choice for processing images with intricate patterns and details. In this model, DenseNet121 serves as the initial feature extractor. It takes as input grayscale images of size 224×224 pixels. While the original DenseNet 121 was designed for color images (three channels), we have adapted it to work seamlessly with grayscale images (one channel) by modifying the input shape accordingly. By leveraging the pretrained weights of DenseNet121, our model gains the ability to identify high-level features and nuanced characteristics specific to malware types.

2.3.2.2 *LSTM for sequential analysis*

Following the feature extraction layers, our model incorporates LSTM units. The LSTM architecture excels in processing sequential data, making it a valuable addition to malware classification tasks. Malware often exhibits complex behaviors that unfold over time, making sequential analysis essential. The feature representations extracted by DenseNet 121 are reshaped and passed into an LSTM layer. This sequential analysis allows our model to capture temporal dependencies and patterns within the feature representations. By doing so, it becomes proficient at discerning the subtle nuances that differentiate malware categories from one another.

2.3.3 Classification and deployment

After the LSTM layer, our model includes a dense layer with softmax activation. This final layer assigns one of six possible class labels to each input image, effectively categorizing it into the appropriate malware type. The softmax activation ensures that the model's output is a probability distribution over the classes, enabling confident and accurate predictions. Once trained on a labeled dataset containing grayscale malware images, the model is ready for deployment in various cybersecurity applications. Whether it is real-time network traffic analysis, endpoint security, or email scanning, the model's ability to automatically detect and classify malware contributes significantly to enhancing the security posture of computing environments.

In summary, our malware classification model, built upon the DenseNet121 and LSTM architecture, represents an innovative and effective approach to classifying malware as images. By capitalizing on deep learning and sequential analysis, this model empowers cybersecurity professionals to identify and mitigate malicious threats in the ever-evolving landscape of cybersecurity [17]. Figure 2.5 showcases the flow of the DenseNet and LSTMs model.

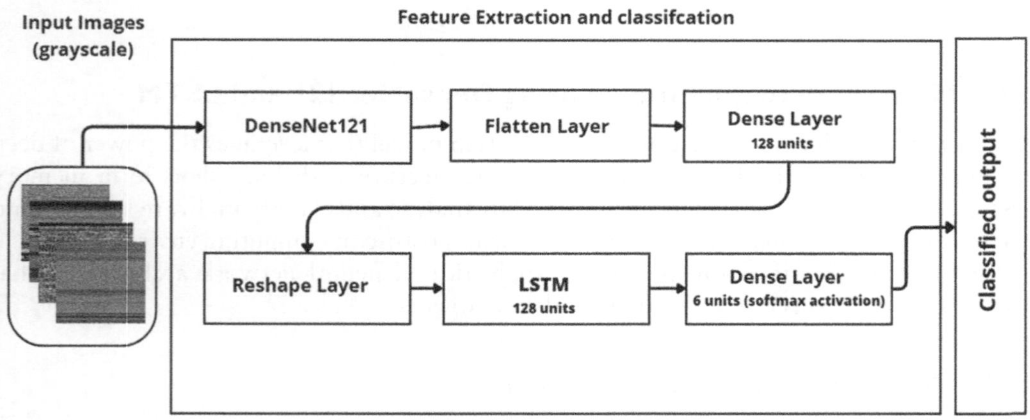

Figure 2.5 Flow of the DenseNet + LSTM model.

2.4 EXPERIMENTAL RESULTS

2.4.1 Dataset

The performance of the proposed model was assessed malware dataset named Malimg. The Malimg dataset consists of 9,339 grayscale images of malicious samples, where each sample belongs to 1 of the 25 malware classes. But in our research, we created a new dataset based on Table 2.2, we merged the families of malware by their type, and we got a dataset that contains six classes of type malware. Figure 2.6 shows the distribution of samples across classes for the new dataset.

2.4.2 Results and discussion

In this section, we delve into the outcomes of our experiments and engage in a comprehensive discussion. To begin, we partitioned the created dataset, allocating 80% for training and the remaining 20% for testing. This allocation resulted in 7,471 samples for training and 1,868 samples designated for validation. Our experimental setup was executed on a Windows system equipped with an Intel(R) Core(TM) i7-8750H CPU @ 2.20 GHz, 16 GB of RAM, and an NVIDIA GeForce RTX 2060 with 6,144 MB of dedicated memory.

Table 2.2 Malimg dataset – malware families and types

Class ID	*Family*	*Type*
0	Adialer.C	Dialer
1	Agent.FYI	Backdoor
2	Allaple.L	Worm
3	Allaple.A	Worm
4	Alueron.gen!J	Worm
5	Autorun.K	Worm
6	C2LOP.P	Trojan
7	C2LOP.gen!g	Trojan
8	Dialplatform.B	Dialer
9	Dontovo.A	Trojan
10	Fakerean	Rogue
11	Instantaccess	Dialer
12	Lolyda.AA1	PWS
13	Lolyda.AA2	PWS
14	Lolyda.AA3	PWS
15	Lolyda.AT	PWS
16	Malex.gen!J	Trojan
17	Obfuscator.AD	Trojan
18	Rbot!gen	Backdoor
19	Skintrim.N	Trojan
20	Swizzor.gen!E	Trojan
21	Swizzor.gen!L	Trojan
22	VB.AT	Worm
23	Wintrim.BX	Trojan
24	Yuner.A	Worm

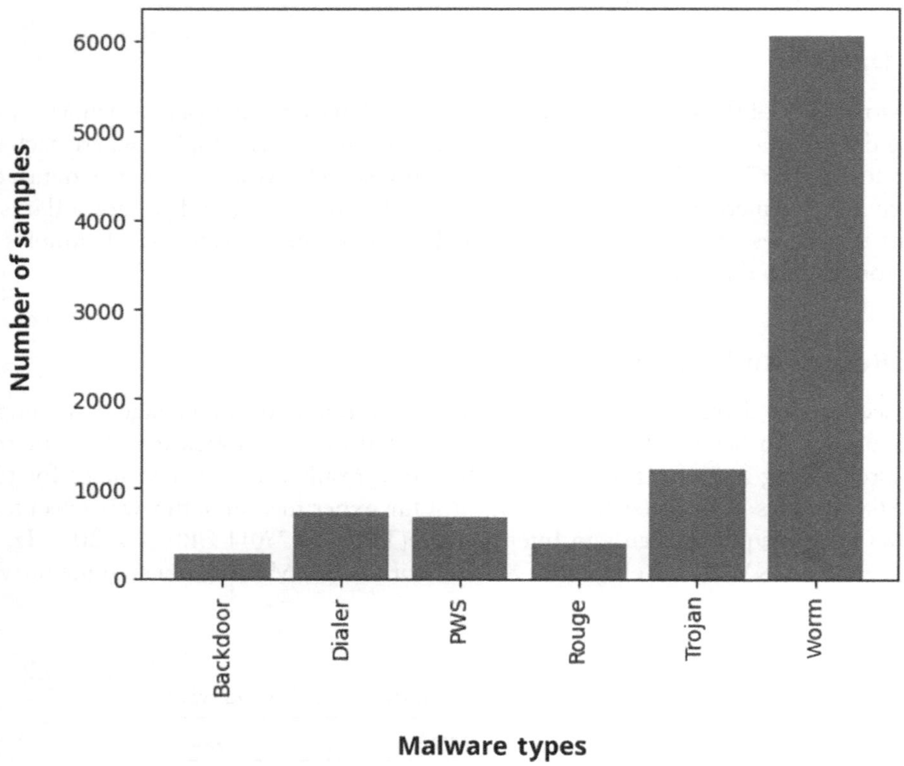

Figure 2.6 Distribution of type malware in the Malimg dataset.

The crux of our investigation revolves around the development of a machine learning model geared toward malware classification, employing advanced deep learning methodologies. Specifically, our model amalgamates the capabilities of the DenseNet121 CNN architecture and the robust LSTM layers.

We have started our experimentation by initializing our model with a pretrained DenseNet121 variant, initially primed on the ImageNet dataset for a multitude of image recognition tasks. This pretrained model forms the cornerstone for our feature extraction process. Our model's architectural configuration is sequential in nature, commencing with the DenseNet121 layer, which adeptly extracts pertinent features from input grayscale images skillfully resized to dimensions of 224×224 pixels. It is noteworthy that our model gracefully adapts to grayscale images (single channel), harmoniously accommodating the distinctive requirements of our dataset while benefiting from the knowledge ingrained in the pretrained model. Subsequent to feature extraction, we introduce a flattened layer, skillfully reshaping the data to render it amenable for further analysis. Following this, we incorporate a dense layer characterized by 128 units, significantly enhancing the model's aptitude to discern intricate high-level patterns within the data. Consecutively, we introduce a reshape layer, diligently preparing the output for the ensuing LSTM layer, which excels in scrutinizing sequential data. This LSTM layer plays a pivotal role in comprehending temporal dependencies and nuanced characteristics latent within the feature representations extracted from the malware images [18]. Last, our model culminates with a dense layer featuring a softmax activation function, enabling the multiclass classification of malware into six distinct categories. To fine-tune our model, we meticulously compile it, harnessing categorical cross-entropy as the chosen loss function and the Adam optimizer, thoughtfully configured with a learning rate of 0.001. To gauge its performance, we rely on accuracy as the principal evaluation metric.

To gain deeper insights into the efficacy of our model, we compute and analyze four key metrics integral to classification predictions: True Positive (TP), True Negative (TN), False Positive (FP), and False Negative (FN). These metrics offer a granular perspective on our model's ability to accurately classify instances within the dataset.

- **Accuracy:** The proportion of correctly classified samples to the total number of samples.

$$\text{Accuracy} = \frac{\text{TP} + \text{TN}}{\text{TP} + \text{TN} + \text{FP} + \text{FN}}$$

- **Precision:** The ratio of TPs to the sum of TPs and FPs.

$$\text{Re call} = \frac{\text{TP}}{\text{TP} + \text{FN}}$$

- **Recall:** The ratio of TPs to the sum of TPs and FNs.

$$\text{Precision} = \frac{\text{TP}}{\text{TP} + \text{FP}}$$

- **F1-Score:** The harmonic mean of precision and recall.

$$\text{F1} - \text{Score} = 2 \cdot \frac{\text{Precision} - \text{Recall}}{\text{Precision} + \text{Recall}}$$

To provide a comprehensive summary of our model's performance, we employ a confusion matrix. This matrix serves as a visual representation of our model's classification results, organizing various evaluation metrics into an $N \times N$ matrix. Here, N represents the number of classes present within our dataset. The matrix delineates true class labels along the left axis and the assigned class labels along the top axis. Each cell within the confusion matrix signifies a specific combination of true and predicted classes. The values within each cell denote the count or frequency of elements that belong to the true class on the left axis and are classified as the class indicated on the corresponding element along the top axis.

Through a meticulous examination of these values, we gain a profound understanding of our model's performance across different classes. The confusion matrix serves as the foundation for calculating critical evaluation metrics, including accuracy, precision, recall, and the F1-Score. Collectively, these metrics furnish a comprehensive assessment of our model's effectiveness in classifying diverse instances within our dataset. Table 2.3 provides a comparison of other models of deep learning on different datasets.

Moreover, Figures 2.7–2.10 illustrate the confusion matrix, ROC curve, training, and loss curve of DenseNet+LSTM on the created dataset based on Malimg, which contains six classes. As we can show, from our extensive experimentation, the proposed combination of DenseNet and LSTM model emerged as the undisputed leader in our quest for effective malware detection. This hybrid architecture exhibited unparalleled performance, boasting the lowest loss, highest accuracy, and exceptional precision, recall, and F1-Score values among all models examined.

Table 2.3 Malimg dataset – compared models

Model	*Loss*	*Accuracy*	*Precision*	*Recall*	*F1-Score*
DenseNet	0.0117	0.9967	0.9943	0.9991	0.9967
VGG16	0.0162	0.9957	0.9959	0.9920	0.9939
VGG19	0.0163	0.9967	0.9949	0.9950	0.9949
AlexNet	0.0732	0.9801	0.9755	0.9694	0.9724
ResNet-50	0.0422	0.9866	0.9808	0.9875	0.9841
SqueezeNet	0.1524	0.9646	0.9610	0.9518	0.9564
VGG16+LSTM	0.1782	0.9930	0.9863	0.9875	0.9869
DenseNet+LSTM	0.0077	0.9973	0.9953	0.9965	0.9959

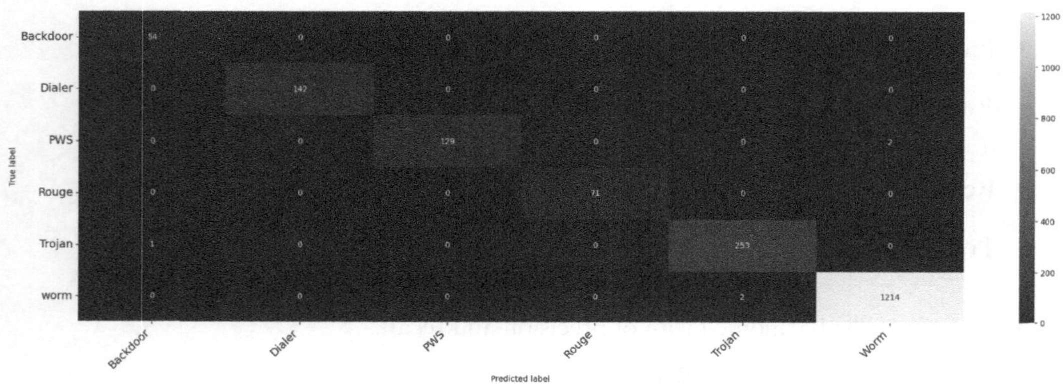

Figure 2.7 Confusion matrix – Malimg (six classes) – DensNet+LSTM.

2.5 CONCLUSION

In this comprehensive research on malware detection utilizing deep learning techniques, a diverse range of models were meticulously explored and evaluated. The study encompassed well-established architectures, including DenseNet, VGG16, VGG19, AlexNet, ResNet-50, SqueezeNet, VGG16+LSTM, and DenseNet+LSTM, all applied to the challenging task of classifying malware into six distinct categories.

Hence, the incorporation of LSTM in conjunction with DenseNet has proved to be particularly effective at capturing temporal dependencies and intricate patterns inherent in malware sequences. This synergy significantly enhanced the model's capacity for accurate classification of malicious samples. While the other models delivered commendable results, it is unequivocal that the fusion of DenseNet and LSTM stands as the optimal choice for robust and precise malware detection. This research underscores the pivotal role played by hybrid deep learning architectures and their potential to furnish effective solutions in the ever-evolving cybersecurity landscape.

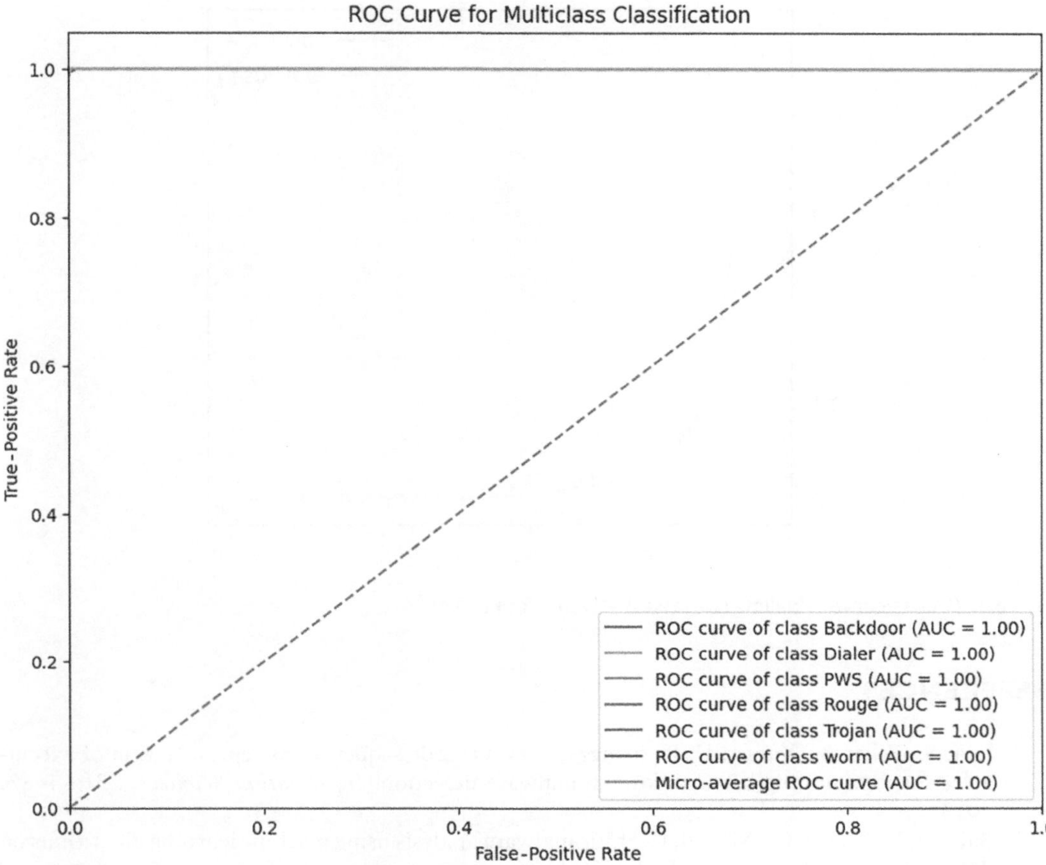

Figure 2.8 ROC curve – Malimg (six classes) – DensNet + LSTM.

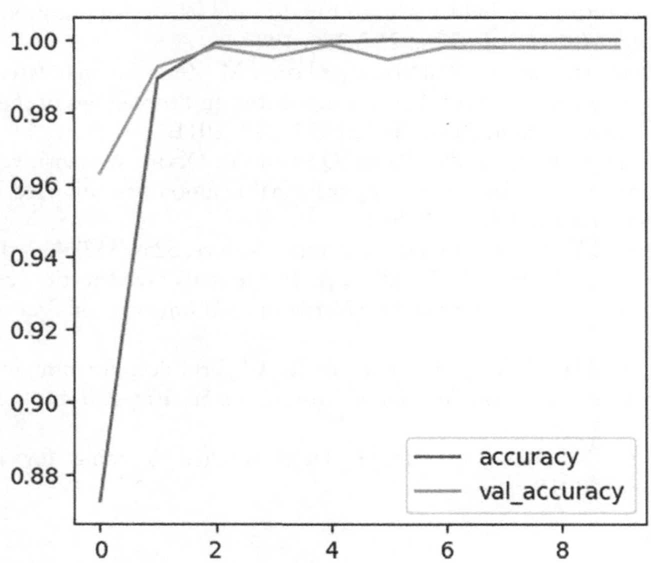

Figure 2.9 Train curve – Malimg (six classes) – DensNet + LSTM.

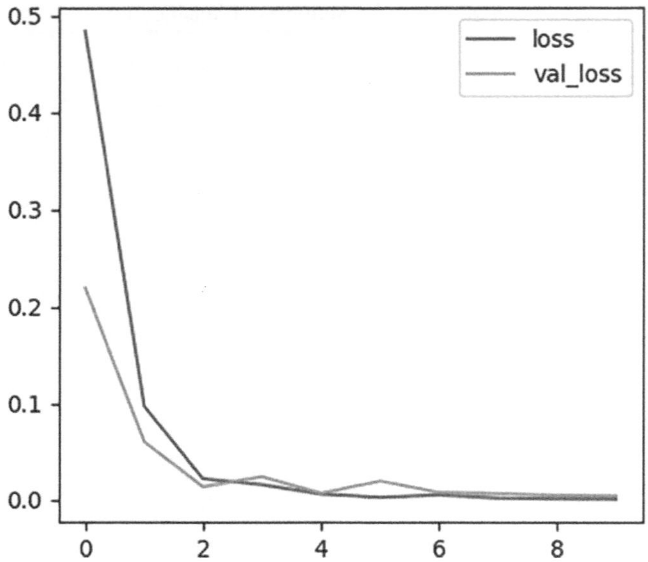

Figure 2.10 Loss curve – Malimg (six lasses) – DensNet+LSTM.

REFERENCES

1. Igor, S., Felix, B., Xabier, U.-P., Bringas, P. G. Opcode sequences as representation of executables for data-mining-based unknown malware detection. *Information Sciences*, 231:64–82, 2013.
2. Balram, N., Hsieh, G., McFall, C. Static malware analysis using machine learning algorithms on APT1 dataset with string and PE header features. *International Conference on Computational Science and Computational Intelligence (CSCI)*, Las Vegas, NV, USA, pp. 90–95, 2019.
3. Shijo, P.V., Salim, A. Integrated static and dynamic analysis for malware detection. *Procedia Computer Science*, 46:804–811, 2015.
4. Rabiner, L. R. A tutorial on hidden Markov models and selected applications in speech recognition. *Proceedings of the IEEE*, 77(2):257–286, 1989.
5. Alazab, M., Venkatraman, S., Watters, P., Alazab, M. Zero-day malware detection based on supervised learning algorithms of API call signatures. In *Proceedings of the Australasian Data Mining Conference*, Ballarat, Australia 121:171–182, 2011.
6. Zhang, J., Zhang, J., Kehuan, Z., Zheng, Q., Hui, Y., Qixin, W. Sensitive system calls based packed malware variants detection using principal component initialized MultiLayers neural networks. *Cybersecurity*, 1(1):1–13, 2018.
7. LeCun, Y., Bengio, Y., Hinton, G. Deep learning. *Nature*, 521(7553):436–444, 2015.
8. Abdelkhalki, J. E., Ahmed, M., Boudhir, A. Image malware detection using deep learning. *International Journal of Communication Networks and Information Security (IJCNIS)*, 12(2): 180–189, 2020.
9. Venkatraman, S., Alazab, M., Vinayakumar, R. A hybrid deep learning image-based analysis for effective malware detection. *Journal of Information Security and Applications*, 47:377–389, 2019.
10. He, K., Zhang, X., Ren, S., Sun, J. Deep residual learning for image recognition. arXiv:1512.03385, 2015.

11. Hemalatha, J., Roseline, S. A., Geetha, S., Kadry, S., Damas̆evic̆ius, R. An efficient densenet-based deep learning model for malware detection. *MDPI Entropy Journal*, 23(3):344, 2021.
12. Cui, Z., Xue, F., Cai, X., Cao, Y., Wang, G.-G., Chen, J. Detection of malicious code variants based on deep learning. *IEEE Transactions on Industrial Informatics*, 14(7):3187–3196, 2018.
13. Bozkir, A. S., Cankaya, A. O., Aydos, M. Utilization and comparision of convolutional neural networks in malware recognition. *27th Signal Processing and Communications Applications Conference (SIU)*, Sivas, Turkey, 1–4, 2019.
14. Nataraj, L., Karthikeyan, S., Jacob, G., Manjunath, B. S. Malware images: Visualization and automatic classification. *Proceedings of the 8th International Symposium on Visualization for Cyber Security*, Pittsburgh Pennsylvania, USA, 1–7, 2011.
15. Jocher, G., Chaurasia, A., and Qiu, J. YOLO by ultralytics. https://github.com/ultralytics/ultralytics, 2023.
16. Bochkovskiy, A., Wang, C.-Y., Liao, H.-Y. M. YOLOv4: Optimal speed and accuracy of object detection. arXiv:2004.10934, 2020.
17. Terven, J. R., Cordova-Esparaza, D. M. A comprehensive review of YOLO: From YOLOv1 and beyond. arxiv:2304.00501, 2023.
18. Ultralytics: https://github.com/ultralytics/ultralytics/blob/main/ultralytics/cfg/models/v8/yolov8-cls.yaml

Chapter 3

A comprehensive overview of AI-driven behavioral analysis for security in Internet of Things

Hicham Yzzogh, Hiba Kandil, and Hafssa Benaboud

3.1 INTRODUCTION

IoT encompasses a wide range of devices, from smart home appliances and wearable gadgets to industrial sensors and critical infrastructure components. The diverse nature of IoT devices introduces various vulnerabilities, and potential threats include unauthorized access, data breaches, device manipulation, and denial-of-service attacks. Some key points regarding cybersecurity in IoT are below:

- **Authentication and authorization:** Weak credentials and inadequate access controls can lead to unauthorized access to IoT devices and networks.
- **Data encryption:** Encrypting data in transit and at rest helps protect sensitive information exchanged between devices and stored on IoT platforms.

3.1.1 A comprehensive overview of AI-driven behavioral analysis for security in IoT

- **Firmware and software security:** Regular updates to device firmware and software are crucial for patching security vulnerabilities. Manufacturers and users should prioritize security in the development and maintenance processes.
- **Network security:** Securing IoT networks involves implementing robust firewalls, intrusion detection systems (IDS), and network segmentation to isolate critical devices from potential threats.
- **Privacy concerns:** The vast amount of data generated by IoT devices raises privacy concerns. Implementing privacy-by-design principles and complying with regulations help address these issues.
- **Standards and regulations:** Adhering to cybersecurity standards and regulations helps establish a baseline for secure IoT implementations.
- **End-user awareness:** Educating end-users about cybersecurity best practices, such as updating device passwords, recognizing phishing attempts, and understanding device permissions, is crucial.
- **Continuous monitoring:** Continuous monitoring of IoT devices and networks is essential for identifying and responding to evolving cybersecurity threats.
- **Behavioral analysis:** AI-driven behavioral analysis plays a significant role in identifying anomalies in device behavior, helping detect potential security threats by analyzing patterns and deviations.

This chapter explores the final aspect and behavioral analysis in IoT, and offers an overview of this concept. It is structured as follows. Section 3.2 delves into Behavioral Analysis in IoT, discussing

DOI: 10.1201/9781032714806-4

the understanding of Device Behavior in IoT Networks and the Importance of Behavioral Analysis for Cybersecurity in IoT. Section 3.3 explores the Applications of Behavioral Analysis by AI in IoT, focusing on two key subsections: Anomaly Detection and Predictive Analysis. In Section 3.4, we give a comprehensive exploration of AI Techniques for Behavioral Analysis, structured into three key subsections: Machine Learning Algorithms, Deep Learning Models, and Federated Learning (FL) Frameworks. In addition, the section dedicates a separate segment to review Related Work in the field. Section 3.5 focuses on real-world applications of AI-driven Behavioral Analysis in IoT Security. We conclude our chapter in Section 3.5.

3.2 BEHAVIORAL ANALYSIS IN IOT

3.2.1 Understanding device behavior in IoT networks

The interconnected nature of IoT networks presents a tapestry of diverse devices, each with its unique functionalities, communication protocols, and data transmission patterns. Understanding the behavior of these devices within the network becomes paramount for effective security measures.

1. **Device interactions and communication protocols:** IoT devices communicate through various protocols exhibiting distinct interaction patterns. Understanding how devices exchange information, the frequency of communication, and the protocols they utilize form the foundation for behavioral analysis.
2. **Data patterns and transactions:** Analysis of data transactions between devices unveils crucial insights into the normal flow of information within the IoT ecosystem. Examining data patterns aids in discerning regular communication versus anomalous or potentially malicious data exchanges.
3. **Behavioral profiles of devices:** Establishing behavioral profiles for various devices based on their operational patterns, including data transmission intervals, types of exchanged data, and typical usage scenarios, is crucial in identifying potential security threats or anomalies by detecting deviations from these behavioral norms.
4. **Contextual understanding of device behavior:** Context plays a vital role in determining the legitimacy of device behavior. For instance, a sudden spike in data transmission during specific operational hours might be standard for certain devices but anomalous for others. Considering contextual information enhances the accuracy of behavioral analysis and anomaly detection.

Understanding how devices interact, communicate, and transact data within IoT networks lays the groundwork for effective behavioral analysis. It forms the basis for AI-driven systems to discern normal behavior from potential security risks or anomalies, contributing significantly to IoT security frameworks.

3.2.2 Importance of behavioral analysis for cybersecurity in IoT

The behavioral analysis in the realm of IoT cybersecurity serves as a linchpin for proactive threat detection, anomaly identification, and preemptive mitigation strategies within interconnected ecosystems.

1. **Proactive threat detection:** Traditional cybersecurity approaches often rely on signature-based detection, which might be insufficient in detecting novel threats in

IoT networks. Behavioral analysis enables the identification of anomalies or deviations from normal device behavior, signaling potential security threats even without predefined signatures.

2. **Anomaly identification and response:** Rapidly evolving attack vectors demand a dynamic defense mechanism. Behavioral analysis detects aberrations in device behavior, indicating potential attacks or compromised devices. Early anomaly identification allows for prompt responses, such as isolating compromised devices or triggering preemptive security measures.
3. **Predictive risk mitigation:** Analyzing historical behavior and patterns equips AI systems with the capability to predict potential security risks or vulnerabilities. Predictive analytics based on behavioral analysis empowers organizations to proactively fortify IoT networks against emerging threats.
4. **Reducing false positives and enhancing accuracy:** Behavioral analysis minimizes false positives by distinguishing normal variations in device behavior from actual security threats. By creating behavioral baselines and context-aware analysis, false alarms are reduced, allowing for more accurate threat identification.
5. **Continuous monitoring and adaptation:** The dynamic nature of IoT ecosystems demands continuous monitoring. Behavioral analysis, supported by AI, continuously adapts to evolving device behaviors and emerging threats. Continuous monitoring ensures that security measures remain aligned with the current threat landscape.
6. **Comprehensive security posture:** Integrating behavioral analysis with other security measures, such as access control, encryption, and vulnerability management, strengthens the overall security posture of IoT networks. It provides a comprehensive defense mechanism against diverse cyber threats targeting interconnected devices.

Behavioral analysis can constitute robust cybersecurity in IoT environments. Its ability to identify anomalies, predict threats, and enable proactive responses safeguards the integrity, confidentiality, and availability of IoT ecosystems, fostering resilience against evolving cyber threats. In the upcoming section, we will discuss AI-based behavioral analysis applied to anomaly detection, as well as the role of AI-powered predictive analysis in addressing security issues in the IoT environment.

3.3 AI-BASED BEHAVIORAL ANALYSIS IN IOT

3.3.1 Anomaly detection by AI-based behavioral analysis

This section highlights the critical role of AI-based behavioral analysis in anomaly detection within IoT environments. It emphasizes the significance of identifying abnormal behaviors and their implications for early threat detection and preemptive security measures.

3.3.1.1 Identifying abnormal behavior in devices

In IoT environments, where devices have diverse functionalities and operate within complex networks, detecting anomalies is crucial. Identifying abnormal behavior allows for preemptive actions to prevent potential security breaches or system compromises. AI-driven behavioral analysis establishes baseline patterns of normal device behavior based on historical data. Deviations from these established baselines are flagged as anomalies, indicating potential security risks or irregularities. Anomalies can manifest in various ways, such as unusual data transmission patterns, atypical access times, or unexpected device interactions. AI algorithms, trained to recognize deviations from normal behavior, flag such activities as potential anomalies for further investigation.

3.3.1.2 Early detection of security threats

In IoT ecosystems, early detection of security threats contributes to fortifying the overall security posture. It minimizes the impact of potential attacks, protecting sensitive data and ensuring the integrity of interconnected devices. AI-driven behavioral analysis systems continuously monitor IoT networks in real time. Immediate alerts are generated upon detecting anomalies, enabling security teams to investigate and respond promptly.

3.3.2 Predictive analysis

This section highlights the role of AI-powered predictive analysis in anticipating and addressing potential security issues within IoT environments. It underscores the significance of leveraging historical behavioral data to forecast and mitigate future threats, fostering a proactive security posture.

3.3.2.1 Forecasting potential security issues based on behavior patterns

In an IoT environment, predictive analysis helps organizations stay ahead of cyber threats by anticipating and preparing for potential security issues. It provides a proactive stance, allowing for preemptive measures to be taken, thereby reducing the impact of potential security incidents. AI-powered systems analyze historical behavioral patterns of devices within IoT networks. By learning from these patterns, the AI models predict potential security issues based on deviations or trends identified in the data. Predictive analysis goes beyond current anomalies to forecast potential future security threats. It identifies evolving patterns or subtle deviations that might indicate emerging threats before they materialize into actual security breaches.

3.3.2.2 Proactive threat mitigation

Proactive measures based on predictive analysis bolster the resilience of IoT ecosystems against evolving cyber threats. This proactive approach might involve updating security protocols, patching vulnerabilities, or isolating potentially compromised devices. Proactive threat mitigation strategies minimize the window of vulnerability, protecting IoT networks from potential breaches or attacks.

3.4 AI TECHNIQUES FOR BEHAVIORAL ANALYSIS

Within the realm of IoT cybersecurity, the utilization of AI techniques stands as a pivotal asset. These techniques offer innovative avenues for behavioral analysis, empowering systems to comprehend and respond to dynamic device interactions within interconnected networks. In this section, we discuss various AI techniques (Figure 3.1), such as machine learning and deep learning algorithms, along with their applications in IoT security. We provide recent research related to each technique in this field.

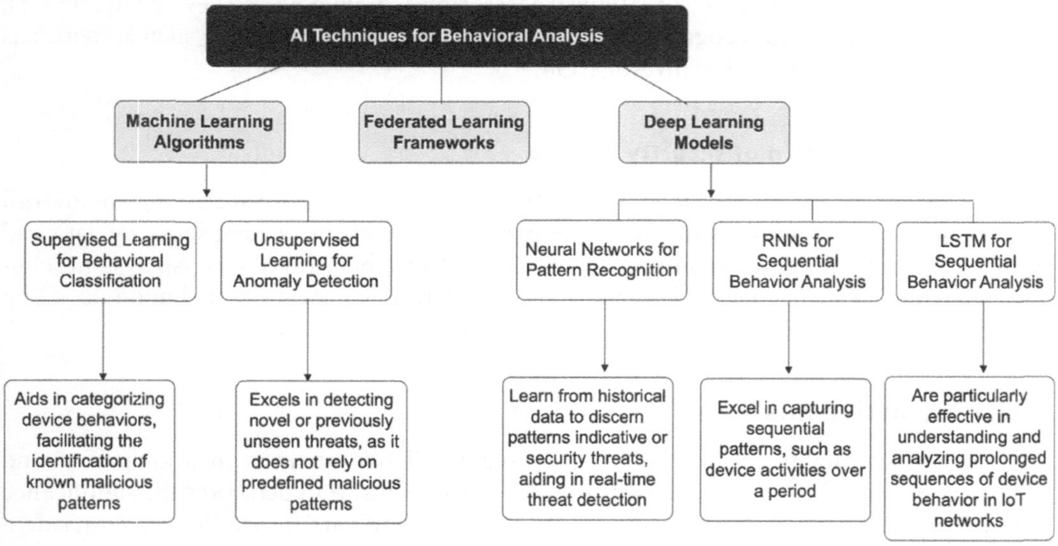

Figure 3.1 AI techniques for behavioral analysis.

3.4.1 Machine learning algorithms

Machine learning forms the backbone of AI-driven behavioral analysis in IoT, offering diverse methodologies to comprehend and interpret device behavior within interconnected networks.

3.4.1.1 Supervised learning for behavior classification

Supervised learning involves training models on labeled datasets, where devices' behaviors are classified into predefined categories based on known patterns. Classification models, such as decision trees, random forests, or neural networks, learn from labeled data to classify behaviors into predefined classes (e.g., normal behavior, suspicious behavior).

Devices' historical behavioral data labeled with known outcomes (normal or malicious) serve as training datasets. The model learns patterns from this data and can subsequently classify new behaviors based on the learned patterns. Evaluation metrics, such as accuracy, precision, and recall, gauge the model's effectiveness in accurately classifying device behavior.

3.4.1.2 Unsupervised learning for anomaly detection

Unsupervised learning techniques, such as clustering or autoencoders, discern patterns in unlabeled data without predefined categories. These models identify anomalies by detecting deviations from established normal behavior without prior labeled information. Unsupervised models create a representation of normal behavior based on the majority of observed data points. Any deviation or outlier from this learned normal behavior is flagged as an anomaly, potentially indicating a security threat.

3.4.1.3 Online learning

Online learning, also known as incremental learning, is a machine learning paradigm where a model is updated continuously as new data becomes available. This approach is particularly relevant in scenarios where data arrives in a stream or where the distribution of the data may change over time, which is the case in an IoT environment. Online learning is advantageous for tasks that require adaptability to changing conditions and the ability to process data in real time.

Online learning is well-suited for applications that demand real-time processing, such as intrusion detection, fraud detection, or anomaly detection in behavioral analysis. Since the model is updated continuously, it can utilize computational resources more efficiently. It doesn't need to process the entire dataset each time, making it scalable for large and evolving datasets.

3.4.2 Deep learning models

Deep learning, a subset of machine learning, leverages neural networks with multiple layers to extract intricate patterns and representations from data, making it highly effective for behavioral analysis in IoT.

3.4.2.1 Neural networks for pattern recognition

Neural networks, consisting of interconnected layers of nodes (neurons), learn complex patterns from data. Through forward and backward propagations, these networks extract hierarchical repress, aiding in understanding intricate behavior patterns.

3.4.2.2 RNNs for sequential behavior analysis

Recurrent neural networks (RNNs) are specialized neural networks designed to process sequential data by retaining memory of previous inputs. This memory capability enables RNNs to analyze temporal dependencies in sequences, making them well-suited for analyzing time-series data, prevalent in IoT contexts.

A specialized form of RNN, long short-term memory (LSTM) networks, addresses the vanishing or exploding gradient problems and are adept at capturing long-range dependencies in sequences. LSTMs are particularly effective in understanding and analyzing prolonged sequences of device behavior in IoT networks.

3.4.3 Federated learning frameworks

FL [1] is a machine learning approach designed to train models across decentralized devices or servers holding local data samples without exchanging them. Instead of sending raw data to a centralized server, where a model is trained, the model is sent to the decentralized devices. These devices then compute an update based on their local data and send only the model updates back to the central server. The server aggregates these updates to improve the global model, and this process is iteratively repeated. Behavioral analysis can be performed using FL [2]. FL enables the training of behavioral analysis models without centrally aggregating raw data. Local models are trained on individual devices, preserving privacy. Multiple entities contributing to the behavioral analysis model can do so without sharing detailed data. This collaborative approach enhances the accuracy and robustness of the model.

3.4.4 Related work

In this section, we discuss related work by grouping papers according to the AI techniques they employ.

Supervised learning models aid in categorizing device behaviors, facilitating the identification of known malicious patterns. For instance, these models can identify known malware behaviors or deviations from normal behavior patterns detected during the training phase. The proposal in [3] focuses on using Support Vector Machines (SVM) to detect anomalies within by creating a hyperplane from benign and malicious sensor activity. Unlike generic datasets, the authors use real IoT network traffic with specific attacks, achieving up to 100% accuracy within the trained network and 81% in unknown topologies. The authors in [4] introduce TRICA, a security framework designed for smart homes to ensure only legitimate users control their IoT devices through controlling apps. Utilizing User Behavior Analytics and Anomaly Detection techniques, TRICA builds a One-Class SVM model based on historical cyber and physical user activities and smart home system states. This model distinguishes anomalous commands as outliers, rejecting them, while allowing the execution of normal commands. Experiments on real-world data demonstrate TRICA's feasibility with low false accept and false reject rates, ensuring both security and user convenience. Paper [5] introduces an IDS using a novel feature set from BoT–IoT data, swiftly distinguishing benign and malicious traffic. Supervised machine learning algorithms (KNN, LR, SVM, MLP, DT, RF) validate system efficiency, where Decision Tree and Random Forest show 99.9% accuracy, yet Random Forest demonstrates better training and testing times. Metrics such as precision, recall, F1-Score, and ROC validate system performance. The paper [6] predicts different anomalies based on various dataset features, utilizing Logistic Regression and Artificial Neural Network (ANN) models. Two experimental approaches are explored due to the dataset's size of over 350,000. The first involves applying classification algorithms to the entire dataset, while the second excludes the "value" feature with data as 0 and 1. Data is split into training and test sets (75% and 25%, respectively). The ANN achieves 99.4% accuracy in the first case and 99.99% in the second, indicating potential for identifying threats in smart devices and IoT solutions to prevent attacks. The study in [7] introduces a detection framework utilizing device behavioral fingerprinting and machine learning to identify anomalies and classify various malware affecting real IoT spectrum sensors. Analyzing kernel events, the framework achieved promising results in detecting and classifying malicious behaviors, with a true positive rate of 0.88–0.90 for unseen or zero-day attacks and an F1-Score of 0.94–0.96. Testing on an infected IoT spectrum sensor demonstrated the effectiveness of both supervised and semi-supervised approaches in different network configurations.

Unsupervised learning excels in detecting novel or previously unseen threats, as it does not rely on predefined malicious patterns. Anomalies in device behavior, such as sudden spikes in data transmission or unusual access patterns, can be detected as potential security threats. Authors in [8] propose a behavior-learning anomaly detection for vulnerable environments using network-centric methods. Their unsupervised ML-based approach leverages TCP traffic predictability from IoT devices to detect various DDoS attacks in real time. By distinguishing normal and anomalous traffic with a small feature set, it allows identifying compromised endpoints despite IP spoofing. Compared to supervised ML, their unsupervised methods are easier to implement and better at detecting new attacks. The paper in [9] explores unsupervised anomaly detection in IoT's multidimensional time-series data through a GRU-based Gaussian Mixture VAE, GGM–VAE. Using GRU cells, authors uncover data correlations and employ Gaussian Mixture priors to handle multimodal data. They introduce a model selection mechanism guided by Bayesian Inference Criterion (BIC) during training. This approach effectively estimates distributions in the Gaussian Mixture latent space. Simulations across four datasets demonstrate the proposed scheme's superiority, achieving an average improvement of up to 47.88% in F1 scores compared to state-of-the-art anomaly detection methods.

In IoT, neural networks excel in recognizing patterns within device behavior, such as identifying normal data transmission patterns or distinguishing between authorized and unauthorized access attempts. These networks learn from historical data to discern patterns indicative of security threats, aiding in real-time threat detection. In the same context, RNNs excel in capturing sequential patterns, such as device activities over a period. RNNs can detect abnormalities or deviations in sequential behavior, flagging potential security threats like unexpected sequences of commands or data transmissions. The paper [10] introduces DAIMD, a dynamic analysis for IoT malware detection using a convolutional neural network (CNN) model. DAIMD performs dynamic analysis in a nested cloud environment, extracting behaviors related to memory, network, virtual file system, process, and system call [11]. By converting the analyzed behavior data into images and employing CNN, DAIMD classifies and trains on IoT malware behaviors, aiming to reduce infection damage to IoT devices. In their paper [12], the authors introduce an automated IDS tailored for Fog security against cyber attacks. The model employs multilayered RNNs specifically designed for Fog computing security, targeting proximity to end-users and IoT devices. Using a balanced version of the NSL–KDD dataset, they showcased the model's performance across various metrics, including Mathew correlation and Cohen's Kappa coefficients, revealing its stability and robustness against cyber threats. The paper [13] focuses on leveraging sequential models, proposing novel methods to extract network and application layer features for intrusion detection. Employing Text-CNN and GRU models enables treating sequential data akin to language models, extracting richer features. Results indicate deep learning's higher F1-Score, advocating for sequential model-based IDS to bolster IoT server security. Authors in paper [14] introduce IoT-DeepSense, a security detection system for IoT devices utilizing firmware virtualization and deep learning. This system captures fine-grained system behaviors, employs an LSTM-based abnormality detection approach, and achieves a 92% behavioral detection rate without significant impact on IoT device performance. Implemented independently on a behavior detection server, IoT-DeepSense ensures scalability without modifying the limited resources of IoT devices. The paper [15] presents an Anomaly Behavior Analysis Methodology utilizing ANNs for an adaptive IDS targeting compromised Fog nodes. Experimental results demonstrate the method's effectiveness in characterizing normal Fog Node behavior adaptively and detecting anomalies with high accuracy and low false alarms, showcasing its robustness against various sources of anomalies.

FL is a framework that can be applied to various machine learning models and algorithms. The choice of models and algorithms in FL depends on the specific requirements of the task and the characteristics of the data. The paper [16] discusses the challenges of AI-based behavioral fingerprinting. It proposes an FL approach to build a global behavioral model while addressing privacy concerns through secure strategies such as homomorphic encryption and Blockchain. The proposed solution, evaluated through a security analysis and experiments, demonstrates satisfactory performance with an average accuracy of 0.85 in distinguishing between normal and anomalous behaviors. In [17], the authors introduce DÏoT, an autonomous distributed system utilizing FL for detecting compromised IoT devices. DÏoT relies on device-type-specific communication profiles, autonomously learning without human intervention or labeled data. It achieves a high detection rate (95.6%) and fast response (257 ms) in identifying compromised devices, such as those affected by Mirai malware. The paper [18] proposes a decentralized FL approach for anomaly detection in IoT using neural networks, aiming to enhance accuracy while ensuring local data security. The decentralized model mitigates issues associated with traditional FL, eliminating a single point of failure. Simulation experiments on the IoT23 dataset validate the system's performance.

The paper [19] presents IoTCom, a compositional threat detection system that automatically identifies unsafe interactions between IoT apps and devices. Utilizing static analysis, IoTCom infers app behaviors and employs a unique strategy to reduce the complexity of formal analysis.

Experimental results demonstrate IoTCom's effectiveness in identifying various interaction threats across cyber and physical channels, outperforming existing techniques in computational time. The system maintains its analysis capability across different IoT platforms.

The following chapters provide reviews and surveys on recent advances in IoT security. The paper [20] reviews anomaly detection techniques using machine learning in IoT, analyzing existing literature and datasets. It identifies key dataset issues and outlines future research directions. The paper [21] presents a comprehensive survey on the applications of deep learning in enhancing security for IoT, examining system architecture, methodologies, and performance evaluation. The paper [22] offers an overview of physical security for IoT equipment, highlighting research opportunities. Among the solutions covered in this paper, behavior analysis is included.

3.5 APPLICATIONS OF AI-DRIVEN BEHAVIORAL ANALYSIS IN IOT SECURITY

This section illustrates the diverse real-world applications and the effectiveness of AI-driven behavioral analysis in securing IoT environments across various sectors. We summarize these applications in Table 3.1 and provide recent research references for each. These applications include IDSs, Predictive Maintenance in Industrial IoT, Healthcare IoT Security, Smart Grid Security, Connected Vehicle Security, Smart Home, and also Smart City Security.

Table 3.1 Applications of AI-driven behavioral analysis in IoT security

Real-word applications	*Description*	*References*
Intrusion detection systems (IDS)	IDS leveraging AI-driven behavioral analysis monitors network traffic, flagging suspicious activities or anomalies in real time. For instance, anomaly-based IDS can identify unusual data patterns or access attempts across IoT devices in smart buildings, alerting security teams of potential breaches.	[17,23,24]
Predictive maintenance in Industrial IoT (IIoT)	AI-powered predictive maintenance systems analyze equipment behavior, detecting anomalies indicative of potential breakdowns or malfunctions. By monitoring changes in device behavior, these systems forecast equipment failures, allowing preemptive maintenance, reducing downtime, and improving operational efficiency.	[25]
Healthcare IoT security	AI-based behavioral analysis in healthcare IoT ensures the security and integrity of medical devices and patient data. Behavioral analysis systems detect abnormal patterns in device behavior, preventing unauthorized access or tampering with critical medical equipment.	[26]
Smart grid security	AI-driven behavioral analysis in smart grids identifies anomalous behavior, such as unusual power consumption patterns or cyber attacks targeting energy infrastructure. Early detection allows for prompt responses, safeguarding the integrity and reliability of the power grid.	[27]

(Continued)

Table 3.1 (Continued) Applications of AI-driven behavioral analysis in IoT security

Real-word applications	*Description*	*References*
Connected vehicle security	Behavioral analysis in connected vehicles detects deviations from standard driving behavior or anomalous communications within vehicular networks. Identifying potential cyber threats enables proactive security measures to protect vehicle systems from unauthorized access or manipulations.	[28]
Smart home security	AI-based behavioral analysis in smart homes monitors device interactions, detecting unusual behaviors indicating potential security risks, such as unauthorized access attempts or irregular device activities. Swift responses to anomalies safeguard home automation systems and user privacy.	[4,29]
Smart city security	AI-based behavioral analysis in smart city security involves leveraging artificial intelligence techniques to analyze the behavior of interconnected devices, sensors, and systems within urban environments. This approach aims to enhance security, efficiency, and responsiveness across various aspects of city operations.	[30]

3.6 CONCLUSION

The exploration of AI-driven Behavioral Analysis emerges as a pivotal paradigm, with a specific focus on enhancing IoT security through anomaly detection and predictive analysis. The comprehensive examination of AI techniques, including machine learning algorithms, deep learning models, and FL frameworks, highlights the multifaceted approach needed to address the evolving challenges in securing IoT ecosystems. By reviewing related work, this chapter contributes to the ongoing discourse in the field, emphasizing the importance of continuous research and innovation to fortify the integrity, confidentiality, and resilience of IoT environments against emerging cybersecurity threats.

REFERENCES

1. C. Zhang, Y. Xie, H. Bai, B. Yu, W. Li, and Y. Gao, "A survey on federated learning", *Knowledge-Based Systems*, 216, 2021, 106775, https://doi.org/10.1016/j.knosys.2021.106775
2. B. McMahan, E. Moore, D. Ramage, S. Hampson, and B. Aguera y Arcas, "Communication-efficient learning of deep networks from decentralized data", *Artificial Intelligence and Statistics*, Fort Lauderdale, Florida, USA. PMLR, 2017, pp. 1273–1282.
3. C. Ioannou and V. Vassiliou, "Classifying security attacks in IoT networks using supervised learning", *2019 15th International Conference on Distributed Computing in Sensor Systems (DCOSS)*, Santorini, Greece, 2019, pp. 652–658, https://doi.org/10.1109/DCOSS.2019.00118
4. N. Amraoui, A. Besrour, R. Ksantini, and B. Zouari, "Securing smart homes using a behavior analysis based authentication approach", *2020 IEEE Eighth International Conference on Communications and Networking (ComNet)*, Hammamet, Tunisia, 2020, pp. 1–5, https://doi.org/10.1109/ComNet47917.2020.9306081.
5. T. Himani and K. Rajendra, "Attack and anomaly detection in IoT networks using supervised machine learning approaches", *Revue d'Intelligence Artificielle*, 35(1), 2021, 11–21, https://doi.org/10.18280/ria.350102

6. N. K. Sahu and I. Mukherjee, “Machine learning based anomaly detection for IoT network: (Anomaly detection in IoT network)”, *2020 4th International Conference on Trends in Electronics and Informatics (ICOEI)(48184)*, Tirunelveli, India, 2020, pp. 787–794, https://doi.org/10.1109/ICOEI48184.2020.9142921
7. A. H. Celdrán, P. M. S. Sánchez, M. A. Castillo, G. Bovet, G. M. Pérez, and B. Stiller, “Intelligent and behavioral-based detection of malware in IoT spectrum sensors”, *International Journal of Information Security*, 22, 2023, 541–561, https://doi.org/10.1007/s10207-022-00602-w
8. R. Bhatia, S. Benno, J. Esteban, T. V. Lakshman, and J. Grogan, “Unsupervised machine learning for network-centric anomaly detection in IoT”, *Big-DAMA’19: Proceedings of the 3rd ACM CoNEXT Workshop on Big DAta*, December 2012, pp. 42–48, https://doi.org/10.1145/3359992.3366641
9. Y. Guo, T. Ji, Q. Wang, L. Yu, G. Min, and P. Li, “Unsupervised anomaly detection in IoT systems for smart cities”, *IEEE Transactions on Network Science and Engineering*, 7(4), 2020, 2231–2242, https://doi.org/10.1109/TNSE.2020.3027543
10. J. Jeon, J. H. Park, and Y.-S. Jeong, “Dynamic analysis for IoT malware detection with convolution neural network model”, *IEEE Access*, 8, 2020, 96899–96911, https://doi.org/10.1109/ACCESS.2020.2995887
11. S. M. Ali, A. S. Elameer, and M. M. Jaber, “IoT network security using autoencoder deep neural network and channel access algorithm”, *Journal of Intelligent Systems*, 31(1), 2022, 95–103, https://doi.org/10.1515/jisys-2021-0173
12. M. Almiani, A. AbuGhazleh, A. Al-Rahayfeh, S. Atiewi, and A. Razaque, “Deep recurrent neural net- work for IoT intrusion detection system”, *Simulation Modelling Practice and Theory*, 101, 2020, 102031, https://doi.org/10.1016/j.simpat.2019.102031
13. M. Zhong, Y. Zhou, and G. Chen, “Sequential model based intrusion detection system for IoT servers using deep learning methods”, *Sensors*, 21(4), 2021, 1113, https://doi.org/10.3390/s21041113
14. J. Wang, C. Liu, J. Xu, J. Wang, S. Hao, W. Yi, and J. Zhong, “IoT-DeepSense: Behavioral security detection of IoT devices based on firmware virtualization and deep learning”, *Security and Communication Networks*, 2022, https://doi.org/10.1155/2022/1443978
15. J. Pacheco, V. H. Benitez, L. C. Félix-Herrán, and P. Satam, “Artificial neural networks-based intrusion de- tection system for internet of things fog nodes”, *IEEE Access*, 8, 2020, 73907–73918, https://doi.org/10.1109/ACCESS.2020.2988055
16. M. Arazzi, S. Nicolazzo, and A. Nocera, “A fully privacy-preserving solution for anomaly detection in IoT using federated learning and homomorphic encryption”, *Information Systems Frontiers*, 2023, https://doi.org/10.1007/s10796-023-10443-0
17. T. D. Nguyen, S. Marchal, M. Miettinen, H. Fereidooni, N. Asokan, and A.-R. Sadeghi, “DÏoT: A federated self-learning anomaly detection system for IoT”, *2019 IEEE 39th International Conference on Distributed Computing Systems (ICDCS)*, Dallas, TX, 2019, pp. 756–767, https://doi.org/10.1109/ICDCS.2019.00080
18. Z. Lian and C. Su, “Decentralized federated learning for internet of things anomaly detection”, *Proceedings of the 2022 ACM on Asia Conference on Computer and Communications Security*, 2022, pp. 1249–1251, https://doi.org/10.1145/3488932.3527285
19. M. Alhanahnah, C. Stevens, B. Chen, Q. Yan, and H. Bagheri, “IoTCom: Dissecting interaction threats in IoT systems”, *IEEE Transactions on Software Engineering*, 49(4), 2023, 1523–1539, https://doi.org/10.1109/TSE.2022.3179294
20. N. Alghanmi, R. Alotaibi, and S. M. Buhari, “Machine learning approaches for anomaly detection in IoT: An overview and future research directions”, *Wireless Personal Communications*, 122, 2022, 2309–2324. https://doi.org/10.1007/s11277-021-08994-z
21. Y. Yue, S. Li, P. Legg and Fuzhong Li, “Deep learning-based security behaviour analysis in IoT environments: A survey”, *Security and Communication Networks*, 2021, 2021, https://doi.org/10.1155/2021/8873195
22. X. Yang, L. Shu, Y. Liu, G. P. Hancke, M. A. Ferrag, and K. Huang, “Physical security and safety of IoT equipment: A survey of recent advances and opportunities”, *IEEE Transactions on Industrial Informatics*, 18(7), 2022, 4319–4330, https://doi.org/10.1109/TII.2022.3141408

23. S. Ahn, H. Yi, Y. Lee, W. R. Ha, G. Kim, and Y. Paek, "Hawkware: Network intrusion detection based on behavior analysis with ANNs on an IoT device", *2020 57th ACM/IEEE Design Automation Conference (DAC)*, San Francisco, CA, 2020, pp. 1–6, https://doi.org/10.1109/DAC18072.2020.9218559
24. F. Abusafat, T. Pereira, H. Santos, "Proposing a behavior-based IDS model for IoT environment", In Wrycza, S., Mas´lankowski, J. (eds) *Information Systems: Research, Development, Applications, Educa- tion. SIGSAND/PLAIS 2018*. Lecture Notes in Business Information Processing, vol. 333, 2018. Springer, Cham, https://doi.org/10.1007/978-3-030-00060-8_9
25. N. Chander and M. Upendra Kumar, "Metaheuristic feature selection with deep learning enabled cascaded recurrent neural network for anomaly detection in Industrial Internet of Things environment", *Cluster Computing*, 26, 2023, 1801–1819. https://doi.org/10.1007/s10586-022-03719-8
26. A. Manocha, G. Kumar, M. Bhatia, et al., "IoT-inspired machine learning-assisted sedentary behav- ior analysis in smart healthcare industry", *Journal of Ambient Intelligence and Humanized Computing*, 14, 2023, 5179–5192, https://doi.org/10.1007/s12652-021-03371-x
27. A. Habtamu and B. Svetlana, "Anticipatory adaptive security for IoT-based smart grids infrastructure and value-added services", *Research Report*, 248113/O70, NR–notat / Norsk Regnesentral-notater (NR Notes), 2020, https://hdl.handle.net/11250/2732254
28. B. Huber and F. Kandah, "Behavioral model based trust management design for IoT at scale", *2020 International Conferences on Internet of Things (iThings) and IEEE Green Computing and Communications (GreenCom) and IEEE Cyber, Physical and Social Computing (CPSCom) and IEEE Smart Data (SmartData) and IEEE Congress on Cybermatics (Cybermatics)*, Rhodes, Greece, 2020, pp. 9–17, https://doi.org/10.1109/iThings-GreenCom-CPSCom-SmartData-Cybermatics50389.2020.00022
29. I. Priyadarshini, A. Alkhayyat, A. Gehlot, and R. Kumar, "Time series analysis and anomaly detection for trustworthy smart homes", *Computers and Electrical Engineering*, 102, 2022, 108193, https://doi.org/10.1016/j.compeleceng.2022.108193
30. N. Amraoui and B. Zouari, "Anomalous behavior detection-based approach for authenticating smart home system users", *International Journal of Information Security*, 21, 2022, 611–636. https://doi.org/10.1007/s10207-021-00571-6

Chapter 4

A deep dive into IoT security

Machine learning solutions and research perspectives

Ahmad Anwar Zainuddin, Muhammad Ziyad Fathullah Mohd Yazid, Nur Alya Aqilah Razak Ratne, Nur Fatnin Izzati Sidik, Nur Adila Ahmad Faizul, Aliah Maisarah Roslee, and Nuramiratul Aisyah Ruzaidi

4.1 INTRODUCTION

4.1.1 Critical analysis of security challenges in Internet of Things (IoT) systems with a focus on MQTT protocol

It has been discovered that many weaknesses in IoT systems expose them to several cyberattacks [1]. Apart from the potential loss of important information, there is also a threat from additional security concerns, including privacy, confidentiality, and accessibility. It has been stated that most of the attacks target cheap IoT devices with weak resources [2].

The security of the Internet will become a problem due to so many devices and so much data connected to it. This is demonstrated by the enormous expansion of IoT gadgets in our residences and everyday lives in recent years. Technology is rapidly progressing, and many end devices are linked to the Internet.

This total is thought to go up a lot in the next little time and be more than it is now [1]. This shows a significant chance for hackers to exploit these devices by sending malicious emails, launching "denial-of-service" attacks, and installing other dangerous worms or Trojan horses.

This study uses the message queue telemetry transport protocol (MQTT) protocol as an example to address the problems related to embedded systems and the IoT since it is exposed to many forms of attack [3]. MQTT protocol enables the publish/subscribe concept and supports low-bandwidth data transmission, authentication, communication, and termination [4]. The problem happens when the MQTT protocol gets requests from close nodes in the same network section, especially on networks where security checks have not been done correctly.

4.1.2 Enhancing IoT security through advanced ML techniques: a comprehensive analysis and application of high-level security-based intelligent systems

Machine learning (ML) has greatly improved in the last several years as artificial intelligence has evolved from a lab curiosity to a useful device that serves many important applications [1]. IoT device intelligence, which allows these devices to be monitored, provides strong defenses against new or zero-day assaults. Using strong ML techniques, the "normal" and "abnormal" behaviors of IoT components and devices in their environment are determinable [5]. Therefore, these techniques are essential for turning IoT system security into a high-level

 DOI: 10.1201/9781032714806-5

security-based intelligent system [1]. In addition, this research improves the knowledge of cybersecurity threats and weaknesses in organizations that use technology and systems.

4.1.3 Impact of IoT on technology and businesses: revolutionizing connectivity, security, and trust in the modern era

These days, the IoT has completely changed technology. The idea of the IoT concept aims to continually link people all around the world to everything [1]. It is a rapidly growing system that could change people's lives and is the next major advancement in Internet technology. Attaching embedded sensors or small electronics to common objects or things to make them smart objects or things is the basic concept of the IoT [6]. A three-layer design with physical, network, and application levels is typically utilized to define the IoT. Security protocols must be used in every part of the IoT. This helps keep things safe and stable [7].

IoT has also quickly become an invention in businesses and is becoming quite popular because of features like rapid interaction and its capacity to improve business operations. This is due to global interaction made possible by the IoT capacity to connect "things" worldwide via sensor networks and tiny systems [6]. Given this, IoT is expected to be the spark that drives subsequent technological advancements, and its usage is predicted to grow significantly in the upcoming years.

4.1.4 Methodology

The research study "IoT Security Intelligence: Leveraging Machine Learning Solution and Research Perspective" adopts a qualitative approach, conducting a comprehensive literature review from 2018 to analyze state of the art in IoT security and ML. This review forms the basis for identifying research gaps and crafting a framework for utilizing ML to enhance IoT security. With IoT rapidly permeating enterprises and promising technological advancements, ensuring embedded security becomes paramount. ML offers robust defense mechanisms by predicting normal and abnormal behaviors of IoT components, thereby elevating system security to an intelligent level.

This chapter provides a methodical approach to strategic inquiry for managing the IoT ecosystem's complexity and complex relationship to security. The idea and goals of ML, applications, difficulties, and future approaches are all covered in detail in various sections. In addition, discussed are ML and the role of the IoT, how security attacks are classified, and suggestions for improving IoT security. The study concludes with a summary and conclusion emphasizing how critical it is to handle cybersecurity threats in the constantly changing IoT world.

4.2 LITERATURE REVIEWS

Table 4.1 indicates interesting prospective uses, discusses present obstacles, and suggests future paths for the discipline.

4.3 DISCUSSION

The IoT connects various objects and has grown rapidly in recent years [10]. It links real-world items with online spaces using tools such as RFID tags, GPS devices, sensors, NFC detectors, and voice alarms for emergencies. ML is crucial for IoT security, analyzing

Table 4.1 Literature reviews

Article	*Potential applications*	*Challenges*	*Future directions*
[1]	The article delves into using machine learning (ML) to predict and counter threats in IoT, such as DDoS attacks and basic computer programs. Techniques like LR, brain-modeled networks, and MLP are employed for detection and mitigation.	For the safety of Internet of Things (IoT), like using special codes to get in, protecting online space, keeping stuff hidden with strong jumbled words and controlling who can use them, or making sure apps stay safe are not well done. Also, these weaknesses go deep inside and they are hard to fix.	IoT uses communication protocols like Wi-Fi, Zigbee, Bluetooth, and message queue telemetry transport protocol (MQTT). MQTT is widely popular for its simplicity and TCP/IP utilization. Monitoring protocol behavior is crucial for IoT network security.
[7]	The article discusses IoT's role in smart cities for sustainable urban development, emphasizing integrated technologies to monitor and control smart devices and sensor networks in city infrastructure.	Maintaining and protecting these networks with many devices talking in different ways is difficult. As many groups can use Bluetooth apps and have different user controls, it needs a safety setup that lets people trust the data and services shared through its system.	Using AI and ML for threat prediction is essential for IoT security. Enhancing privacy tools and collaborating with stakeholders is vital for establishing standardized safety protocols for broad IoT adoption.
[6]	The article highlights how integrating fog computing and IoT can revolutionize smart cities by enabling efficient traffic management, environmental monitoring, and resource optimization, ultimately enhancing urban living standards with intelligent systems.	The article addresses ensuring data privacy and security in IoT systems. Safeguarding the integrity and confidentiality of sensitive data generated by numerous IoT devices is crucial. In addition, managing the large volume of data and addressing device heterogeneity are additional challenges.	The future entails advanced fog computing for handling growing IoT data. This includes exploring distributed approaches and employing ML for data analysis at the fog layer.
[2]	Recognizing flaws in LPWAN-based IoT systems improves security. Implementing risk mitigation measures necessitates stakeholders' understanding of these flaws and potential attacks.	LPWAN are appealing for IoT applications because of their low consumption, long-range connection, and affordability.	Integrating blockchain and AI into LPWAN-based IoT systems boosts security. Collaborative efforts between industry and academia are vital for establishing standardized security protocols and ensuring safe LPWAN IoT network operation.
[3]	The study by C. Patel and N. Doshi introduces a new security framework for MQTT within the IoT model, discussing applications, challenges, and future directions for enhancing MQTT security in IoT frameworks.	Securing MQTT for diverse IoT devices is complex due to integration challenges from device heterogeneity. Preventing exploits needs careful design and continuous security updates. Ensuring scalability and efficiency in IoT environments requires minimizing resource overheads.	Future MQTT protocol enhancements target improved data integrity and confidentiality via encryption, authentication, and access control. Standardization is crucial for uniform security across IoT devices while refining MQTT security for scalability, efficiency, and resilience against cyber threats remains essential.

(*Continued*)

Table 4.1 (Continued) Literature reviews

Article	*Potential applications*	*Challenges*	*Future directions*
[4]	The research focuses on energy-efficient cryptographic algorithms for threat mitigation in IoT security. It suggests using elliptic curve cryptography to counter replay attacks in IoT networks and identifies areas for current and future research in energy-aware IoT security systems.	The complexity of defending systems and restraining IoT expansion. Ensuring reliable wireless communication is fundamental; security must cover all IoT architecture layers. New machine and deep learning-based security solutions are crucial for addressing IoT security effectively.	Merging edge computing and ML for real-time IoT traffic analysis. It stresses privacy-preserving algorithms and distributed analysis. It also recommends researching ML risks, countermeasures in IoT, and energy-efficient algorithms for constrained IoT devices.
[5]	The article explores using ML to boost IoT network security by analyzing traffic, focusing on anomaly and intrusion detection. Emphasizing traffic analysis for risk identification, it contributes to discussions on ML in IoT security.	Applying ML to IoT network traffic analysis challenges include real-time processing, resource constraints, and security concerns. Adapting models to dynamic IoT contexts adds complexity, emphasizing the need for innovative security approaches.	Integrating edge computing with ML for real-time IoT monitoring, prioritizing privacy. Further research includes studying ML threats, deploying federated learning, and developing energy-efficient IoT algorithms, driving ongoing IoT security research.
[8]	The study investigates how ML boosts IoT security, detecting anomalies and threats to safeguard sensitive data across industries like healthcare and smart homes.	Applying ML to IoT security, including acquiring labeled and diverse datasets, developing robust models, and designing efficient algorithms for resource-constrained IoT devices.	The study recommends exploring federated learning, training ML models on multiple IoT devices without sharing private data, and adapting algorithms to evolving security threats.
[10]	The study illustrates IoT's impact across sectors like automotive, construction, healthcare, and transportation, enabling real-time communication and collaboration among equipment as part of Industry 4.0.	IoT security concerns, citing vulnerabilities in device construction that make them targets for cyberattacks. Challenges include managing multiple devices, lacking security standards, and difficulty with software updates, increasing the risk of cyber assaults.	The essay advocates for a comprehensive IoT security model to address system-wide risks, aiming to advance security measures amidst rising cyberattacks targeting critical infrastructure such as nuclear power plants, steel mills, and healthcare systems.
[11]	The paper underscores the rapid growth of IoT, projecting 50 billion devices by 2020, and its impact on industries like healthcare and transportation. However, it raises security concerns, noting that traditional methods like encryption fall short in safeguarding IoT devices.	The complexity and widespread adoption of IoT lead to security challenges. Current protections are inadequate because of IoT's diverse applications and vulnerable attack surfaces. The paper stresses the necessity for enhanced cybersecurity measures against advanced threats.	The study explores merging ML and deep learning to address issues, particularly in monitoring anomalous activity and mitigating risks. It analyzes these algorithms, categorizes attack surfaces, and considers future directions, underscoring the importance of reliable IoT security solutions.

(Continued)

Table 4.1 (Continued) Literature reviews

Article	*Potential applications*	*Challenges*	*Future directions*
[14]	The paper foresees IoT's significant impacts on daily life, economy, and social interactions while noting device vulnerability to cyber threats due to resource limitations. It explores ML and DL as solutions to enhance device intelligence and security.	IoT is a network of connected systems, highlighting challenges in device management, data handling, and security. Given the network scale, it emphasizes the critical importance of security and privacy for successful IoT implementation across sectors.	The research sees ML as vital for managing IoT data and improving security and privacy. It calls for further exploration to maximize ML and DL for IoT network security.
[9]	The study highlights the importance of addressing security challenges specific to software-defined networks, particularly in detecting and mitigating DDoS attacks. It underscores the value of employing anomaly detection and real-time network traffic analysis for identifying such attacks.	Challenges in feature selection and ML for practical effectiveness, focusing on security issues in software-defined networks, especially in detecting and mitigating DDoS attacks. Key strategies include leveraging programmability, real-time anomaly detection, and tailored intrusion detection systems.	Future research could focus on advanced feature selection techniques to improve DDoS attack detection in software-defined networks, enhancing accuracy and robustness. Subsequent inquiries might prioritize real-time DDoS mitigation strategies, emphasizing dynamic and flexible defense mechanisms for prompt responses.
[21]	The article discusses multilayer self-defense systems for securing enterprise cloud environments, emphasizing comprehensive security measures and addressing cybersecurity challenges in cloud computing.	The study addresses challenges in establishing a multilayer self-defense system for enterprise cloud security, emphasizing integrated infrastructure layers, early threat detection, dynamic security adaptation, regulatory compliance, and scalability for large deployments.	Future advancements in enterprise cloud security may prioritize early risk detection and mitigation, dynamic security adaptation, and scalability improvements. Integration of emerging technologies like blockchain and AI could further enhance security.
[20]	IoT exhibits versatility in enhancing efficiency, user experiences, and addressing challenges across industries. As technology progresses, new and innovative IoT applications are expected to arise.	During pilgrimage sessions, large crowds often lead to lost items and people, causing distress and challenges for authorities. An IoT-based system for monitoring pilgrims helps identify and reconnect lost or harmed individuals, effectively managing crowd control.	The applications of IoT span various fields. IoT also exhibits significant potential for widespread application in numerous other domains.
[15]	The main objective of IoT is to provide advanced services while ensuring security. However, creating lightweight security models for IoT devices, which often have limited resources, is challenging. This vulnerability has made IoT devices prone to cyberattacks.	Security is often overlooked in IoT product launches despite rising devices and threats. Existing smart systems often lack security measures, such as automated updates and encryption in communication channels, making them vulnerable to cyberattacks.	IoT faces security and privacy challenges. Manufacturers must assess threats and device exposure carefully. Users' security efforts may fall short, revealing device vulnerabilities. Until robust defense solutions are developed, companies should strengthen attack detection mechanisms to prevent societal attacks.

data to prevent hacking attacks by detecting unusual behaviors and adapting to emerging threats. Its flexibility makes it essential for safeguarding IoT devices and networks in the evolving IoT landscape [11].

4.3.1 Roles of ML in IoT security

ML is a powerful tool for IoT security, efficiently detecting patterns and anomalies in connected devices. ML acts as a vigilant assistant, promptly assessing device behavior for abnormalities. ML uses mathematical modeling of big data to identify abnormal activities in cyberspace ahead of time [12]. In addition, ML optimizes IoT services by autonomously discovering patterns in vast data, reducing the need for human intervention in problem-solving [13]. ML enables computers and smart devices to learn from human or device-generated data, aiding in complex tasks such as robotics and voice recognition. Google utilizes ML for risk assessment in mobile phones and Android apps [11].

To reach this aim, ML and deep learning (DL) are attractive methods for IoT networks for many reasons. One reason is that these types of networks produce a huge amount of data daily, which ML and DL need to make computers smarter [14]. Moreover, using ML and DL ways with IoT data is more helpful. This lets IoT systems make smart decisions when needed. Security, privacy, finding attacks, and looking at bad software are all usual uses for ML or DL. Some of the security-related real-world uses of ML include:

- Forensic face recognition: viewpoint, light use, hiding parts (like glasses and beard), applying make-up, and hair styling.
- Different ways of writing letters for recognizing characters in security codes.

4.3.2 Limit of ML in IoT and the solution

4.3.2.1 Developing ML models for IoT networks

Developing models for IoT data processing, such as in healthcare, requires reliable training for extracting relevant features efficiently. The reliability of ML models is crucial, especially in mission-critical IoT applications affected by abnormalities from ML algorithms. Efficient tuning of ML models is necessary to address specific requirements based on deployed IoT technology and unique case features.

The solution for this limit is to pay special attention to developing lightweight models that are adapted to IoT limitations like edge computing and scarce resources. Moreover, the solution is to efficiently extract pertinent features from IoT data and improve model performance without excessive computational overhead, applying domain-specific feature engineering techniques.

4.3.2.2 Retrain of ML models

However, when ML algorithms are used for IoT applications and resource management in IoT networks, the models' need to be updated based on newly collected data. This mechanism creates many essential issues like what and how much data needs to be supplied for retraining ML models if the environment changes, and whether such a quantity is sufficient or even excessive.

The solution for retraining of ML model limitation is to create trigger-based systems that, upon detection of predetermined thresholds or appreciable shifts in the distribution of data, start the retraining of the model. Using active learning techniques to maintain model accuracy and optimize resource utilization by carefully choosing and prioritizing data samples for retraining the model is also one solution for this limitation.

4.3.3 IoT security attacks classification

Security is a serious concern with the IoT. Most major security efforts are currently focused on protecting/securing traditional servers, workstations, and smartphones, with less concern for IoT security [15]. Figure 4.1 illustrates various cybersecurity threats and where they may occur within the network.

Malicious code injection attacks: A person who wants to cause harm puts harmful code into a part of an IoT system. They do this to fully take over the IoT system [16].

Wormhole attack: An attacker routes all the traffic through a malicious tunnel between two incorrect nodes to mess up how the network and movement work [17].

Ransomware: The virus hijacks a system, locks up its data, and prevents the user from using it. The attacker demands for ransom for restoring control; otherwise, they threaten to remove all information on the computer [18].

Denial-of-Service (**DoS**)/Distributed Denial-of-Service (**DDoS**): Service denial happens when attackers try to take up user resources or bandwidth so they cannot use it properly. Attacks from many infected nodes are called DDoS [19].

Man in the Middle (**MitM**): When wrong people sneakily catch messages between two or more devices, they can secretly change, alter, or replace shared information without anyone knowing [16].

4.3.4 Application of IoT

IoT is being used in many fields, such as agriculture, environment sensing, healthcare, and home monitoring. An overview of smart living, smart products, smart health, and smart cities is given in this chapter, along with how IoT is being used to create self-awareness, autonomous devices and smart settings. IoT is becoming a key component of smart objects and items, helping to create persistent, smart cyber–physical systems. The use of IoT in

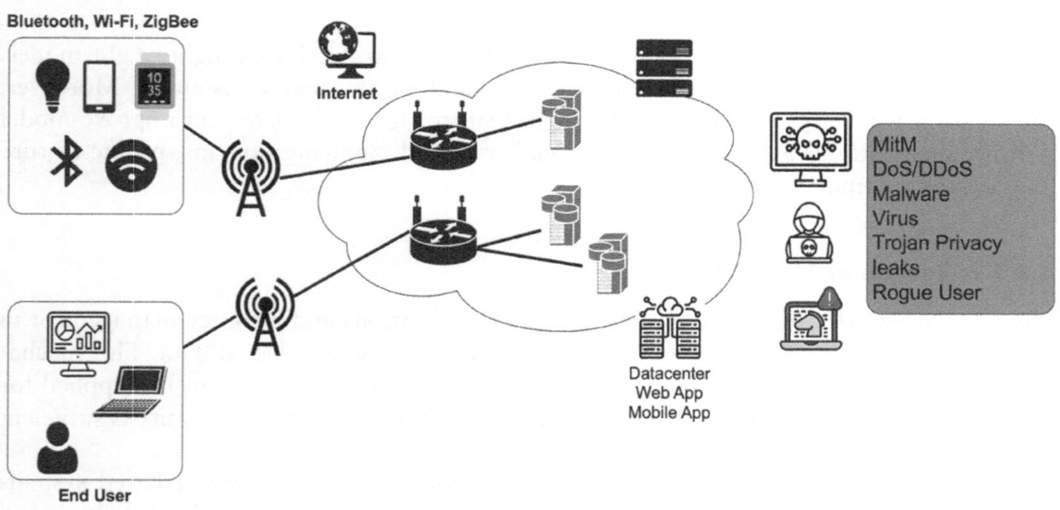

Figure 4.1 Threat model.

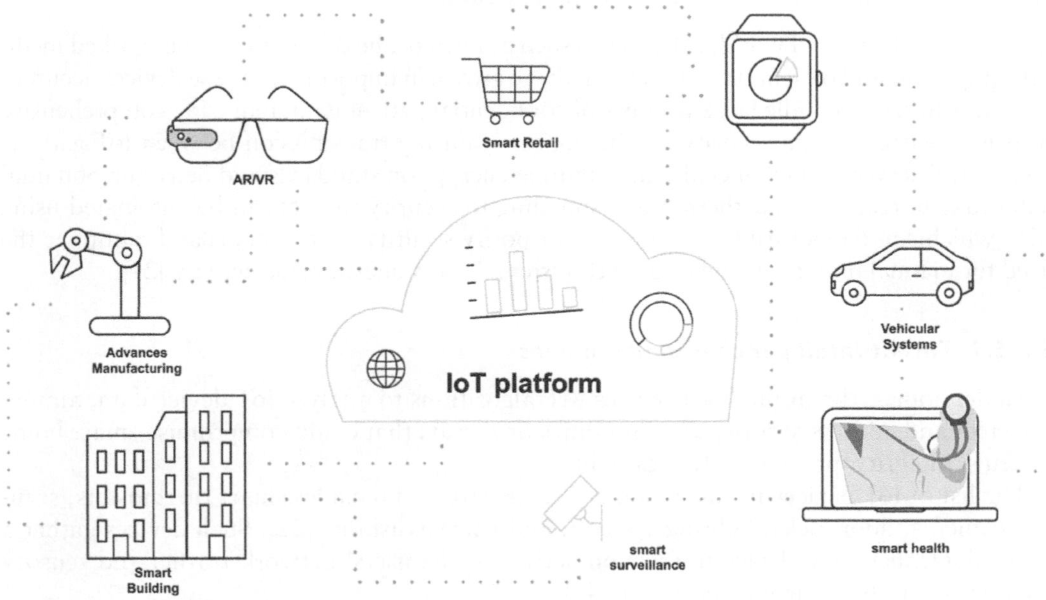

Figure 4.2 Applications of IoT.

smart living could significantly raise people's standards of living [6]. Figure 4.2 describes the applications of IoT in various sectors nowadays.

4.3.4.1 Smart hospital

IoT is essential for modernizing the hospital and enhancing its intelligence. The first step toward creating a smart hospital is underway, which includes mobile application-based inquiries and test result reporting, appointment registration via smartphone based on doctors' availability, etc. Further advancements have been made in designing medical equipment capable of transmitting real-time signals, such as blood pressure, blood sugar, and ECG, to an information platform for diagnosis. In addition, to establish their tracking, very costly healthcare equipment can have an RFID chip implanted [20].

4.3.4.2 Smart metering

Gas, electricity, and water are essential for every household, which are typically billed monthly via mail. IoT-based smart meters revolutionize the process by centrally tracking daily resource consumption. These meters collect usage data and transmit it to a central server over the Internet, where it is processed. The processed data is then made available to the users through central monitoring stations, where they can access their consumption data, receive paperless bills online, and manage electronic accounts for convenient payments.

4.3.5 Case study of ML in IoT security

ML in IoT enables devices to learn from data and make decisions without explicit programming. Through training, devices recognize patterns and relationships in gathered information, enhancing operational efficacy, flexibility, and adaptability to changing environments.

4.3.5.1 Enhancing the security of IoT in healthcare

In today's sophisticated world, IoT gadgets such as low-cost health monitors, networked medical equipment, and patient tracking systems have grown in importance. These devices accumulate, transmit, and handle large volumes of confidential patient data, requiring comprehensive security measures. Data security can be improved during transmission between IoT gadgets and central servers, which includes maintaining encryption standards and detecting potential data leaks in real time. Furthermore, responding to security threats can be automated using ML, which can temporarily restrict access or notify security professionals and minimize the need for manual monitoring when an ML system detects questionable activity [21].

4.3.5.2 Threat intelligence in smart homes

In smart homes, threat intelligence uses ML algorithms to analyze IoT device data, aiming to detect and address security vulnerabilities or threats that could compromise smart home systems' integrity, privacy, or functionality.

The list of IoT devices in smart homes is extensive and includes smart thermostats, security cameras, door locks, lighting systems, and voice assistants [22]. Such devices gather a lot of data, including device interaction activities, the users' network traffic, and sensory readings from the environment (Figure 4.3).

ML algorithms are crucial in analyzing data, particularly identifying patterns that may indicate malicious activities in smart homes. They can efficiently handle large volumes of real-time data, enabling anomaly detection during abnormal behavior [23].

Threat intelligence in smart homes involves analyzing user behavior, device interactions, and network traffic to establish regular activity [24]. Any deviations trigger alerts or automatic responses, such as detecting intrusion attempts or malware infections. ML algorithms monitor interactions to identify emerging threats, enabling preemptive actions such as alerting residents or blocking suspicious traffic to minimize risks.

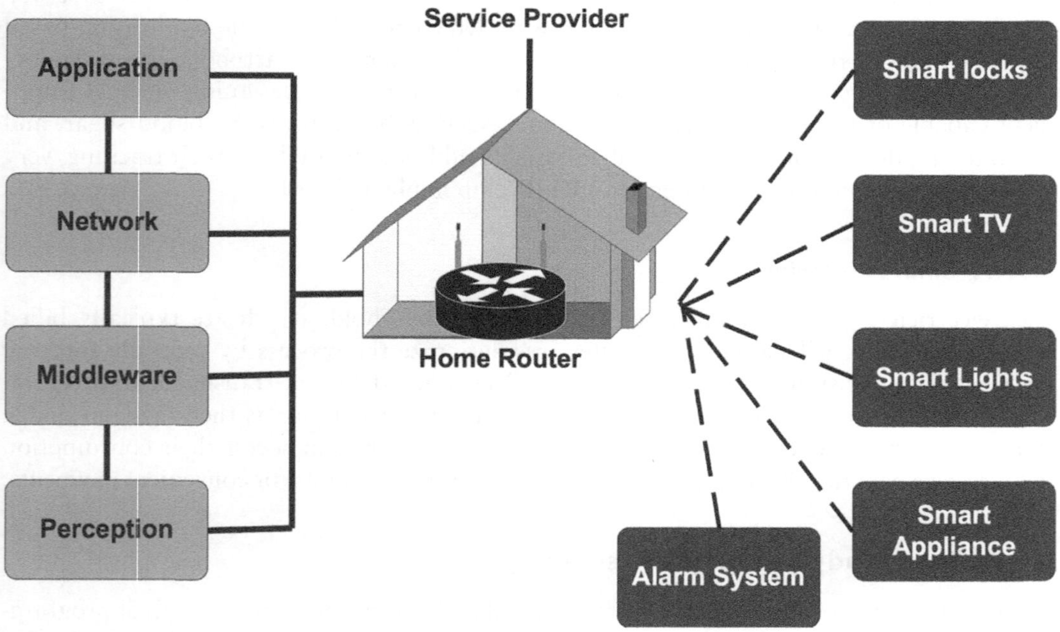

Figure 4.3 Smart home system.

4.3.6 Challenges posed by security threats to IoT

IoT aims to provide both security and advanced services, but finding lightweight security solutions is challenging due to device resource limitations. This vulnerability makes IoT devices susceptible to cyberattacks. Common threats include denial-of-service (DoS) attacks, data breaches, botnets, and malware attacks [15].

DoS attacks can occur at different layers of a network. In such attacks, the perpetrator floods the network with useless data to overwhelm its resources, preventing legitimate users from accessing it. These attacks have evolved to target sensitive data at the application level, allowing the attacker to seize control after breaching the network's defenses. This can lead to the entire network becoming inaccessible [6].

Due to inadequate security, IoT devices are prime targets for hackers. They can launch DoS attacks, exploit open ports such as FTP, SSH, or Telnet, and use brute-force attacks to access devices. With numerous connected devices in IoT (smart meters, medical equipment, etc.), malware can exploit open ports and default credentials. Over the years, IoT malware has evolved, targeting new victims with varied architectures [25].

4.3.6.1 Privacy challenges in IoT

Privacy concerns in IoT arise from the extensive data generated by IoT devices, raising fears of heightened surveillance and data collection. Protecting user privacy is vital for fostering trust in IoT devices and services. Efforts, including frameworks and legislation, aim to grant consumers control over their data, dictating its collection and sharing. However, privacy safeguards vary geographically and across private enterprises, underscoring the necessity for comprehensive legislation and governance to ensure effective privacy protection [26].

It is critical for businesses and legislators to examine the privacy implications of the IoT and to implement mechanisms that provide users more control over their data while also protecting their privacy [27].

4.3.6.2 Security risks and challenges in IoT

IoT devices and services must be secure, especially as they become more intertwined with human life. Cyberattacks and user data leakage grow more possible with less secure IoT devices and services. The usage of insecure IoT devices can have a global impact on Internet security and resilience because these devices give attackers a large and accessible attack surface [28].

In a common IoT attack, a third party hijacks a communication channel to conceal participants' identities, facilitating a man-in-the-middle scenario. This allows the attacker to intercept communication between genuine participants, potentially tricking the bank's servers into accepting the transaction as legitimate without knowing the victim's name, effectively compromising IoT device and network security [29].

4.3.7 Impact of emerging technologies on IoT security

4.3.7.1 Data privacy, confidentiality, and integrity

According to [10], IoT data must be appropriately encrypted to maintain its confidentiality as it passes over several network hops. Due to the varied integration of devices, services, and networks, device data is vulnerable to privacy breaches by breached nodes inside an IoT network. This is because IoT devices are vulnerable to hacking, an attacker may alter stored data maliciously to breach data integrity.

4.3.7.2 *Authentication and secure communication*

Key management systems are necessary for user and device authentication in IoT [30]. Any breach in the network layer or the large costs related to safeguarding communication could leave the network vulnerable to several threats. For example, the overhead of Datagram Transport Level is limited by insufficient resources. Thus, security should be minimized while using it (i.e., DTLS), and the cryptographic methods ensuring safe data transmission in IoT should also incorporate efficiency and the limited availability of resources [10].

4.4 ETHICAL IMPLICATION AND PRIVACY CONCERNS

4.4.1 Potential risk and mitigation strategies

Ethical concerns arise in IoT, demanding transparency and ethical frameworks to guide practices and adapt to diverse ecosystems. Upholding strong ethics encourages better solutions, fostering global connectivity. Prioritizing ethical IoT practices is crucial for maintaining consumer trust and data access. Meanwhile, ML detects anomalies in IoT networks, adapting to new threats with historical data for more resilient defenses against cyber threats [15].

4.4.2 Mitigation strategies

- **Privacy problems:**
  One common problem in IoT setups is privacy. Users are not entirely aware of where and how their personal information is being utilized in IoT environments, where smart devices like wearables and sensors are used to transmit data and information shared using these gadgets. Smart IoT devices save users' and clients' sensitive information, which could be exploited. All IoT devices use security protocols, such as authentication and encryption, to communicate with one another. Users are hesitant to utilize these technologies due to significant obstacles such as threats, privacy exposures, and leaks [20].
- **Data security and integrity:**
  The accuracy of ML hinges on high-quality datasets, especially in IoT network protection. Authorized training data is essential for ML-based IoT security. However, hackers may exploit vulnerabilities to manipulate these datasets, complicating safeguarding efforts. This vulnerability makes it challenging to identify diverse attacks and their frequency in IoT. In conclusion, ensuring data security and integrity remains a challenging aspect of ongoing IoT security research [20].

4.5 IMPLEMENTATION WITH SMART DOOR

4.5.1 Function of the proposed system

The primary goal of this chapter is to improve the safety of the locking door mechanism. The hardware and software requirements for the system that is proposed have been shown (Table 4.2). The mobile will be broadcasting a signal through Bluetooth to the Arduino circuit, of course, as a connection. When implemented properly, the authentication is done using the database from the smartphone to the servo motor. The use of smartphones with Bluetooth is to facilitate accessibility. More security than the regular key.

Table 4.2 The hardware and software for the proposed system

Category	*Components*	*Function*
Software	MIT App inventor	To create an Android app
	Arduino IDE	To write and upload Arduino code/sketch
	Firebase database	To store and retrieve user credentials
	Wireshark	To capture network packets

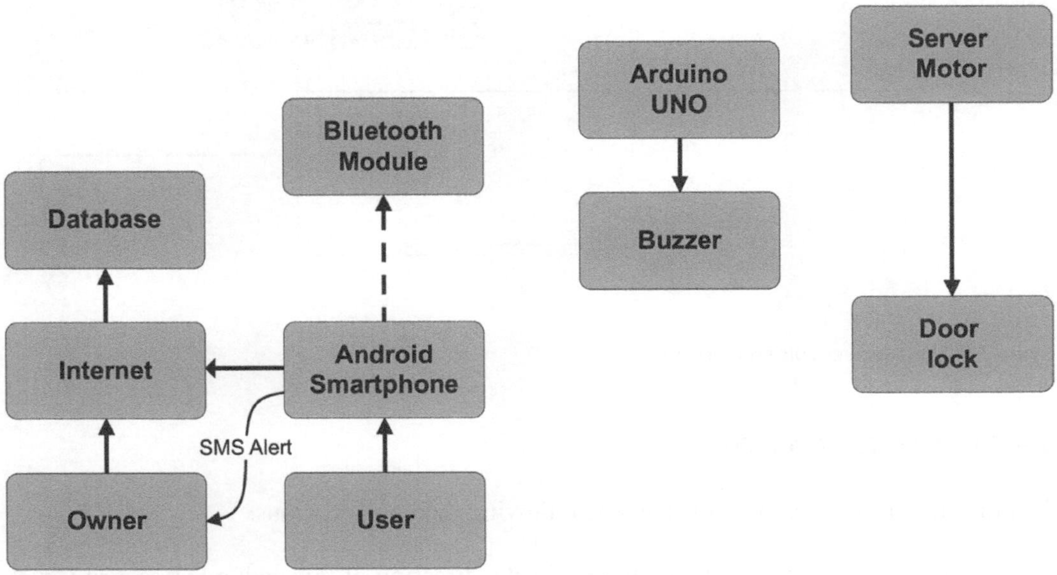

Figure 4.4 Block diagram.

4.5.2 Modeling of the proposed system

Figure 4.4 represents an automated door security block diagram controlled via their Android app. Figure 4.5 demonstrates the connections of the suggested system. Connect the board to the computer using a universal serial bus (USB) cable. The USB interface connecting to the PC is required to program the board as well, but also provides power. The UNO board connected to the USB is automatically powered and can also be powered by an external power source.

Figure 4.6 represents the flowchart model. It is a step approach that was adopted in the writing of the automatic door security development, with the execution of commands that come from the Android app that was developed.

The smart home automation system, referred to as a digital door lock, is suggested for security. The system does not need physical keys to lock or unlock the door. Instead, an Android application is loaded for smart devices, credentials, etc. The database validates the user credentials. If invalid credentials are in the application, a buzzer alarm is offered by a text message to the householder accompanied by a warning indicator notification to the user. This improves the security of the proposed method. It is a dynamic system that is easy to install cheaply with no overhead, such as drawing construction works.

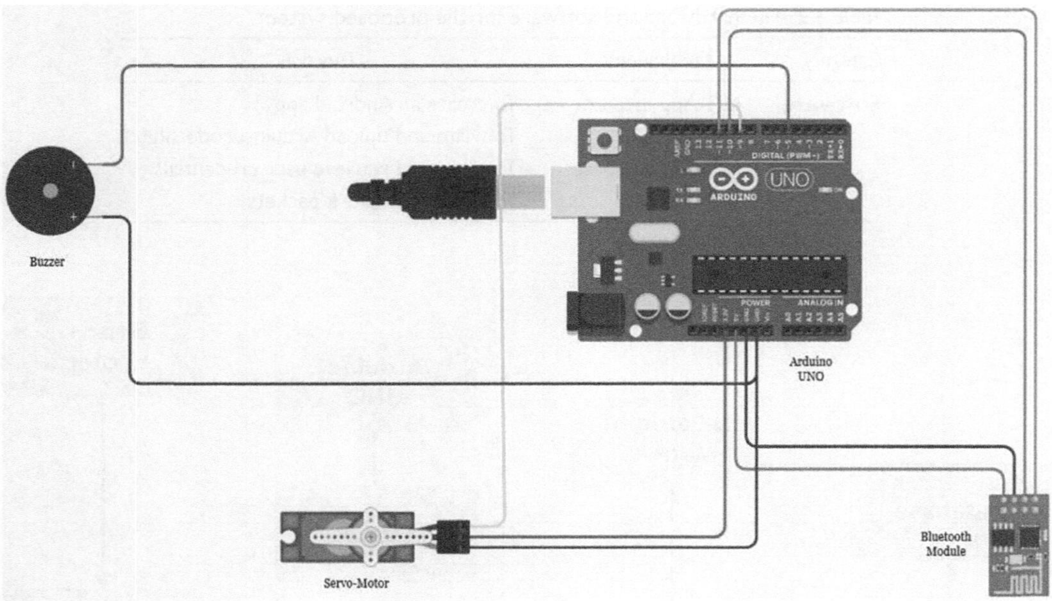

Figure 4.5 Hardware circuit connection.

4.6 RECOMMENDATION

To innovate in this area, we consider the following recommendations:

- **Exploration of advanced techniques:** In the direction of improving IoT security intelligence, look at advanced DL and ML approaches, including transfer learning, generative adversarial networks (GANs), and reinforcement learning. Following the study by Sarker et al. [31], clustering approaches are useful in addressing several security issues on the IoT, including outlier detection, detection of anomalies, pattern of the extraction process, identification of fraud, and cyberattack detection. These methods can provide fresh perspectives on security feature optimization, anomaly identification, and threat detection.
- **Real-time security solutions:** Create real-time security solutions that can adjust to changing IoT settings by utilizing ML models. To enable prompt answers to security concerns, this could entail utilizing edge computing, streaming data processing, and adaptive learning algorithms. According to the study by Kebande et al. [32], real-time monitoring (RTM) is essential for handling security-related issues. It emphasizes the understanding of the complexity of network environments through knowledge-based, behavior-based, and rule-based techniques.
- **Privacy protection techniques:** Explore methods such as differential privacy and federated learning to safeguard sensitive IoT data, ensuring efficient security analysis. These techniques mitigate risks of private data leaks and address evolving legal frameworks restricting data access and usage, posing obstacles to maximizing ML's potential for data-driven applications [33]. By concentrating on these suggestions, scholars and professionals might propel advancements in IoT security intelligence, harnessing the potency of ML to tackle the constantly changing security quandaries in networked IoT settings.

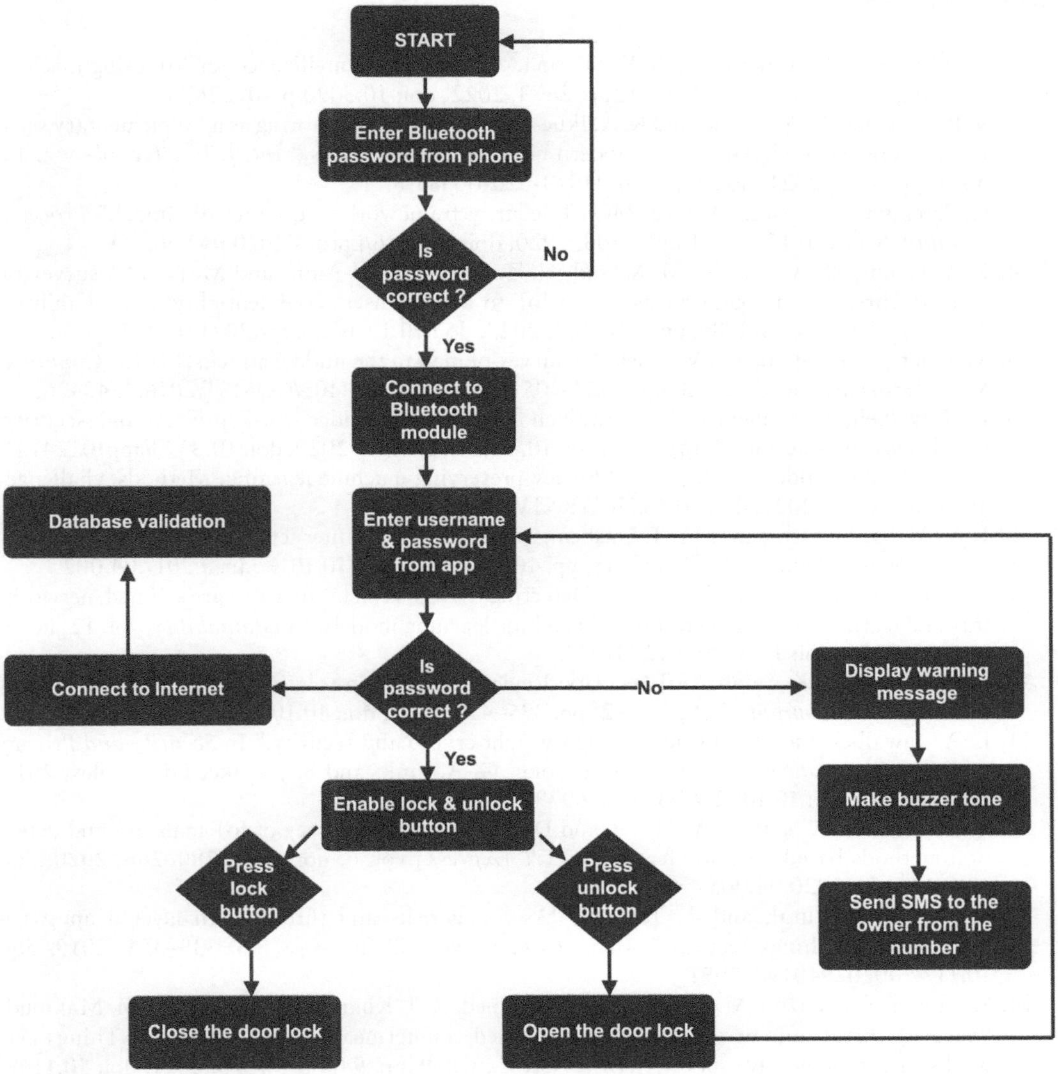

Figure 4.6 Flow chart.

4.7 CONCLUSION

We acknowledge its potential and vulnerabilities after exploring the IoT. While IoT advancements revolutionize connectivity across sectors, security remains a pressing concern due to potential exploitation by attackers. Software-defined networking (SDN) offers scalability solutions, serving as a unified management entity for IoT devices. Implementing comprehensive security protocols and integrating ML are vital for proactive defense, especially in healthcare and smart cities. Collaboration among stakeholders is crucial for pioneering security methods and standardized protocols to combat emerging threats. ML and IoT security underscore the importance of vigilance in addressing evolving challenges. Thus, the security of IoT should be a collective goal. Together we should strive toward making the digital era a safe place for future generations.

REFERENCES

1. S. Mishra, A. Albarakati, and S. K. Sharma, "Cyber threat intelligence for IoT using machine learning," *Processes*, vol. 10, no. 12, p. 2673, 2022, doi: 10.3390/pr10122673.
2. V. R. Kebande, N. M. Karie, and R. A. Ikuesan, "Real-time monitoring as a supplementary security component of vigilantism in modern network environments," *Int. J. Inf. Tecnol.*, vol. 13, no. 1, pp. 5–17, 2021, doi: 10.1007/s41870-020-00585-8.
3. C. Patel and N. Doshi, "A novel MQTT security framework in generic IoT model," *Procedia Comput.* Sci., vol. 171, pp. 1399–1408, 2020, doi: 10.1016/j.procs.2020.04.150.
4. K. Y. Najmi, M. A. AlZain, M. Masud, N. Z. Jhanjhi, J. Al-Amri, and M. Baz, "A survey on security threats and countermeasures in IoT to achieve users confidentiality and reliability," *Mater. Today Proc.*, vol. 81, pp. 377–382, 2023, doi: 10.1016/j.matpr.2021.03.417.
5. M. Conti, N. Dragoni, and V. Lesyk, "A survey of man in the middle attacks," *IEEE Commun. Surv. Tutorials*, vol. 18, no. 3, pp. 2027–2051, 2016, doi: 10.1109/COMST.2016.2548426.
6. L. Tawalbeh, F. Muheidat, M. Tawalbeh, and M. Quwaider, "IoT privacy and security: Challenges and solutions," *Appl. Sci.*, vol. 10, no. 12, p. 4102, 2020, doi: 10.3390/app10124102.
7. R. Xu, N. Baracaldo, and J. Joshi, "Privacy-preserving machine learning: Methods, challenges and directions," 2021, doi: 10.48550/ARXIV.2108.04417.
8. F. A. Alaba, M. Othman, I. A. T. Hashem, and F. Alotaibi, "Internet of Things security: A survey," *J. Netw. Comput. Appl.*, vol. 88, pp. 10–28, 2017, doi: 10.1016/j.jnca.2017.04.002.
9. H. Polat, O. Polat, and A. Cetin, "Detecting DDoS attacks in software-defined networks through feature selection methods and machine learning models," *Sustainability*, vol. 12, no. 3, p. 1035, 2020, doi: 10.3390/su12031035.
10. M. A. Khan and K. Salah, "IoT security: Review, blockchain solutions, and open challenges," *Future Gener. Comput. Syst.*, vol. 82, pp. 395–411, 2018, doi: 10.1016/j.future.2017.11.022.
11. L. A. Tawalbeh and H. Tawalbeh, "Lightweight crypto and security," In *Security and Privacy in Cyber-Physical Systems*, 1st ed., H. Song, G. A. Fink, and S. Jeschke, Eds., Wiley, 2017, pp. 243–261. doi: 10.1002/9781119226079.ch12.
12. Q.-D. Ngo, H.-T. Nguyen, V.-H. Le, and D.-H. Nguyen, "A survey of IoT malware and detection methods based on static features," *ICT Express.*, vol. 6, no. 4, pp. 280–286, 2020, doi: 10.1016/j.icte.2020.04.005.
13. M. Aydos, Y. Vural, and A. Tekerek, "Assessing risks and threats with layered approach to Internet of Things security," *Meas. Control*, vol. 52, no. 5–6, pp. 338–353, 2019, doi: 10.1177/0020294019837991.
14. S. Zaman, K. Alhazmi, M. A. Aseeri, M. R. Ahmed, R. T. Khan, M. S. Kaiser, and M. Mahmud, "Security threats and artificial intelligence based countermeasures for Internet of Things networks: A comprehensive survey," *IEEE Access.*, vol. 9, pp. 94668–94690, 2021, doi: 10.1109/ACCESS.2021.3089681.
15. A. Al Hayajneh, M. Z. A. Bhuiyan, and I. McAndrew, "Improving Internet of Things (IoT) security with software-defined networking (SDN)," *Computers*, vol. 9, no. 1, p. 8, 2020, doi: 10.3390/computers9010008.
16. S. Singh, R. Sulthana, T. Shewale, V. Chamola, A. Benslimane, and B. Sikdar, "Machine-learning-assisted security and privacy provisioning for edge computing: A survey," *IEEE Internet Things J.*, vol. 9, no. 1, pp. 236–260, 2022, doi: 10.1109/JIOT.2021.3098051.
17. I. H. Sarker, A. I. Khan, Y. B. Abushark, and F. Alsolami, "Internet of Things (IoT) security intelligence: A comprehensive overview, machine learning solutions and research directions," *J. Math. Comp. Sci.*, preprint, 2022. doi: 10.20944/preprints202203.0087.v1.
18. O. G. Dorobantu and S. Halunga, "Security threats in IoT," In *2020 International Symposium on Electronics and Telecommunications (ISETC)*, Timisoara, Romania: IEEE, Nov. 2020, pp. 1–4. doi: 10.1109/ISETC50328.2020.9301127.
19. I. Idrissi, M. Azizi, and O. Moussaoui, "IoT security with deep learning-based intrusion detection systems: A systematic literature review," In *2020 Fourth International Conference On Intelligent Computing in Data Sciences (ICDS)*, Fez, Morocco: IEEE, Oct. 2020, pp. 1–10. doi: 10.1109/ICDS50568.2020.9268713.

20. M. A. Ferrag and L. Shu, "The performance evaluation of blockchain-based security and privacy systems for the Internet of Things: A tutorial," *IEEE Internet Things J.*, vol. 8, no. 24, pp. 17236–17260, 2021, doi: 10.1109/JIOT.2021.3078072.
21. F. Alwahedi, A. Aldhaheri, M. A. Ferrag, A. Battah, and N. Tihanyi, "Machine learning techniques for IoT security: Current research and future vision with generative AI and large language models," *Internet of Things Cyber-Phys. Syst.*, vol. 4, pp. 167–185, 2024, doi: 10.1016/j.iotcps.2023.12.003.
22. T. A. Ahanger and A. Aljumah, "Internet of Things: A comprehensive study of security issues and defense mechanisms," *IEEE Access.*, vol. 7, pp. 11020–11028, 2019, doi: 10.1109/ACCESS.2018.2876939.
23. Maleh, Y., Sahid, A., & Belaissaoui, M. (2021). Optimized machine learning techniques for IoT 6LoWPAN cyber attacks detection. In *Proceedings of the 12th International Conference on Soft Computing and Pattern Recognition (SoCPaR 2020) 12* (pp. 669-677). Springer International Publishing.
24. J. C. Sapalo Sicato, P. K. Sharma, V. Loia, and J. H. Park, "VPNFilter malware analysis on cyber threat in smart home network," *Appl. Sci.*, vol. 9, no. 13, p. 2763, 2019, doi: 10.3390/app9132763.
25. F. Hussain, S. A. Hassan, R. Hussain, and E. Hossain, "Machine learning for resource management in cellular and IoT networks: Potentials, current solutions, and open challenges," *IEEE Commun. Surv. Tutorials*, vol. 22, no. 2, pp. 1251–1275, 2020, doi: 10.1109/COMST.2020.2964534.
26. S. R. J. Ramson, S. Vishnu, and M. Shanmugam, "Applications of Internet of Things (IoT)—An overview," In *2020 5th International Conference on Devices, Circuits and Systems (ICDCS)*, Coimbatore, India: IEEE, Mar. 2020, pp. 92–95. doi: 10.1109/ICDCS48716.2020.243556.
27. S. Mishra, S. K. Sharma, and M. A. Alowaidi, "Multilayer self-defense system to protect enterprise cloud," *Comput. Mater. Contin.*, vol. 66, no. 1, pp. 71–85, 2020, doi: 10.32604/cmc.2020.012475.
28. F. Hussain, R. Hussain, S. A. Hassan, and E. Hossain, "Machine learning in IoT security: Current solutions and future challenges," *IEEE Commun. Surv. Tutorials*, vol. 22, no. 3, pp. 1686–1721, 2020, doi: 10.1109/COMST.2020.2986444.
29. M. A. Al-Garadi, A. Mohamed, A. K. Al-Ali, X. Du, I. Ali, and M. Guizani, "A survey of machine and deep learning methods for Internet of Things (IoT) security," *IEEE Commun. Surv. Tutorials*, vol. 22, no. 3, pp. 1646–1685, 2020, doi: 10.1109/COMST.2020.2988293.
30. R. Ahmad and I. Alsmadi, "Machine learning approaches to IoT security: A systematic literature review," *Internet of Things*, vol. 14, p. 100365, 2021, doi: 10.1016/j.iot.2021.100365.
31. G. Kornaros, "Hardware-assisted machine learning in resource-constrained IoT environments for security: Review and future prospective," *IEEE Access.*, vol. 10, pp. 58603–58622, 2022, doi: 10.1109/ACCESS.2022.3179047.
32. N. Torres, P. Pinto, and S. I. Lopes, "Security vulnerabilities in LPWANs—An attack vector analysis for the IoT ecosystem," *Appl. Sci.*, vol. 11, no. 7, p. 3176, 2021, doi: 10.3390/app11073176.
33. M. Park, H. Oh, and K. Lee, "Security risk measurement for information leakage in IoT-based smart homes from a situational awareness perspective," *Sensors*, vol. 19, no. 9, p. 2148, 2019, doi: 10.3390/s19092148.

Chapter 5

Exploring Blockchain techniques for enhancing IoT security and privacy

A comprehensive analysis

Ahmad Anwar Zainuddin, Muhammad Aizzul Izzuddin Zulhazizi, Muhammad Firdaus Darmawan, Shahmie Abd Jalil, Muhammad Hafizudin Jamhari, and Muhammad Syafiq

5.1 INTRODUCTION

Blockchain-based approaches offer a means to enhance security and privacy within Internet of Things (IoT) systems by providing decentralized solutions. The Blockchain application spans various layers of IoT security and privacy [1]. Security measures such as authentication, authorization, and secure firmware updates are upheld at the device level using unique IDs and secure channels. Blockchain ensures secure interactions among IoT devices and other entities through secure protocols and cryptography, establishing a decentralized ledger for managing IoT data, guaranteeing integrity and privacy, and detecting unauthorized access. Smart contracts automate security policies, restricting access to authorized users. As exemplified in supply chain management, decentralized apps at the application layer utilize Blockchain for security and privacy. Challenges include energy consumption, delays, and computational overhead, rendering Blockchain less suitable for IoT applications with stringent performance requirements or constrained resources [2–5].

Organization of paper

The remainder of this chapter is organized as follows. Section 5.1 discusses the introduction of IoT Security and Privacy by utilizing Blockchain. Section 5.2 outlines the analysis of Blockchain. Blockchain integration in IoT is covered in Section 5.3. Section 5.4 discusses difficulties, open issues, and potential areas for future direction. Section 5.5 outlines the methodology. Section 5.6 concludes this chapter with a recommendation.

5.1.1 IoT safety issues

The rise of IoT devices has raised safety concerns, with manufacturers often neglecting security measures during development, leading to exploitable vulnerabilities for cybercriminals.

5.1.1.1 Access control systems of IoT

Access management systems within the domain of IoT provide a safe and user-friendly method for managing access to networked objects and systems.

- **Authentication**: Authentication is a fundamental aspect of access control, ensuring the user or application is who or what it claims to be [6]. By implementing robust authentication measures, organizations can effectively manage access rights and enhance their overall security posture [7].

 DOI: 10.1201/9781032714806-6

- **Encryption**: Encryption in access control employs encryption techniques to safeguard data and uphold secrecy, especially from untrusted parties. It helps organizations reduce risks linked to unauthorized access and potential data leaks.
- **Accountability**: Accountability in access control is crucial for tracing actions back to individuals or entities, ensuring integrity and governance of systems. Organizations establish a strong layer of accountability by linking each activity to a specific person [8].

5.1.1.2 Different types of cyberattack

- **Trojan horse malware:** Trojan attacks utilize malicious software disguised as legitimate code or content to deceive users, leading them to activate the malware on their devices unknowingly. This enables hackers to perform actions such as uploading, downloading, or executing files without permission.
- **BotNets attack:** A botnet is a collection of hacked computers and devices infected with malicious software, sometimes called bot malware. This malware allows a hacker to control and manipulate the infected devices from a distant location [9].
- **Distributed Denial of Service (DDoS):** A DDoS strike directed at IoT devices is a form of cyber assault aimed at disrupting the availability of network resources and servers by inundating the communication channels from various locations [10].

Most of these attacks are categorized into one of the four common types of security attacks in IoT, as shown in Figure 5.1, and these technologies can be applied to avoid these attacks in Blockchain technology.

5.1.2 Blockchain of IoT security

Lately, Blockchain technology has gained recognition for its effectiveness in mitigating security risks [11,12]. Figure 5.2 illustrates the challenges of applying Blockchain to the IoT, which are discussed in detail in the following sections.

Connectivity issue: IoT devices are expected to connect across different networks and share data with stakeholders. However, their limited storage capacity poses a challenge when integrating with Blockchain for new services and economic opportunities across various applications [13].

- **Transparency and privacy issue:** Blockchain ensures that all its transactions are transparent. Creating an efficient Blockchain-based access control system for IoT is crucial to striking the right balance between privacy and transparency.
- **Heterogeneity of IoT devices:** IoT devices employ various communication media to communicate with other devices in a system of small sensors and connected objects. The diversity of IoT devices poses a challenge to enabling connectivity between Blockchain and IoT.
- **Trade between the performance, security, and power consumptions:** Blockchain algorithms' demanding computational and power needs often make them unsuitable for resource-limited applications like IoT. This prompts questions about Blockchain's effectiveness with IoT data. Researchers suggest improving Blockchain consensus algorithms to enhance transaction speed and throughput [14,15].
- **Comparing between scalability, throughput, and reliability**: Blockchain-based IoT apps face scalability hurdles due to massive data volumes. Although Blockchain ensures transparent and secure data handling, scalability issues, bandwidth overhead,

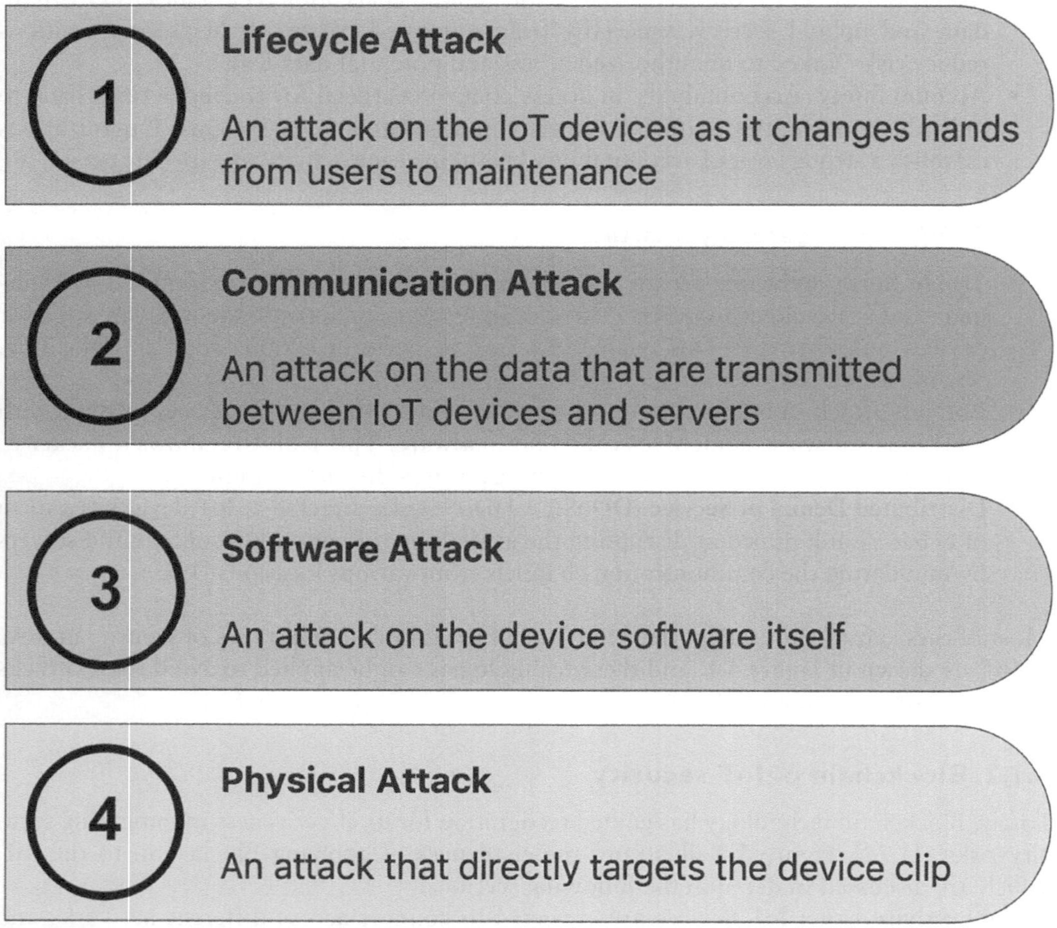

Figure 5.1 Types of security attacks in IoT.

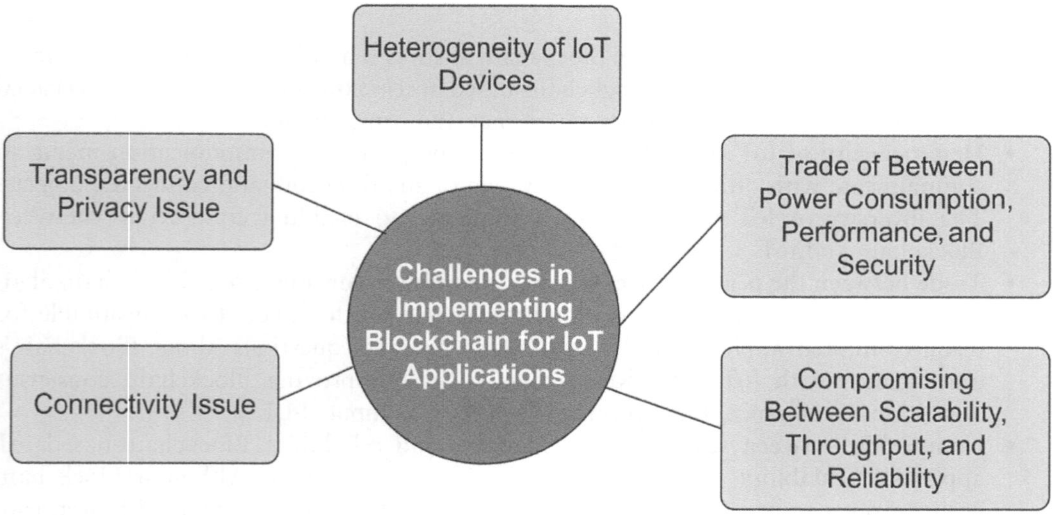

Figure 5.2 Challenges in implementing Blockchain for IoT application.

and computational complexity impede its use for large-scale IoT projects. Traditional encryption methods provide robust security but may lack scalability and trust. Deciding between Blockchain and nonBlockchain solutions for IoT security hinges on the application's needs, balancing efficiency, scalability, privacy, and security. Ultimately, the choice depends on the inherent trade-offs of each IoT scenario [16].

5.2 METHODOLOGY

The study adopts a systematic approach, reviewing academic papers, articles, and online resources to explore how Blockchain technology enhances security and privacy in IoT systems. It focuses on understanding cybersecurity, IoT security, and privacy challenges while investigating Blockchain integration. Extensive literature analysis from databases such as IEEE Xplore, ScienceDirect, and Springer identifies key concepts and trends, emphasizing IoT security threats and Blockchain potential. It examines security threats in IoT, such as access control issues and denial-of-service attacks, and explores Blockchain's role in addressing them, including different Blockchain types and consensus algorithms. The study also identifies future research areas and technological advancements impacting Blockchain integration with IoT security, employing a structured and comprehensive methodology drawing from diverse scholarly sources.

5.3 ANALYSIS OF BLOCKCHAIN

In the rapidly evolving landscape of digital innovation, Blockchain has garnered significant attention for its transformative potential. The capacity of Blockchain technology to augment operational efficiency, fortify integrity, and foster more transparency.

5.3.1 How Blockchain works

Blockchain, a revolutionary digital ledger, has emerged as a cornerstone of decentralized systems, providing a peer-to-peer network that redefines how transactions are recorded. The Blockchain's core architecture consists of interconnected blocks, forming an immutable ledger. Each block contains transactions and a block header, which includes metadata such as a pointer to the previous block's hash, target difficulty for Proof-of-Work (PoW), and creation time with a nonce for security. The Merkle Root, derived from the Merkle Tree, is also included [17]. Transactions are securely embedded within each block, representing value transfers [17]. This structure requires substantial computational effort to add blocks, enhancing security. The diagram illustrates how each block references the previous block's hash, highlighting the interconnected nature of the Blockchain's design (Figure 5.3).

5.3.2 Blockchain platform classification

Blockchain platforms are classified into different categories based on their fundamental characteristics and usage scenarios. Blockchain can be classified into three types of platforms:

5.3.2.1 Public Blockchain

A public Blockchain is the epitome of decentralization, functioning as a permissionless and open environment where anyone with internet access can join and contribute to the network [18].

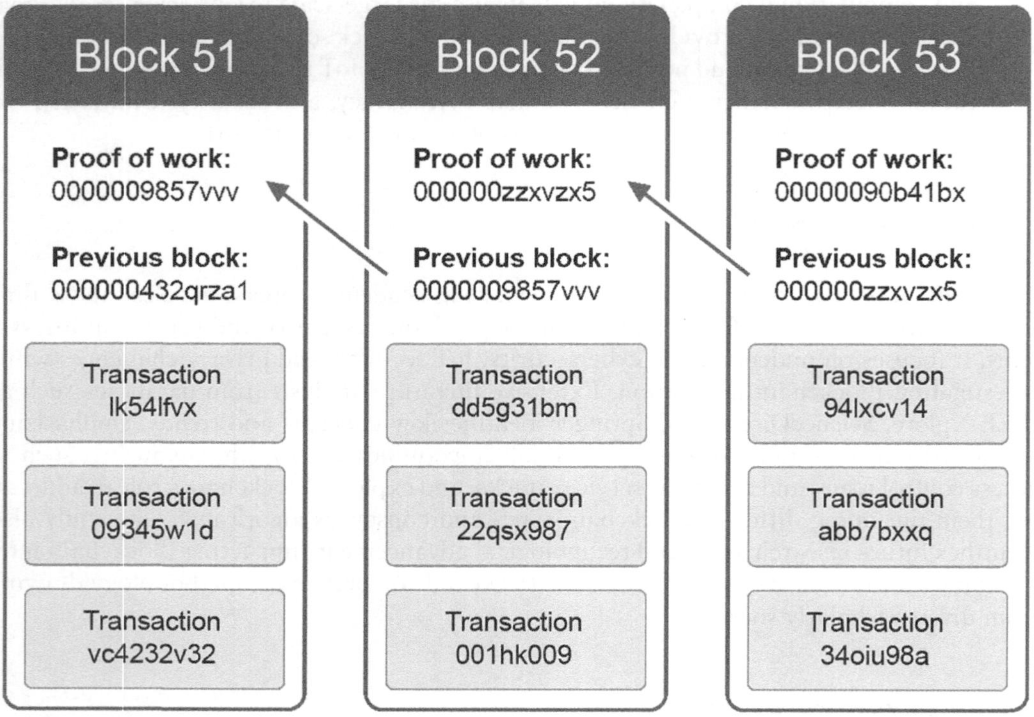

Figure 5.3 General Blockchain framework.

Security is enhanced through decentralization, making it difficult for attackers to compromise the network. Public Blockchains may struggle to handle many transactions from IoT devices, leading to higher costs and delays. Public Blockchain usually uses energy-heavy mechanisms such as PoW, which might not work well for IoT devices with limited resources.

5.3.2.2 Private Blockchain

Private Blockchains, unlike public ones, are controlled by a central organization, allowing for strict governance and privacy, which is beneficial in enterprise settings [17]. They are commonly used for organizations' supply chain management, digital identity, and asset ownership. However, integrating them with IoT devices owned by untrusted parties can be challenging due to the need for permission to join and use the network [18].

5.3.2.3 Hybrid Blockchain

A hybrid Blockchain integrates public and private Blockchains, combining decentralization with privacy benefits. It enables selective data disclosure, keeping sensitive information within the private network while making certain data publicly accessible [18]. Using encryption and access controls can help strike this balance. Hybrid Blockchains might use various consensus methods for their public and private parts, so they need careful coordination to ensure that transactions are verified consistently and data stays intact across the network.

5.3.3 Consensus mechanism

In a Blockchain network, where participants may be anonymous, consensus mechanisms are employed to validate transactions and agree on adding new blocks to the Blockchain.

5.3.3.1 Proof of work (PoW)

PoW is a consensus algorithm where nodes or miners solve complex mathematical problems to validate new blocks on the Blockchain. This ensures transaction trustworthiness and prevents malicious activities in the decentralized ecosystem. The competitive aspect arises from multiple nodes vying to append the next block to the ledger [17].

5.3.3.2 Proof of stake (PoS)

Block verifiers are chosen based on the amount of cryptocurrency they stake, not through competitive computational work like in PoW. Staking involves users committing cryptocurrency to the network, and those who stake more have a greater chance of being selected as validators [17].

5.4 BLOCKCHAIN INTEGRATION IN IOT

Blockchains and IoT can be combined to create a secure and tamper-resistant data-sharing system. Private networks based on Blockchain technology can receive data from IoT devices, creating unchangeable records of shared transactions. Combining Blockchain with IoT creates a new strategy that improves IoT network security, transparency, and trust [19]. There are potential drawbacks of using Blockchain for IoT security as mentioned in Figure 5.4. Integrating Blockchain and IoT holds significant promise, namely, improving security, transparency, and reliability in IoT. Developing IoT applications cannot reach a large audience and may lose all their potential without a reliable and compatible IoT ecosystem [20].

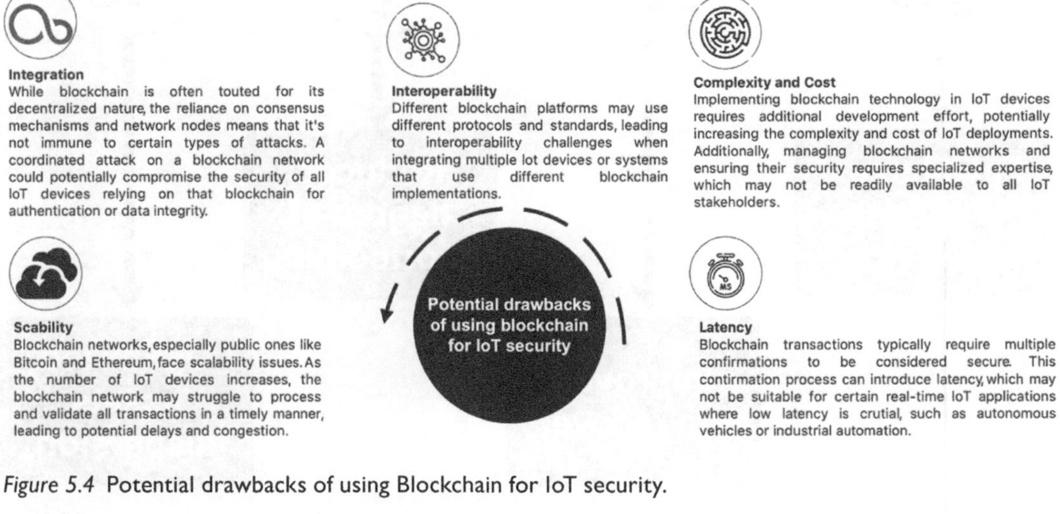

Figure 5.4 Potential drawbacks of using Blockchain for IoT security.

5.4.1 Need for integration

5.4.1.1 Transparency and reliability

Due to Blockchain's transparency and immutability, users can trust that their data and transactions are secure and tamper-proof. Cryptography algorithms are utilized to protect information on the Blockchain, making it difficult for hackers to alter or manipulate recorded data.

5.4.1.2 Decentralization

Figure 5.5 emphasizes the benefits of Blockchain's decentralized traits. The decentralized nature of Blockchain can aid in data storage and sharing, eliminating dependency on centralized systems. It was created as the underpinning technology for the cryptocurrency Bitcoin, but it has now spread to other areas such as banking, healthcare, and supply chain management [21].

While decentralized Blockchain systems emphasize security above performance, they provide transparent and complete management and the possibility of physical decentralization through the dispersion of Blockchain servers across several regions.

5.4.1.3 Smart contracts

Smart contracts are programmable scripts operating on the Blockchain that enable, perform, and maintain a contract between untrusted parties without a reliable intermediary [22]. Figure 5.6 demonstrates how the smart contract in Blockchain integration in IoT works. They facilitate the formation of multiparty digital contracts that are visible, unchangeable, and enforced through programming, increasing confidence and effectiveness in transactions. Table 5.1 shows the challenges in Blockchain integration in IoT.

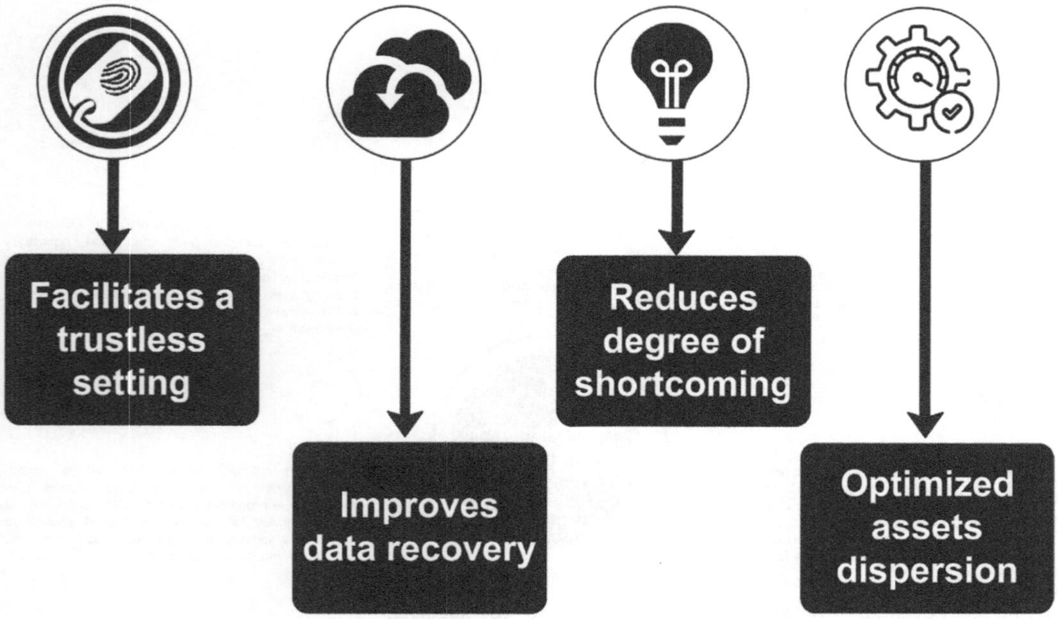

Figure 5.5 Benefits of decentralization.

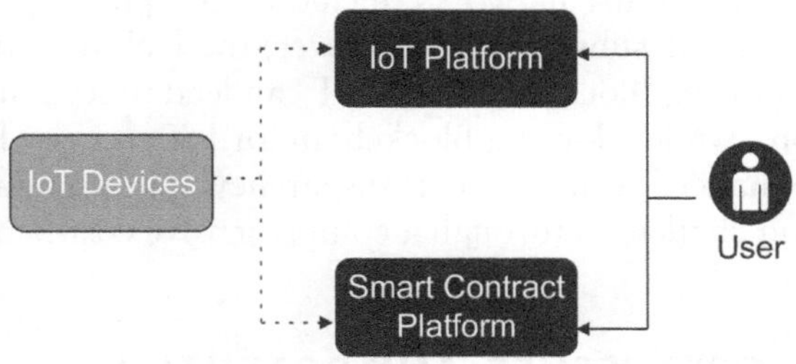

Figure 5.6 Smart contract in Blockchain integration in IoT.

Table 5.1 Challenges in Blockchain integration in IoT

Area of challenges	*Details*
Scalability issues	Decentralized Blockchain networks can face scalability issues, making it challenging to integrate them with IoT systems that have devices with limited disk space, RAM, and slow internet connections
Computing power and time	The time required to create a block in a Blockchain network can be significant, which can be a challenge when integrating it with IoT devices that have limited computing power
Storage space, RAM, and internet connection requirements	Integrating decentralized Blockchain with IoT networks can require a lot of storage space, RAM, and internet connection, which can be challenging for devices with limited resources
Reliability of IoT data	Ensuring the reliability of data generated by IoT devices is a main challenge in the integration of Blockchain and IoT
Trade-off between decentralized security and scalability	Organizations need to find a balance between decentralized security and scalability, as seen in the IOTA cryptocurrency, which uses a different technology solution to address these challenges

5.4.2 Security analysis

Several methodologies exist for assessing the security of Blockchain–IoT systems, including the following.

5.4.2.1 Security frameworks

Security frameworks are crucial for assessing the security of Blockchain–IoT systems. These frameworks offer a systematic method for detecting, analyzing, and reducing security threats in integrated settings. They frequently include a collection of security measures, optimal methods, and instructions customized to address the distinct difficulties presented by merging Blockchain with IoT.

5.4.2.2 Cost benefit

The economic implications of adopting Blockchain for IoT security are multifaceted and can be analyzed through a cost-benefit analysis to provide a more rounded perspective. On the one hand, integrating Blockchain with IoT can lead to increased security and transparency in IoT ecosystems, potentially reducing the economic impact of security breaches and data tampering. Blockchain technology is known for its high energy consumption, which can lead to increased operational costs, especially for resource-constrained IoT devices. The cost of integrating Blockchain with IoT networks, including computing power, storage capacity, and energy resources, can be substantial. Moreover, the lack of standardized regulations and guidelines for integrating Blockchain with IoT can lead to legal uncertainties and compliance costs. Therefore, while adopting Blockchain for IoT security holds the potential for economic benefits in enhanced security and transparency, it is essential to carefully evaluate the associated costs and challenges through a comprehensive cost-benefit analysis [23].

5.5 DIFFICULTIES, OPEN ISSUES, AND POTENTIAL AREAS FOR FUTURE DIRECTIONS

This section presents a concise summary of the unresolved problems and potential research prospects.

5.5.1 Difficulties related to privacy and security in the Blockchain–IoT paradigm

5.5.1.1 Energy consumption and initial cost

Energy use in PoW Blockchain systems raises sustainability concerns. High upfront costs and the scarcity of experts pose challenges for organizations considering adoption. This makes it difficult for small- and medium-sized businesses to adopt Blockchain systems, either fully or partially [24].

5.5.1.2 Privacy and security

Ensuring data privacy and security on the Blockchain remains a significant challenge, especially as the technology advances across sectors [25]. To address this, corporations are actively enhancing Blockchain network security, utilizing formal verification to identify smart contract vulnerabilities and implementing multisignature wallets for added transaction security. However, ongoing vigilance and robust security procedures are crucial to safeguard Blockchain platforms from potential risks and challenges associated with incorporating Blockchain technology into the IoT.

5.5.1.3 Reliability of the data

The data produced by IoT devices is frequently extensive and intricate, and guaranteeing its integrity and dependability is essential for the efficient operation of the IoT ecosystem. Blockchain technology can mitigate this difficulty by offering a decentralized and unalterable ledger that guarantees the reliability and trustworthiness of data from the IoT [26].

5.5.1.4 Interoperability and security

The integration also raises concerns related to interoperability, poor security, trust, and confidentiality. Addressing these issues is vital for the seamless integration of Blockchain with IoT devices. Interoperability issues can be addressed by developing common standards and protocols for Blockchain and IoT devices. To address security issues, it is important to install strong security mechanisms, including multifactor authentication, encryption, and access restriction [27].

5.5.1.5 Regulatory challenges

The ever-evolving regulatory landscape surrounding Blockchain technology poses a significant challenge for compliance and legal navigation. Varied regulations across jurisdictions impact Blockchain adoption in IoT systems, given Blockchain's global operation and the differing regulatory approaches to cryptocurrencies, smart contracts, and related activities. This creates complexity and uncertainty for organizations integrating Blockchain with IoT, particularly concerning data privacy and security. Cross-border transactions facilitated by Blockchain face additional regulatory hurdles, compounding the complexity. To tackle these challenges, proactive collaboration with regulators, monitoring regulatory developments, and consulting legal and compliance experts are crucial for compliance and navigating the evolving regulatory environment [28,29].

5.5.2 Open issues and possible avenues for future directions

5.5.2.1 Scalability and performance

Scalability and performance in Blockchain pose significant challenges, especially when integrated with IoT. Research is necessary to improve Blockchain systems, particularly in managing the vast data generated by IoT devices. This approach has shown promising results and holds potential for scalability and efficiency improvements in the future [26].

5.5.2.2 Energy consumption and its implication

The energy-intensive consensus mechanisms like PoW utilized in Blockchain networks can hinder their adoption in IoT systems because of significant energy consumption. Nonetheless, integrating Blockchain with IoT networks can ensure trusted data provision and decentralized operation, eliminating intermediaries and fostering trust within the IoT network [30]. The potential of Blockchain in IoT energy management is substantial, revolutionizing energy management through transparent, secure, and efficient energy transactions. Combining Blockchain technologies with IoT devices allows consumers to trade and purchase energy from the grid directly, offering innovative solutions for renewable energy distribution and enhancing efficiencies for utility providers.

5.6 RECOMMENDATIONS

Various suggestions might be considered while aiming to improve the utilization of decentralized Blockchain methods.

1. **Leverage Blockchain's decentralized nature:** Blockchain operates on a decentralized, peer-to-peer network, facilitating a secure and transparent exchange of information and assets. Its lack of a single point of vulnerability makes it difficult for hackers to target and attack the system. Decentralization creates a trustless environment and enhances data reconciliation across the network.
2. **Employ privacy-preserving techniques:** Zero-knowledge proofs, state channels, and side chains effectively preserve privacy in Blockchain-based IoT systems. They enable secure transactions without disclosing sensitive information. However, the ongoing development of IoT systems utilizing these methods is still under investigation.

5.7 CONCLUSION

This study thoroughly examines how Blockchain technology can enhance IoT systems' security and privacy. It analyzes ongoing projects combining Blockchain and IoT security, categorizes IoT vulnerabilities, and explores Blockchain's operations and security concerns. The findings indicate significant potential for Blockchain in improving IoT data security and confidentiality, though challenges remain. Future research should prioritize developing standardized protocols for seamless Blockchain integration with IoT, addressing interoperability, scalability, and security issues. Focusing on these areas can help build robust and privacy-preserving Blockchain-enabled IoT systems, advancing IoT applications across various domains.

REFERENCES

1. R. Alajlan, N. Alhumam, and M. Frikha, "Cybersecurity for Blockchain-Based IoT Systems: A Review," *Applied Sciences*, vol. 13, no. 13, p. 7432, Jun. 2023, doi: 10.3390/app13137432.
2. M. Picone, S. Cirani, and L. Veltri, "Blockchain Security and Privacy for the Internet of Things," *Sensors*, vol. 21, no. 3, p. 892, Jan. 2021, doi: 10.3390/s21030892.
3. S. Jagdale, "The Impact of Blockchain on IoT Security and Privacy," [Online]. https://iot.eetimes.com/the-impact-of-blockchain-on-iot-security-and-privacy/
4. C. Chakray, "Blockchain and IoT Security: Everything You Need to Know," [Online]. https://www.chakray.com/blockchain-iot-security/
5. trendmicro, "Blockchain: The Missing Link between Security and the IoT?" [Online]. https://www.trendmicro.com/vinfo/us/security/news/internet-of-things/blockchain-the-missing-link-between-security-and-the-iot
6. IBM, "Authentication Versus Access Control," [Online]. https://www.ibm.com/docs/en/wca/3.0.0?topic=security-authentication-versus-access-control
7. G. Mehta, "What Is Access Control? What Are the Different Aspects of Access Control Systems?" [Online]. https://www.oloid.ai/blog/what-is-access-control-what-are-the-different-aspects-of-access-control-systems/
8. C. Cybrary, "CISSP Study Guide: Access Control and Accountability," [Online]. https://www.cybrary.it/blog/access-control-and-accountability
9. KasperSky, "What Is a Botnet?" [Online]. https://usa.kaspersky.com/resource-center/threats/botnet-attacks
10. P. Kumari and A. K. Jain, "A comprehensive study of DDoS attacks over IoT network and their countermeasures," *Computers & Security*, vol. 127, p. 103096, Apr. 2023, doi: 10.1016/j.cose.2023.103096.
11. H.-N. Dai, Z. Zheng, and Y. Zhang, "Blockchain for Internet of Things: A Survey," *IEEE Internet of Things Journal*, vol. 6, no. 5, pp. 8076–8094, Oct. 2019, doi: 10.1109/JIOT.2019.2920987.
12. T. Sharma, S. Satija, and B. Bhushan, "Unifying Blockchian and IoT: Security Requirements, Challenges, Applications and Future Trends," In *2019 International Conference on Computing, Communication, and Intelligent Systems (ICCCIS)*, Greater Noida, India: IEEE, pp. 341–346, Oct. 2019, doi: 10.1109/ICCCIS48478.2019.8974552.
13. H. F. Atlam and G. B. Wills, "Technical Aspects of Blockchain and IoT," *Advances in Computers*, vol. 115, pp. 1–39, 2019. doi: 10.1016/bs.adcom.2018.10.006, Elsevier.
14. S. M. H. Bamakan, A. Motavali, and A. B. Bondarti, "A Survey of Blockchain Consensus Algorithms Performance Evaluation Criteria," *Expert Systems with Applications*, vol. 154, p. 113385, Sep. 2020, doi: 10.1016/j.eswa.2020.113385.
15. C. Fan, S. Ghaemi, H. Khazaei, and P. Musilek, "Performance Evaluation of Blockchain Systems: A Systematic Survey," *IEEE Access*, vol. 8, pp. 126927–126950, 2020, doi: 10.1109/ACCESS.2020.3006078.

16. Q. Zhou, H. Huang, Z. Zheng, and J. Bian, "Solutions to Scalability of Blockchain: A Survey," *IEEE Access*, vol. 8, pp. 16440–16455, 2020, doi: 10.1109/ACCESS.2020.2967218.
17. S. Das, S. Namasudra, and V. H. C. De Albuquerque, "Blockchain Technology: Fundamentals, Applications, and Challenges," In *Blockchain Technology in e-Healthcare Management*, S. Namasudra and V. H. C. De Albuquerque, Eds., Institution of Engineering and Technology, pp. 1–30, 2022, doi: 10.1049/PBHE048E_ch1.
18. A. S. Gaikwad, "Overview of Blockchain," *International Journal for Research in Applied Science and Engineering Technology*, vol. 8, no. 6, pp. 2268–2270, 2020, doi: https://doi.org/10.22214/ijraset.2020.6364.
19. D. H. Prasadhini, H. R, and R. J. Michelle, "Blockchain-Enabled IoT: Enabling Trust and Security in a Connected World," *IJRASET*, vol. 11, no. 5, pp. 2631–2634, May 2023, doi: 10.22214/ijraset.2023.52051.
20. V. Hassija, V. Chamola, V. Saxena, D. Jain, P. Goyal, and B. Sikdar, "A Survey on IoT Security: Application Areas, Security Threats, and Solution Architectures," *IEEE Access*, vol. 7, pp. 82721–82743, 2019, doi: 10.1109/ACCESS.2019.2924045.
21. T. Aditya Sai Srinivas, A. David Donald, I. Dwaraka Srihith, D. Anjali, and A. Chandana, "The Rise of Secure IoT: How Blockchain is Enhancing IoT Security," *IJARSCT*, pp. 32–40, Apr. 2023, doi: 10.48175/IJARSCT-9006.
22. S. N. Khan, F. Loukil, C. Ghedira-Guegan, E. Benkhelifa, and A. Bani-Hani, "Blockchain Smart Contracts: Applications, Challenges, and Future Trends," *Peer-to-Peer Networking and Applications*, vol. 14, no. 5, pp. 2901–2925, Sep. 2021, doi: 10.1007/s12083-021-01127-0.
23. J. Pothuru, "IoT In Blockchain: Benefits, Use Cases, and Challenges," [Online]. https://www.reveation.io/blog/iot-in-blockchain/
24. B. M. Bitcoin Magazine, "Five Challenges Blockchain Technology Must Overcome Before Mainstream Adoption." [Online]. https://www.nasdaq.com/articles/five-challenges-blockchain-technology-must-overcome-before-mainstream-adoption-2018-01-03
25. B. Marr, "The 5 Biggest Problems with Blockchain Technology Everyone Must Know About," [Online]. https://www.forbes.com/sites/bernardmarr/2023/04/14/the-5-biggest-problems-with-blockchain-technology-everyone-must-know-about/?sh=61d8fde455d2
26. A. Reyna, C. Martín, J. Chen, E. Soler, and M. Díaz, "On Blockchain and Its Integration with IoT. Challenges and Opportunities," *Future Generation Computer Systems*, vol. 88, pp. 173–190, Nov. 2018, doi: 10.1016/j.future.2018.05.046.
27. K. Rupareliya, "IoT In Blockchain: Benefits, Use Cases, and Challenges," [Online]. https://www.intuz.com/blog/iot-in-blockchain-benefits-use-cases-and-challenges
28. T. I. Technology Innovators, "Regulatory Challenges in Blockchain: Navigating Compliance and Legal Frameworks," [Online]. https://www.technology-innovators.com/regulatory-challenges-in-blockchain-navigating-compliance-and-legal-frameworks/
29. L. Tao, Y. Lu, X. Ding, Y. Fan, and J. Y. Kim, "Throughput-Oriented Associated Transaction Assignment in Sharding Blockchains for IoT Social Data Storage," *Digital Communications and Networks*, vol. 8, no. 6, pp. 885–899, Dec. 2022, doi: 10.1016/j.dcan.2022.05.024.
30. S. Wadhwa, S. Rani, Kavita, S. Verma, J. Shafi, and M. Wozniak, "Energy Efficient Consensus Approach of Blockchain for IoT Networks with Edge Computing," *Sensors*, vol. 22, no. 10, p. 3733, May 2022, doi: 10.3390/s22103733.

Part II

Analyzing and responding to emerging threats

Chapter 6

Integrating security analysis module for proactive threat intelligence

Yassine Maleh and Abdelekbir Sahid

6.1 INTRODUCTION

In today's rapidly evolving digital landscape, organizations face an ever-expanding array of cybersecurity threats, ranging from malware and phishing attacks to sophisticated cyber espionage campaigns. Traditional security approaches, which rely heavily on reactive measures such as signature-based detection and incident response, are increasingly proving inadequate in the face of these dynamic and persistent threats. As a result, there is a growing recognition of the need for a paradigm shift toward proactive cybersecurity strategies that can anticipate and preemptively mitigate emerging risks (Maleh et al., 2022).

At the core of proactive cybersecurity is integrating advanced technologies and methodologies that enable organizations to detect and respond to threats in real time (Maleh et al., 2021). One such technology is Docker, a leading containerization platform that allows organizations to encapsulate applications and their dependencies into lightweight, portable containers. By leveraging Docker's containerization capabilities, organizations can isolate and secure their applications, reducing the attack surface and minimizing the impact of potential security breaches (Alaoui et al., 2022).

In addition to containerization, proactive cybersecurity relies on the collaborative sharing of threat intelligence and compromise indicators. Malware Information Sharing Platform (MISP) and Threat Sharing is a powerful open-source platform designed for precisely this purpose. With MISP, organizations can securely share and analyze threat intelligence data, enabling them to identify and respond to emerging threats more effectively. By leveraging MISP's capabilities for indicative compromise sharing, organizations can enhance their situational awareness and bolster their defenses against cyber threats (Briliyant et al., 2021; Maleh, 2023).

Furthermore, proactive cybersecurity necessitates robust intrusion detection capabilities to identify and thwart potential threats before they can cause harm. Wazuh is an open-source intrusion detection system with real-time monitoring, threat detection, and response capabilities. By deploying Wazuh agents across their network infrastructure, organizations can continuously monitor for signs of unauthorized access, malware infections, and other security incidents, enabling them to respond promptly and effectively to potential threats (Guarascio et al., 2022).

Proactive cybersecurity requires advanced data analysis capabilities to identify patterns, trends, and anomalies that may indicate potential security breaches. OpenSearch, an open-source search and analytics engine, provides organizations with the tools to analyze large volumes of security data in real time. By leveraging OpenSearch's powerful search and visualization capabilities, organizations can gain deeper insights into their security posture, identify emerging threats, and take proactive measures to mitigate risks before they escalate into significant incidents (Preuveneers & Joosen, 2021).

DOI: 10.1201/9781032714806-8

This chapter is organized into six main sections to explore the integrated security analysis module comprehensively. It begins with the literature review, providing insights into the contemporary cybersecurity landscape and organizations' challenges. The Research Methodology section outlines the approach to developing and assessing the proposed model. The Proposed Model section elucidates the architecture and functionality of the integrated module, detailing its components and synergies. A step-by-step deployment process is presented in implementation, emphasizing the integration of Docker, MISP, Wazuh, and OpenSearch. The Benefits and Limitations section evaluates the model's strengths and weaknesses. Finally, the chapter concludes by reflecting on the findings and suggesting future research directions.

6.2 LITERATURE REVIEW

This section examines existing literature on the challenges associated with comprehending and utilizing threat intelligence across various units, focusing on different frameworks, platforms, services, standards, and terminologies pertinent to the study's research questions. The review identifies and evaluates the diverse elements and challenges of established standards, frameworks, services, and related components.

6.2.1 Cyber Threat Intelligence (CTI)

Cyber Threat Intelligence (CTI) is a repository of detailed information concerning cybersecurity threats directed at an organization, aiding security teams to adopt proactive measures and data-driven actions to forestall potential cyberattacks (Stojkovski et al., 2021). This intelligence facilitates detecting and responding to ongoing threats and enhances an organization's overall cyber resilience. Security analysts compile CTI by collecting and scrutinizing raw threat data from various sources, subsequently correlating and analyzing this data to unveil trends, patterns, and relationships. The resultant intelligence is tailored to the organization, delving into specific vulnerabilities within the attack surface, potential attackers, tactics, techniques, and procedures (TTPs), and indicators of compromise (IoCs), thus furnishing actionable insights for security teams. The CTI lifecycle embodies an iterative and continuous process through which security teams generate, disseminate, and refine threat intelligence. This process typically involves six stages, commencing with planning, where security analysts collaborate with organizational stakeholders to define intelligence requirements. Subsequently, threat data is collected from diverse sources, including threat intelligence feeds, information-sharing communities, and internal security systems, and centralized for analysis. In the treatment stage, security analysts aggregate, standardize, and correlate the raw data to facilitate analysis and glean insights (Wagner et al., 2019). The study marks the transformation of raw threat data into actionable intelligence, enabling security teams to formulate recommendations based on verified trends and patterns. Broadcast entails sharing intelligence and recommendations with relevant stakeholders, prompting actions such as updating detection rules or blocking suspicious IP addresses. Finally, the comments stage enables stakeholders and analysts to reflect on the recent threat intelligence cycle, identifying emerging requirements or intelligence gaps that inform subsequent lifecycle iterations. Through this systematic approach, organizations can effectively harness CTI to fortify their cybersecurity posture and mitigate potential threats (Chantzios et al., 2019; Faiella et al., 2019).

The review highlights various threat intelligence standards, including the Traffic Light Protocol (TLP), Incident Object Description and Exchange Format (IODEF) (de Melo e Silva et al., 2020), Vocabulary for Event Recording and Incident Sharing (VERIS) (Tounsi, 2019), Real-time Inter-network Defense (RID), Open IoC (OpenIOC) (Doerr, 2018), Cyber Observable eXpression (CybOX), Structured Threat Information Expression (STIX), Trusted Automated eXchange of Indicator Information (TAXII), and Open Threat Exchange (OTX) (Faiella et al., 2019). Each standard offers unique features and capabilities for exchanging and sharing threat intelligence data. While some standards excel in certain aspects, such as data exchange format and machine readability, others face challenges in user interface design, data sharing capabilities, and authentication mechanisms. Table 6.1 provides a comparison of threat intelligence standards

Table 6.1 Threat intelligence standards

Standards	*Interoperability challenges*	*Popularity*
TLP	– Lack of standard format for data exchange – Limited user interface – Lack of automation between import/export – Issues with authentication and confidentiality – Difficulty in using data for detective and preventive controls – Challenges in filtering data based on dates – Difficulty in measuring efficacy	High
IODEF	– Limited user interface – Inability to share data based on attached attributes – Challenges in using data for detective and preventive controls – Difficulty in filtering data based on dates – Difficulty in measuring efficacy	Medium
RID	– Limited user interface – Inability to share data based on attached attributes – Challenges in using data for detective and preventive controls – Difficulty in filtering data based on dates – Difficulty in measuring efficacy	Medium
OpenIOC	– Limited user interface – Inability to share data based on attached attributes – Challenges in using data for detective and preventive controls – Difficulty in filtering data based on dates – Difficulty in measuring efficacy – Lack of automation between import/export – Issues with authentication and confidentiality	High
VERIS	– Lack of standard format for data exchange – Limited user interface – Inability to share data based on attached attributes – Challenges in using data for detective and preventive controls – Difficulty in filtering data based on dates – Difficulty in measuring efficacy – Lack of automation between import/export – Issues with authentication and confidentiality	Low

(Continued)

Table 6.1 (Continued) Threat intelligence standards

Standards	*Interoperability challenges*	*Popularity*
OTX	– Lack of standard format for data exchange – Limited user interface – Inability to share data based on attached attributes – Issues with authentication and confidentiality – Challenges in using data for detective and preventive controls – Difficulty in filtering data based on dates – Difficulty in measuring efficacy	Medium

Note: Popularity is categorized as high, medium, or low based on the perceived usage and adoption of each standard in the cybersecurity community.

6.2.2 Cyber threat intelligence platforms

Threat Intelligence Platforms (TIPs) are indispensable tools for security investigators, allowing them to leverage global threat intelligence for comprehensive risk assessment and mitigation strategies (Möller, 2023). These platforms play a critical role in gathering and analyzing data pertaining to an organization's threat landscape, addressing crucial questions such as the identity, frequency, timing, and motivation behind potential. Utilizing threat scores, TIPs aid in identifying patterns and vulnerabilities within the infrastructure, while automated data feeds streamline data collection from diverse external sources, such as the Internet (Cascavilla et al., 2021).

Moreover, TIPs furnish contextual information regarding the nature of threats and participants' responses, facilitating informed decision-making processes (Faiella et al., 2019). They also play a pivotal role in integrating CTI with real-time identification and verification frameworks like Security Information and Event Management (SIEM) systems (Safarzadeh et al., 2019), ensuring the timely retrieval and analysis of structured and unstructured data from external repositories. Figure 6.1 shows a Threat Intelligence Platform Architecture.

In the landscape of proprietary TIPs, several prominent platforms stand out. The Accenture Cyber Intelligence Platform leverages artificial intelligence algorithms and operational analytics to deliver actionable threat information across diverse technical environments, including mobile, IoT, and cloud infrastructures. Similarly, Facebook Threat Exchange facilitates the secure sharing of threat intelligence information among security professionals through its API platform (Chantzios et al., 2019). McAfee Threat Intelligence Exchange, HP Threat Central, and IBM X-Force Exchange are also notable proprietary platforms, each offering unique capabilities in threat detection, aggregation, and response (Hadi et al., 2023).

Conversely, open-source TIPs like the MISP provide collaborative platforms for sharing threat data and IoCs among community members (Chantzios et al., 2019). MISP, in particular, facilitates sharing incident data and IoCs to enhance incident response capabilities across organizations and national Computer Emergency Teams (CERTs). Its flexible data model and sharing mechanisms enable effective collaboration and information exchange among diverse stakeholders, enhancing collective defense against cyber threats.

OpenCTI, initiated by the French National Cyber Security Agency (ANSSI) in collaboration with the EU Computer Emergency Team (CERT-EU) in September 2018, embodies

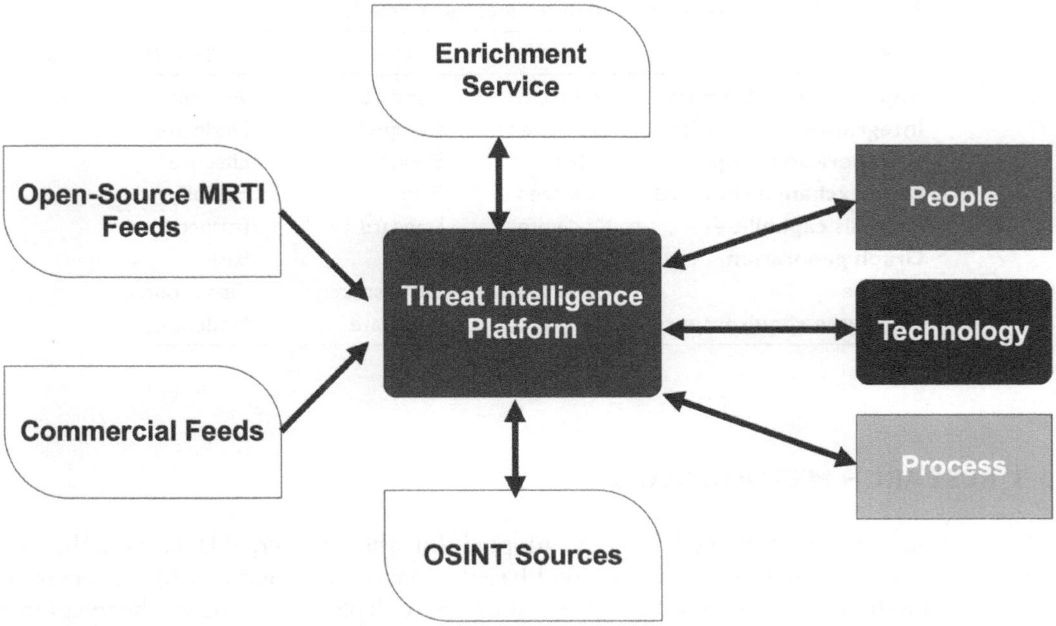

Figure 6.1 Threat intelligence platform architecture.

a significant step toward fostering CTI sharing while fortifying defenses against digital threats. This open-source platform serves as a comprehensive repository for structuring, organizing, visualizing, and disseminating data on cyber threats, catering to diverse levels of sophistication within the information security community. ANSSI continually enhances and shares information through OpenCTI, bolstering its capabilities to anticipate and counter evolving threats (Schlette et al., 2021). By leveraging OpenCTI, ANSSI and its collaborators aim to augment their collective understanding of digital threats, facilitating more effective threat mitigation strategies. OpenCTI adheres to STIX standards for data organization and boasts a modern web application interface, ensuring seamless integration with other cybersecurity measures and frameworks such as MISP, The Hive, and MITRE ATT&CK. Its data model enables the incorporation of specialized and nonspecialized data, with features like data linking, trust levels, and customizable datasets empowering users to derive actionable insights from diverse sources. Another noteworthy platform in threat intelligence is the Collective Intelligence Framework (CIF), which was established within the Research and Education Networks Information Sharing and Analysis Center (REN-ISAC). CIF's primary objective is to aggregate security data from various sources and facilitate its analysis, correlation, and dissemination among stakeholders. Implementing the IODEF standard, CIF ensures comprehensive data collection for incident reporting, generating periodic feeds of recent threat reports based on distinctive identifiers such as IP addresses and URLs. With over 1,000 subscribers worldwide, CIF delivers information in various formats, including STIX, JSON, CSV, and Snort pairs, catering to the diverse needs of formal and informal CERTs, private investigators, and industry security groups. Table 6.2 compares the three TIPs based on their import/export formats, integration capabilities, collaboration support, data exchange standards, analysis capabilities, graph generation, license, and hardware requirements.

Table 6.2 Comparative analysis of threat intelligence platforms

Criteria	*MISP*	*CIF*	*OpenCTI*
Import/export formats	Diverse	Standard	Versatile
Integration capabilities	Extensive	Comprehensive	Moderate
Collaboration support	Robust	Strong	Effective
Data exchange standard	Advanced	Basic	Flexible
Analysis capabilities	Moderate	Standard	Balanced
Graph generation	Basic	Basic	Basic
License	Open source	Open source	Open source
Hardware requirements	High	Moderate	Moderate

6.3 RESEARCH METHODOLOGY

This section outlines the methodology adopted for the Docker, MISP, Wazuh, and OpenSearch implementation within a cloud-based Amazon Web Services (AWS) environment. The methodology encompasses several key steps to deploy and evaluate the integrated security ecosystem systematically.

6.3.1 Infrastructure design

The initial phase of the research involved designing the cloud-based infrastructure on AWS to support the deployment of Docker, MISP, Wazuh, and OpenSearch. This process included selecting appropriate EC2 instances, configuring networking settings, and establishing security groups to ensure a secure and scalable environment.

6.3.2 Component selection

Each component (Docker, MISP, Wazuh, OpenSearch) was carefully selected based on ease of use, flexibility, security features, and cost-effectiveness criteria. Alternative technologies were considered and evaluated to determine the most suitable options for the project's requirements.

6.3.3 Configuration and integration

Following component selection, meticulous configuration and integration efforts were undertaken. Docker containers were deployed using Docker Compose to ensure consistency and efficiency in managing the containerized infrastructure. Configuration files for MISP, Wazuh, and OpenSearch were customized to align with the project's objectives and security requirements.

6.3.4 Testing and validation

Testing was conducted post-deployment to validate the functionality and performance of the integrated system. This encompassed testing for interoperability between components, assessing the system's resilience to cyber threats, and evaluating its scalability under varying workloads.

6.3.5 Feedback and iteration

Feedback was actively solicited from project stakeholders, including IT professionals, security analysts, and system administrators. This feedback guided iterative updates and enhancements to the implementation setup, addressing any identified issues, or shortcomings.

By adhering to this research methodology, the project establishes a robust and effective infrastructure for deploying Docker, MISP, Wazuh, and OpenSearch within a cloud environment. This infrastructure aims to enhance organizational cybersecurity capabilities and resilience against cyber threats.

6.4 THE PROPOSED MODEL

The design of the preventive security model's architecture relies on the synergistic integration of open-source tools such as Docker, MISP, Wazuh, and OpenSearch to enhance the preventive detection of threats. Figure 6.2 presents the proposed architecture.

Integrating IoCs between MISP and Wazuh strengthens preventive detection by providing rich contextual information for each generated alert. This interaction allows in-depth analysis of suspicious activities by associating Wazuh alerts with complementary data from MISP, as shown in Figure 6.3.

- **Data collector:**
  Wazuh Agents are deployed on each system within your infrastructure. These agents collect local security logs and events, capturing detailed information about system activity. Wazuh Agents can be configured to monitor specific configuration files, system logs, registry changes, etc.
- **Interaction and aggregation:**
  Wazuh Agents, with their tailored configuration, aggregate data from local logs, Sysmon alerts, and OSSEC alerts. These aggregated data are then transmitted to the central Wazuh management server, which is stored and indexed in OpenSearch for further analysis. Wazuh and OSSEC detection rules are applied to these aggregated data, generating relevant alerts based on abnormal behaviors or detected threat signatures. Information enriched by MISP is also associated with these alerts, reinforcing their context and understanding.

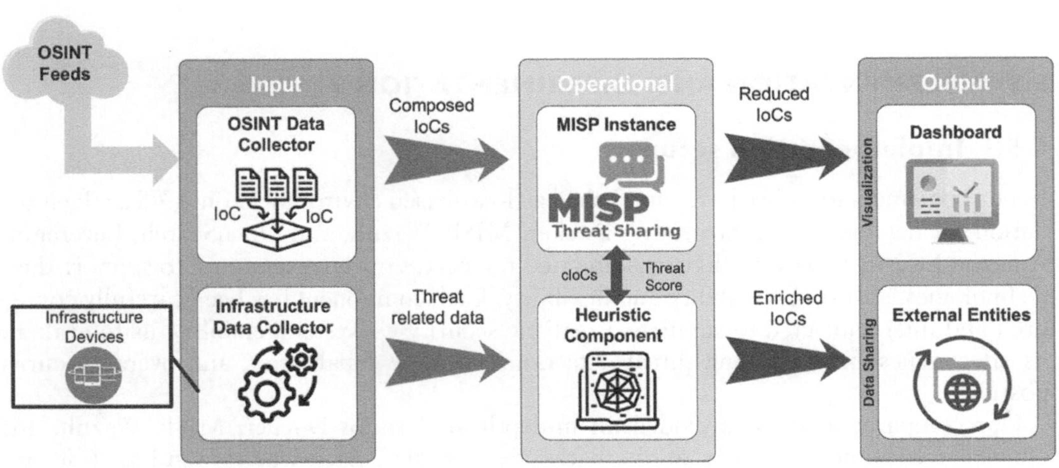

Figure 6.2 The proposed security analysis module architecture.

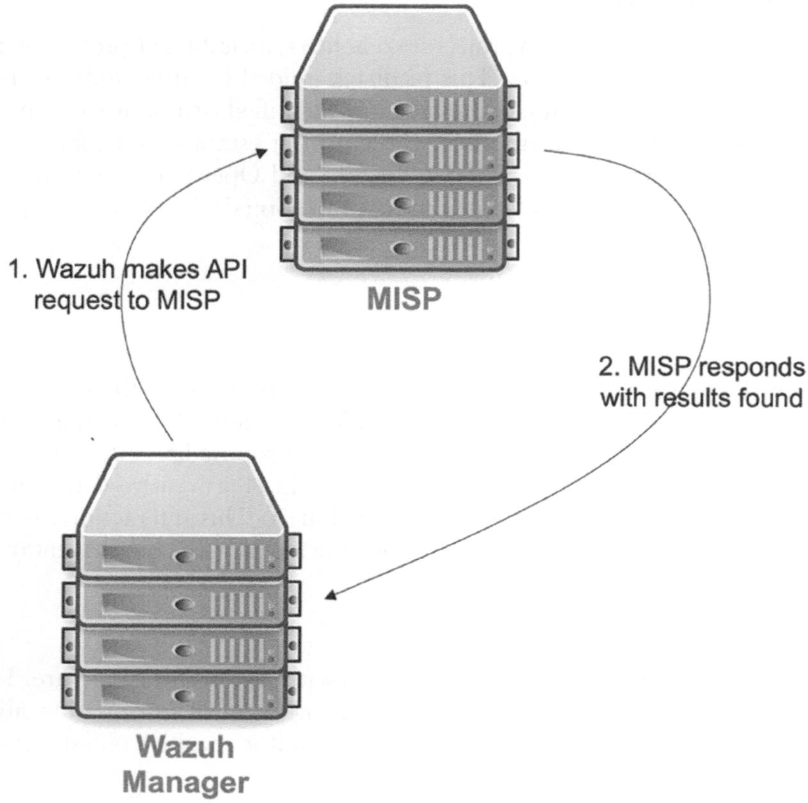

Figure 6.3 Integration of Wazuh and MISP.

- **Visualization in Wazuh with OpenSearch:**
  Data aggregated by Wazuh, from agents and Sysmon, are stored and indexed in OpenSearch. Wazuh utilizes OpenSearch features to generate meaningful visualizations directly within the Wazuh interface. Wazuh Dashboards, powered by OpenSearch data, offer a consolidated view of security activities, highlighting priority alerts and trends.

6.5 IMPLEMENTATION AND EXPERIMENTATION

6.5.1 Implementation setup

In our implementation, we have established a cloud-based environment on AWS as the foundation for deploying and integrating Docker, MISP, Wazuh, and OpenSearch. Leveraging Amazon EC2 instances, we have instantiated the necessary infrastructure to support these technologies, ensuring scalability and flexibility. Each component has been carefully configured and interconnected to simulate a realistic security ecosystem, enabling us to evaluate its effectiveness in enhancing threat detection, response capabilities, and overall security posture.

Our comparative study provided an in-depth analysis of Docker, MISP, Wazuh, and OpenSearch, comparing them to alternative technologies. Each was chosen based on specific criteria like ease of use, flexibility, security, customization, and cost, aligning with our organization's needs and enhancing our security posture. The subsequent sections will cover the implementation setup, evaluation results, and recommendations: Docker Selection: Chosen for ease of use, efficient container management, and lightweight isolation and preferred over alternatives like Podman and Kubernetes due to popularity and community support. MISP Usage: Selected for excellence in sharing IoCs and facilitating open collaboration. Preferred over ThreatConnect and Anomali for its open-source nature and community support. Wazuh Selection: Chosen for robust real-time threat detection and effective log correlation. Preferred over Snort and Suricata for their versatility and customization. OpenSearch Adoption was chosen for its open-source license, performance, and native security integration, preferred over Elasticsearch and Splunk for its open-source nature and cost-effectiveness.

6.5.2 Integration and implementation

6.5.2.1 Component deployment

6.5.2.1.1 Configure and deploy Wazuh SIEM with Docker Compose container

Wazuh containers are deployed via Docker Compose, ensuring consistent security infrastructure implementation. Each Wazuh component, with its specific configurations, helps to strengthen threat monitoring and response.

The Wazuh Manager service is crucial for collecting, analyzing, and managing security data. It uses the Wazuh/Wazuh Manager: 4.9.0 image and exposes several ports, including port 1514 for standard communication, port 1515 for secure connections, port 514 in UDP mode for logs, and port 55000 for internal communication. The Manager is configured to connect to the Wazuh Indexer and uses volumes to store configuration, logs, and other essential data (Figure 6.4).

Wazuh Indexer: The Wazuh Indexer service, based on the Wazuh/Wazuh Indexer: 4.9.0 image, is responsible for indexing security data. It exposes port 9200 and uses volumes to store the indexed data and SSL certificates required for secure communication (Figure 6.5).

Wazuh Dashboard: The Wazuh Dashboard provides a graphical interface for viewing data using the Wazuh/Wazuh Dashboard: 4.9.0 image. It can be accessed via port 443, and SSL certificates are configured to ensure secure communication. The Dashboard is linked to the Wazuh Indexer and Manager for seamless integration (Figure 6.6).

Generating certificates for the Wazuh Indexer: The "dock er-compose -f generate-indexer-certs.yml run - rm generator" command is used to run the "generator" service defined in the "generate-indexer-certs.yml" Docker Compose configuration file. This service generates the certificates required by the Wazuh Indexer. Once the operation is complete, the temporary container is automatically deleted (-rm) (Figure 6.7).

Executing the code, we now execute the "docker- compose up" command. This command is crucial because it orchestrates the launch of all the services defined in the Docker Compose configuration file. When you run this command, Docker Compose will create and start all the containers required according to the parameters specified.

Services such as the Wazuh Manager, Wazuh Indexer, and Wazuh Dashboard will be instantiated and interconnected according to the specifications in the "docker-compose. yml" file. In addition, volumes defined to store data, configurations and SSL certificates will be created and linked to the corresponding containers.

Executing this command marks an essential stage in deploying the infrastructure, ensuring that all components are operational and ready to provide proactive security monitoring. The console will display the start-up logs for the various services, enabling the start-up

```
# Wazuh App Copyright (C) 2017, Wazuh Inc. (License GPLv2)
version: '3.7'

services:
  wazuh.manager:
    image: wazuh/wazuh-manager:5.0.0
    hostname: wazuh.manager
    restart: always
    ulimits:
      memlock:
        soft: -1
        hard: -1
      nofile:
        soft: 655360
        hard: 655360
    ports:
      - "1514:1514"
      - "1515:1515"
      - "514:514/udp"
      - "55000:55000"
    environment:
      - INDEXER_URL=https://wazuh.indexer:9200
      - INDEXER_USERNAME=admin
      - INDEXER_PASSWORD=SecretPassword
      - FILEBEAT_SSL_VERIFICATION_MODE=full
      - SSL_CERTIFICATE_AUTHORITIES=/etc/ssl/root-ca.pem
      - SSL_CERTIFICATE=/etc/ssl/filebeat.pem
      - SSL_KEY=/etc/ssl/filebeat.key
      - API_USERNAME=wazuh-wui
      - API_PASSWORD=MyS3cr37P450r.*-
    volumes:
      - wazuh_api_configuration:/var/ossec/api/configuration
      - wazuh_etc:/var/ossec/etc
      - wazuh_logs:/var/ossec/logs
      - wazuh_queue:/var/ossec/queue
      - wazuh_var_multigroups:/var/ossec/var/multigroups
      - wazuh_integrations:/var/ossec/integrations
      - wazuh_active_response:/var/ossec/active-response/bin
      - wazuh_agentless:/var/ossec/agentless
      - wazuh_wodles:/var/ossec/wodles
      - filebeat_etc:/etc/filebeat
      - filebeat_var:/var/lib/filebeat
      - ./config/wazuh_indexer_ssl_certs/root-ca-manager.pem:/etc/ssl/root-ca.pem
      - ./config/wazuh_indexer_ssl_certs/wazuh.manager.pem:/etc/ssl/filebeat.pem
      - ./config/wazuh_indexer_ssl_certs/wazuh.manager-key.pem:/etc/ssl/filebeat.key
      - ./config/wazuh_cluster/wazuh_manager.conf:/wazuh-config-mount/etc/ossec.conf
```

Figure 6.4 The Wazuh manager part of the Docker compose code.

```
..
  wazuh.indexer:
    image: wazuh/wazuh-indexer:5.0.0
    hostname: wazuh.indexer
    restart: always
    ports:
      - "9200:9200"
    environment:
      - "OPENSEARCH_JAVA_OPTS=-Xms1g -Xmx1g"
    ulimits:
      memlock:
        soft: -1
        hard: -1
      nofile:
        soft: 65536
        hard: 65536
    volumes:
      - wazuh-indexer-data:/var/lib/wazuh-indexer
      - ./config/wazuh_indexer_ssl_certs/root-ca.pem:/usr/share/wazuh-indexer/certs/root-ca.pem
      - ./config/wazuh_indexer_ssl_certs/wazuh.indexer-key.pem:/usr/share/wazuh-indexer/certs/wazuh.indexer.key
      - ./config/wazuh_indexer_ssl_certs/wazuh.indexer.pem:/usr/share/wazuh-indexer/certs/wazuh.indexer.pem
      - ./config/wazuh_indexer_ssl_certs/admin.pem:/usr/share/wazuh-indexer/certs/admin.pem
      - ./config/wazuh_indexer_ssl_certs/admin-key.pem:/usr/share/wazuh-indexer/certs/admin-key.pem
      - ./config/wazuh_indexer/wazuh.indexer.yml:/usr/share/wazuh-indexer/opensearch.yml
      - ./config/wazuh_indexer/internal_users.yml:/usr/share/wazuh-indexer/opensearch-security/internal_users.yml
```

Figure 6.5 Wazuh part indexed in Docker compose code.

```
  wazuh.dashboard:
    image: wazuh/wazuh-dashboard:5.0.0
    hostname: wazuh.dashboard
    restart: always
    ports:
      - 443:5601
    environment:
      - INDEXER_USERNAME=admin
      - INDEXER_PASSWORD=SecretPassword
      - WAZUH_API_URL=https://wazuh.manager
      - DASHBOARD_USERNAME=kibanaserver
      - DASHBOARD_PASSWORD=kibanaserver
      - API_USERNAME=wazuh-wui
      - API_PASSWORD=MyS3cr37P450r.*-
    volumes:
      - ./config/wazuh_indexer_ssl_certs/wazuh.dashboard.pem:/usr/share/wazuh-dashboard/certs/wazuh-dashboard.pem
      - ./config/wazuh_indexer_ssl_certs/wazuh.dashboard-key.pem:/usr/share/wazuh-dashboard/certs/wazuh-dashboard-key.pem
      - ./config/wazuh_indexer_ssl_certs/root-ca.pem:/usr/share/wazuh-dashboard/certs/root-ca.pem
      - ./config/wazuh_dashboard/opensearch_dashboards.yml:/usr/share/wazuh-dashboard/config/opensearch_dashboards.yml
      - ./config/wazuh_dashboard/wazuh.yml:/usr/share/wazuh-dashboard/data/wazuh/config/wazuh.yml
      - wazuh-dashboard-config:/usr/share/wazuh-dashboard/data/wazuh/config
      - wazuh-dashboard-custom:/usr/share/wazuh-dashboard/plugins/wazuh/public/assets/custom
    depends_on:
      - wazuh.indexer
    links:
      - wazuh.indexer:wazuh.indexer
      - wazuh.manager:wazuh.manager
```

Figure 6.6 The Wazuh Dashboard part of the Docker compose code.

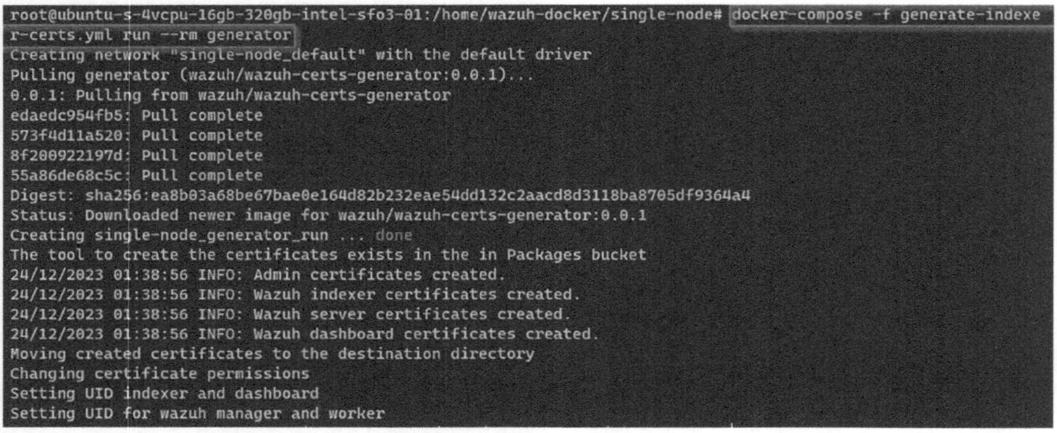

Figure 6.7 Generating SSL certificates for Wazuh Indexer.

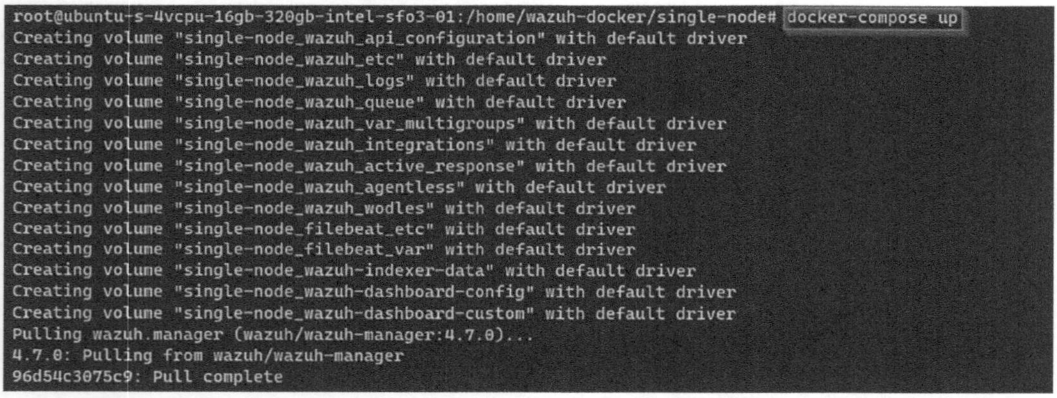

Figure 6.8 Executing the code.

process to be monitored and any potential problems to be identified. Once this stage has been successfully completed, the entire Wazuh system will be operational and ready to contribute to preventive network security (Figure 6.8).

6.5.2.1.2 Discovering the Wazuh Dashboard and launching a client

Wazuh Login page Discovering the Wazuh Dashboard is a crucial step in the implementation of this security solution. Once the infrastructure is deployed and the Wazuh services, such as the manager and indexer, are operational, access to the dashboard becomes possible. By launching a web browser and accessing the URL http://wazuh-ip:5601 specified in the dashboard configuration, users are greeted by an intuitive interface offering a real-time view of safety data, alerts, and events (Figure 6.9).

Once the credentials specified in the YAML code have been entered, access to the Wazuh Dashboard is opened. These credentials, defined in the dashboard configuration, allow authorized users to log in and explore the dashboard's features. Through this interface, administrators and security analysts can view security data in real time, manage system-generated alerts, and perform in-depth analyses of detected events.

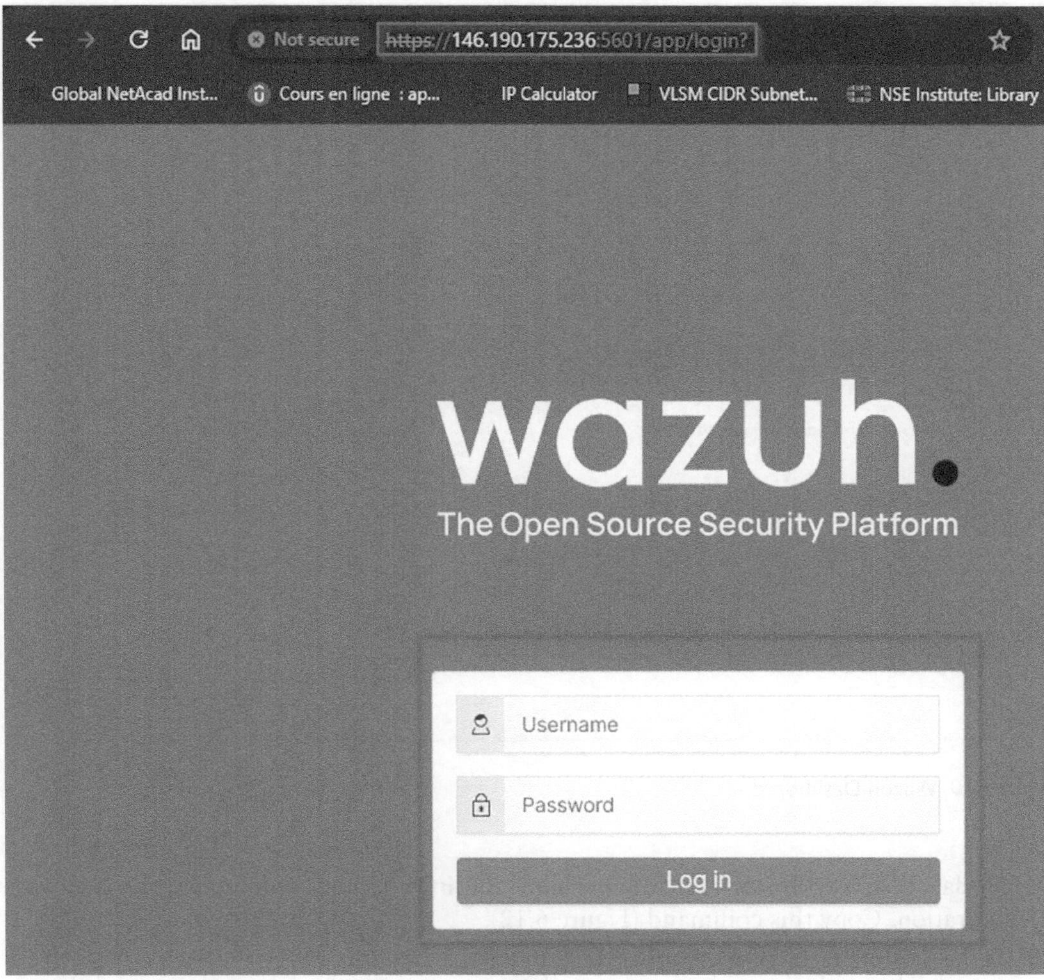

Figure 6.9 Wazuh login page.

- **Wazuh Dashboard:**
  The Wazuh Dashboard offers a user-friendly, customizable interface, making it easy to monitor the network proactively. Secure access to this centralized platform is an essential element of effective security management, strengthening the overall posture of the infrastructure in the face of potential threats. Once users have successfully entered their credentials, they are ready to explore the various features of the Wazuh Dashboard to ensure preventive and reactive security (Figure 6.10).
- **Launch of a Wazuh client:**
  The simplified approach to deploying a Wazuh agent on a Windows system is via the Wazuh Dashboard. Once logged in, go to the "Agents" section in the navigation menu. Then, choose the "Deploy a New Agent" option. This will take you to an interface where you can specify the agent's details (Figure 6.11).

In this interface, select the agent's operating system (Windows), give it a name, and provide the IP address of the Wazuh server to which the agent should connect.

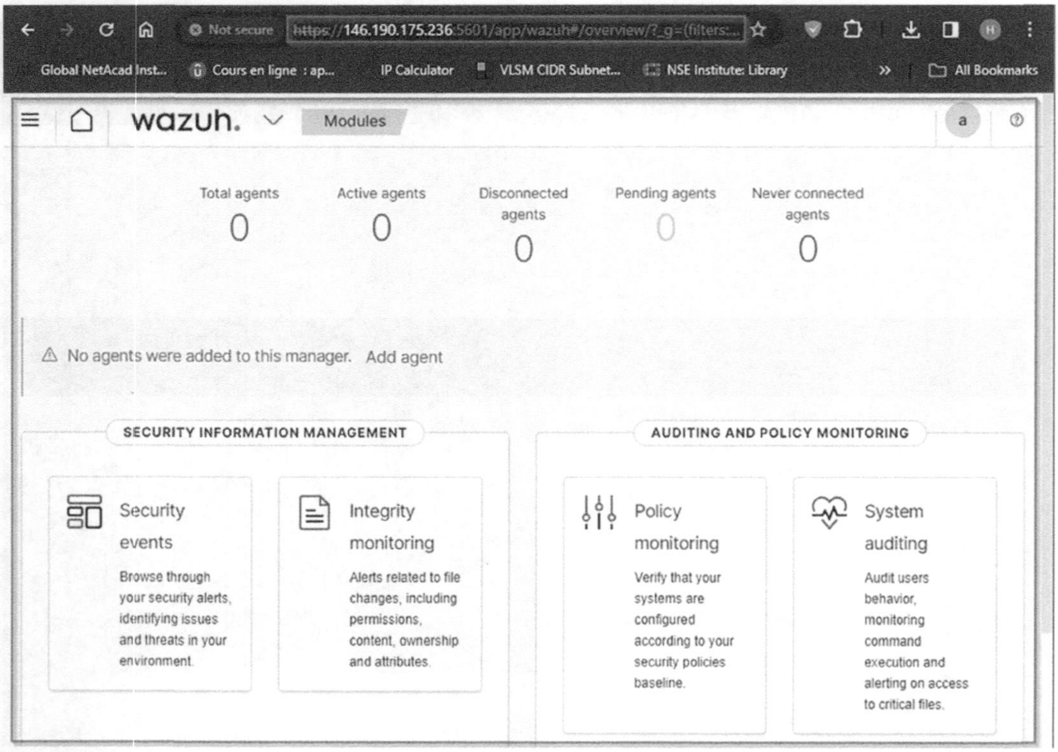

Figure 6.10 Wazuh Dashboard.

The dashboard will automatically generate the installation command specific to your configuration. Copy this command (Figure 6.12).

Open a command window on the target Windows system, paste and run the installation command. This will trigger the agent installation process on the system (Figure 6.13).

This simplified approach is fast and easy to use, enabling users to easily configure and install security agents on their Windows systems without manually manipulating complex configuration files.

6.5.2.1.3 Development of composite Docker code for MISP

MISP containers can also be deployed using Docker Compose, ensuring a structured infrastructure implementation dedicated to threat information management.

Each MISP component, with its specific configurations, enhances the ability to share, and analyze safety data.

Mail, Redis, and DB: These services play auxiliary roles in the MISP ecosystem. The messaging service (Mail) ensures the ability to relay via various platforms, while Redis and the MySQL database (DB) are used for data management (Figure 6.14).

MISP: The MISP service, based on the coolacid/misp-docker:core-latest image, is the system's heart. It is responsible for managing threat information. The deployment exposes ports 80 and 443 for web communication and uses volumes to store the configuration, logs,

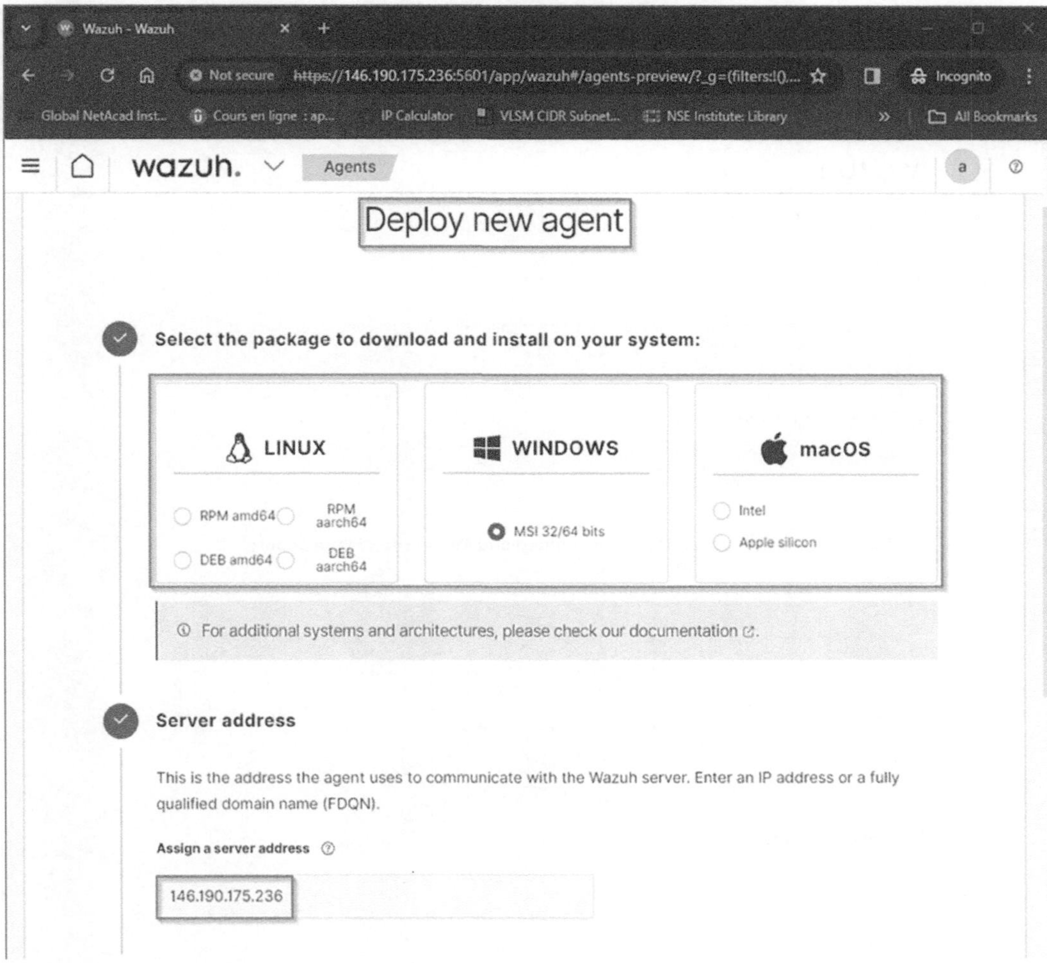

Figure 6.11 First agent deployment stage.

files, and SSL certificates needed to secure communications. Environment variables such as BASEURL and REDIS FQDN are configured to ensure the system operates correctly (Figure 6.15).

Deploying MISP via Docker Compose simplifies the configuration and management of the entire infrastructure, enabling the consistent and efficient implementation of this threat information-sharing system.

6.5.2.1.4 Deployment of containers for MISP

Adjustments to the configuration files are necessary to deploy MISP in an environment that allows access from points other than the local host. Once MISP has been deployed, access the configuration files in the "/var/www/MISP/app/Config/" directory. Open the "config.php" and "server.php" files in a text editor.

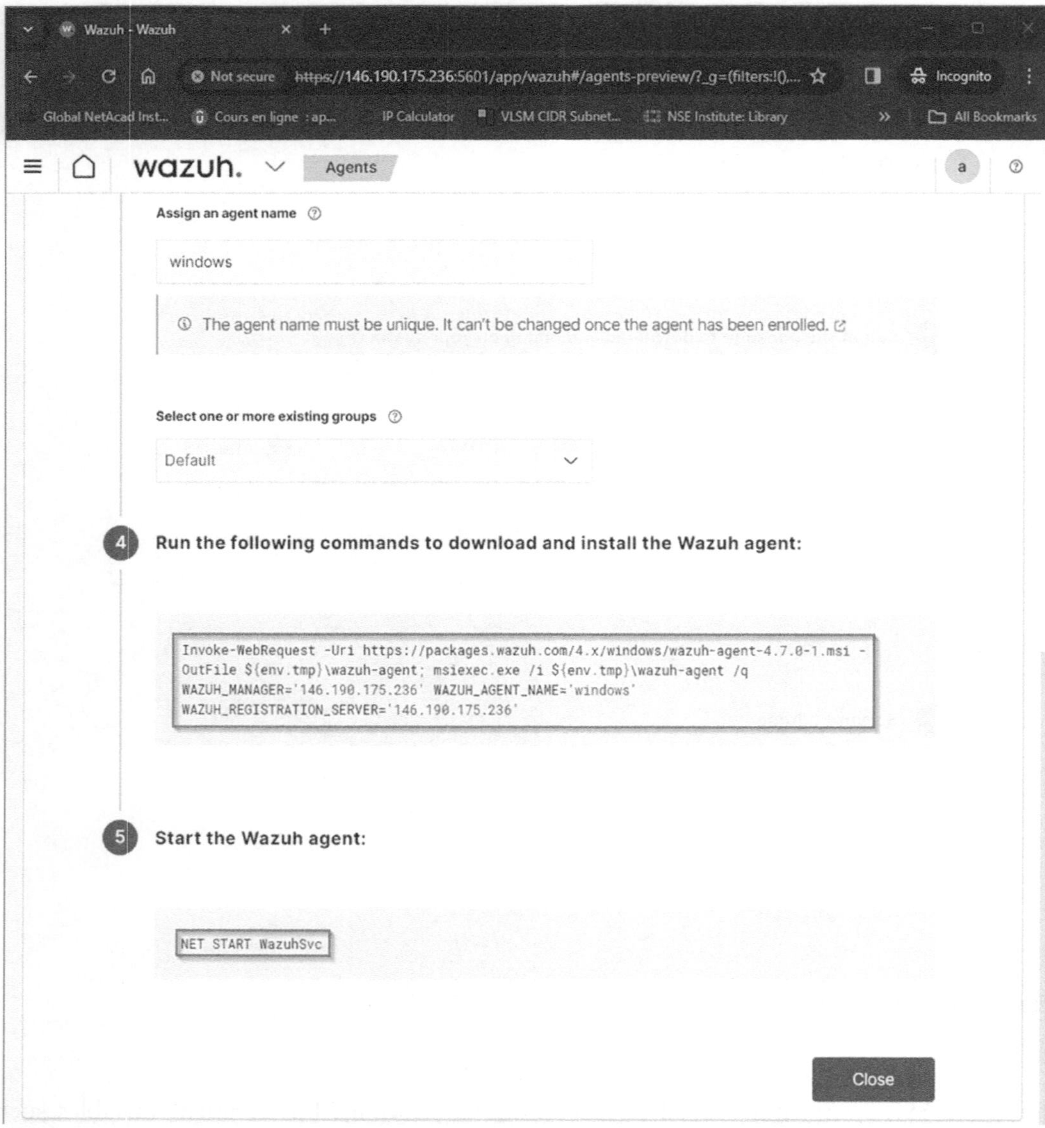

Figure 6.12 Second agent deployment stage.

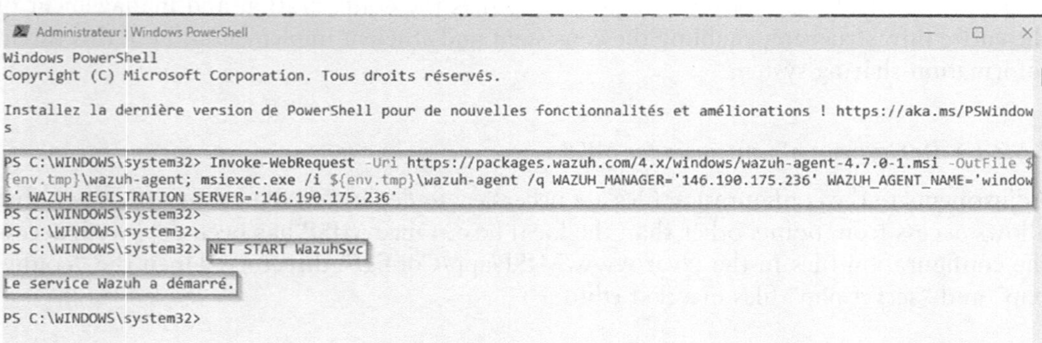

Figure 6.13 Third agent deployment stage.

```
version: '3'
services:
  # This is capable to relay via gmail, Amazon SES, or generic relays
  # See: https://hub.docker.com/r/ixdotai/smtp
  mail:
    image: ixdotai/smtp
    environment:
      - "SMARTHOST_ADDRESS=${SMARTHOST_ADDRESS}"
      - "SMARTHOST_PORT=${SMARTHOST_PORT}"
      - "SMARTHOST_USER=${SMARTHOST_USER}"
      - "SMARTHOST_PASSWORD=${SMARTHOST_PASSWORD}"
      - "SMARTHOST_ALIASES=${SMARTHOST_ALIASES}"

  redis:
    image: redis:7.2

  db:
    # We use MariaDB because it supports ARM and has the expected collations
    image: mariadb:10.11
    restart: always
    environment:
      - "MYSQL_USER=${MYSQL_USER:-misp}"
      - "MYSQL_PASSWORD=${MYSQL_PASSWORD:-example}"
      - "MYSQL_ROOT_PASSWORD=${MYSQL_ROOT_PASSWORD:-password}"
      - "MYSQL_DATABASE=${MYSQL_DATABASE:-misp}"
    volumes:
      - mysql_data:/var/lib/mysql
    cap_add:
      - SYS_NICE  # CAP_SYS_NICE Prevent runaway mysql log
```

Figure 6.14 Mail, Redis, and DB parts in Docker compose code.

In these files, look for occurrences of "localhost" or "127.0.0.1" and replace them with the public IP address of your server. Change parameters such as "baseurl" in "config.php" and "fullBaseUrl" in "server.php." After making these adjustments, restart the MISP service to apply the changes (Figure 6.16).

This modification allows MISP to be accessed outside using the server's public IP address. Also, ensure that the necessary ports are open and properly redirected to your firewall. These adjustments make it easier to access the MISP interface from locations other than localhost, improving the platform's connectivity and usability. The next step in our deployment involves running the Docker Compose configuration files using the following command (Figure 6.17).

Combining the configurations specified in the "docker-compose.yaml" and "build-docker-compose. yaml" files, the "docker-compose up -d" command is used to orchestrate deploying the Docker containers associated with our environment. The "-d" option runs the containers in detached mode, freeing up the console while keeping the services running in the background.

```
  misp-core:
    image: coolacid/misp-docker:modules-latest
    build:
      context: core/.
      args:
          - CORE_TAG=${CORE_TAG}
          - CORE_COMMIT=${CORE_COMMIT}
          - PHP_VER=${PHP_VER}
          - PYPI_REDIS_VERSION=${PYPI_REDIS_VERSION}
          - PYPI_LIEF_VERSION=${PYPI_LIEF_VERSION}
          - PYPI_PYDEEP2_VERSION=${PYPI_PYDEEP2_VERSION}
          - PYPI_PYTHON_MAGIC_VERSION=${PYPI_PYTHON_MAGIC_VERSION}
          - PYPI_MISP_LIB_STIX2_VERSION=${PYPI_MISP_LIB_STIX2_VERSION}
          - PYPI_MAEC_VERSION=${PYPI_MAEC_VERSION}
          - PYPI_MIXBOX_VERSION=${PYPI_MIXBOX_VERSION}
          - PYPI_CYBOX_VERSION=${PYPI_CYBOX_VERSION}
          - PYPI_PYMISP_VERSION=${PYPI_PYMISP_VERSION}
    depends_on:
      - redis
      - db
volumes:
    mysql_data:
```

Figure 6.15 MISP modules in Docker compose code.

```
root@ubuntu-s-4vcpu-16gb-320gb-intel-sfo3-01:/home# grep -r "localhost"
docker-misp/server-configs/email.php:    // to set the return-path header, simply uncomment the line below and change you
@localhost to the desired e-mail address
docker-misp/server/files/entrypoint_nginx.sh:    sed -i "s/'host' => 'localhost'.*/'host' => '$REDIS_FQDN',          \/\
/ Redis server hostname/" "/var/www/MISP/app/Plugin/CakeResque/Config/config.php"
docker-misp/server/files/entrypoint_nginx.sh:    sed -i "s/localhost/$MYSQL_HOST/" $MISP_APP_CONFIG_PATH/database.php
docker-misp/server/files/entrypoint_nginx.sh:        openssl req -x509 -subj '/CN=localhost' -nodes -newkey rsa:4096 -ke
yout key.pem -out cert.pem -days 365
docker-misp/docker-compose.yml:      - "BASEURL=https://localhost"
docker-misp/README.md:-   Login to 'https://localhost'
root@ubuntu-s-4vcpu-16gb-320gb-intel-sfo3-01:/home# grep -r "127.0.0.1"
docker-misp/server/files/entrypoint_nginx.sh:    sed -i "s/'host' => '127.0.0.1'.*/'host' => '$REDIS_FQDN',          \/\
/ Redis server hostname/" "/var/www/MISP/app/Plugin/CakeResque/Config/config.php"
```

Figure 6.16 Occurrences of localhost and 127.0.0.1.

```
root@ubuntu-s-4vcpu-16gb-320gb-intel-sfo3-01:/home/docker-misp# docker-compose -f docker-compose.yml -f build-docker-com
pose.yml up
Creating network "docker-misp_default" with the default driver
Creating volume "docker-misp_mysql_data" with default driver
Pulling db (mysql:8.0.19)...
8.0.19: Pulling from library/mysql
54fec2fa59d0: Pull complete
bcc6c6145912: Pull complete
951c3d959c9d: Pull complete
05de4d0e206e: Pull complete
319f0394ef42: Pull complete
d9185034607b: Pull complete
013a9c64dadc: Pull complete
96d4c3d31f9f: Pull complete
785bc90808da: Pull complete
1339cf094729: Pull complete
beb8f531cc37: Pull complete
```

Figure 6.17 Code execution for MISP.

The "docker-compose. yaml" file defines the basic configuration for deploying services, while the "build- docker-compose. yaml" file can include specific adjustments relating to constructing Docker images or other parameters.

By running the docker ps command, you can check that the deployment has been carried out correctly. This command displays a list of Docker containers currently running on your system, providing visual confirmation that the services specified in your Docker Compose configuration files are operational.

6.5.2.1.5 Discovering MISP functionalities

MISP Login page: The MISP login page is the secure home interface where users can login to access the platform. Once MISP has been successfully deployed, you can access this page by using a web browser and entering the public IP address of your server or the associated domain name, followed by the port (by default, port 80 for HTTP or port 443 for HTTPS).

You will be asked to provide your user details on the login page, usually a username and password. This information is specified during the initial configuration of MISP. Once you have entered the correct information and pressed "Login" (or similar), you will be directed to the main MISP dashboard (Figure 6.18).

MISP Events page: The MISP Events page is an essential interface for incident and threat management. Accessing this page after logging in, users are presented with a list of events containing key information such as ID, creation date, category, and number of associated attributes. This interface provides a convenient overview, allowing users to filter, search, and view each event in detail. Users can access in-depth information by clicking on a specific event, including associated attributes, relationships, and other metadata. The events page is central to collaboration, allowing users to share, comment on, and contextualize threat-related information (Figure 6.19).

To add an organization to MISP, you can follow a simple procedure. After logging into MISP as an administrator, go to the "Administration" section via the top navigation bar. Once in this section, look for the "Organizations" option and click on it. You will be taken to a page where you can add a new organization by providing details such as name,

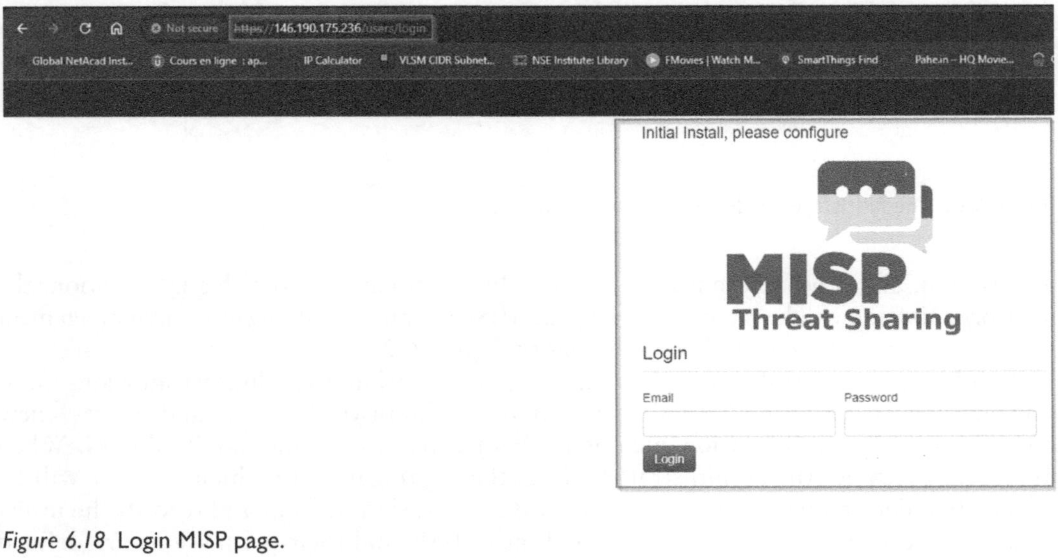

Figure 6.18 Login MISP page.

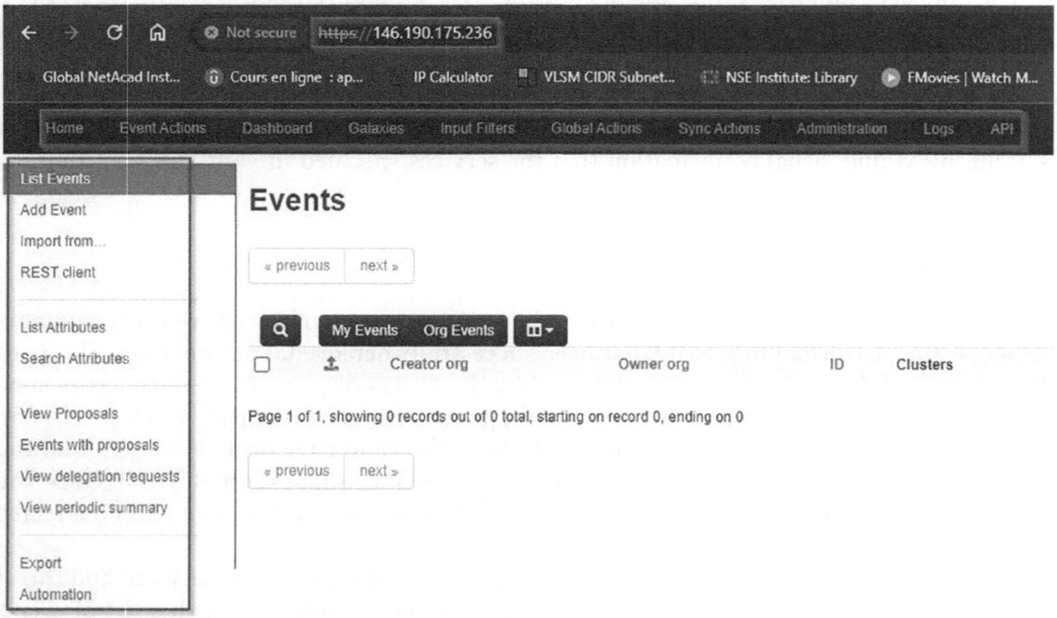

Figure 6.19 Events page MISP.

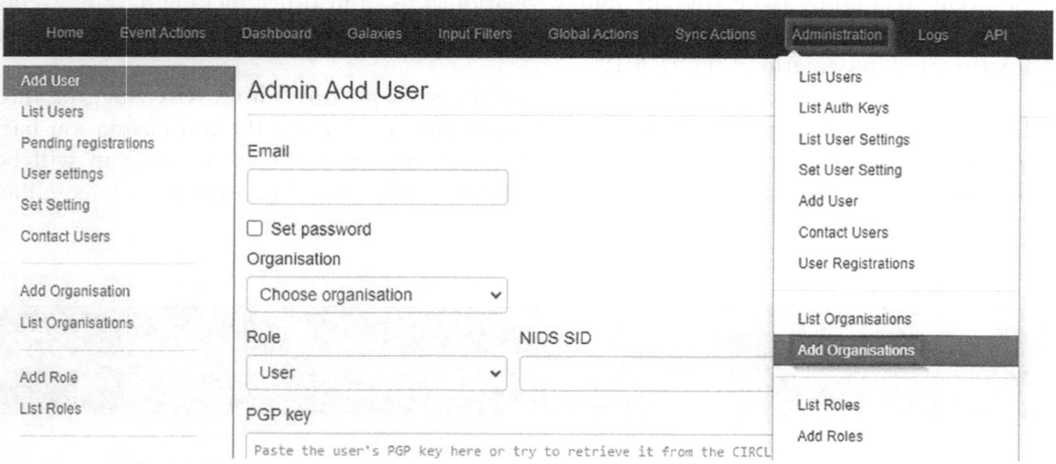

Figure 6.20 The first stage in adding an organization.

description, and other relevant information. Once you have entered this information, the new organization will be integrated into your MISP instance, enabling efficient management of information and events linked to this entity (Figure 6.20).

To add a user to MISP and link them to an organization, the administrator logs on to the platform, accesses the "Users" section in the "Administration" tab, and creates a new user by providing information such as username, password, and email address. When the user is created, the administrator selects the organization to which the user will be associated, determines their role (administrator, user, sharer, etc.), and records the information. In this way, the user is integrated into MISP, and their permissions are defined according to their role within the specified organization, facilitating targeted management of access and responsibilities (Figure 6.21).

To integrate feeds into MISP, users can follow a simple process via the graphical interface. After logging into MISP, access the "Feed Management" section and click on the "Feeds" option. Then, synchronization allows the user to select which feeds to add, such as the default CIRCL OSINT feed and the botvrij.eu data feed. This will enable MISP to connect to the selected feeds and automatically update the DB with threat information from these sources. Adding these feeds enhances the wealth of data available in MISP, improving the ability to detect and respond to threats (Figure 6.22).

6.5.2.2 *Configuration and customization*

In this customization stage, we developed the custom-misp.py file to reinforce the added value of our Wazuh–MISP integration project. This script is essential in proactively searching the MISP DB for IoCs in response to alerts generated by Wazuh.

6.5.2.2.1 *Development of the custom-misp.py integration code*

Initialization section: This section imports the necessary modules, configures the path to the current directory and defines the socket address for communication with Wazuh (Figure 6.23).

Send Events function allows events to be sent to Wazuh via the socket, formatting the message according to the agent concerned (Figure 6.24).

Initialization of MISP Variables: This section initializes the variables needed to interact with the MISP API, specifying the base URL of the MISP server and the API authentication key (Figure 6.25).

Extraction of Sysmon Parameters: These lines extract information specific to the Sysmon event from the alert file, such as the source of the event, the type, and a regex template for SHA256 hashes (Figure 6.26).

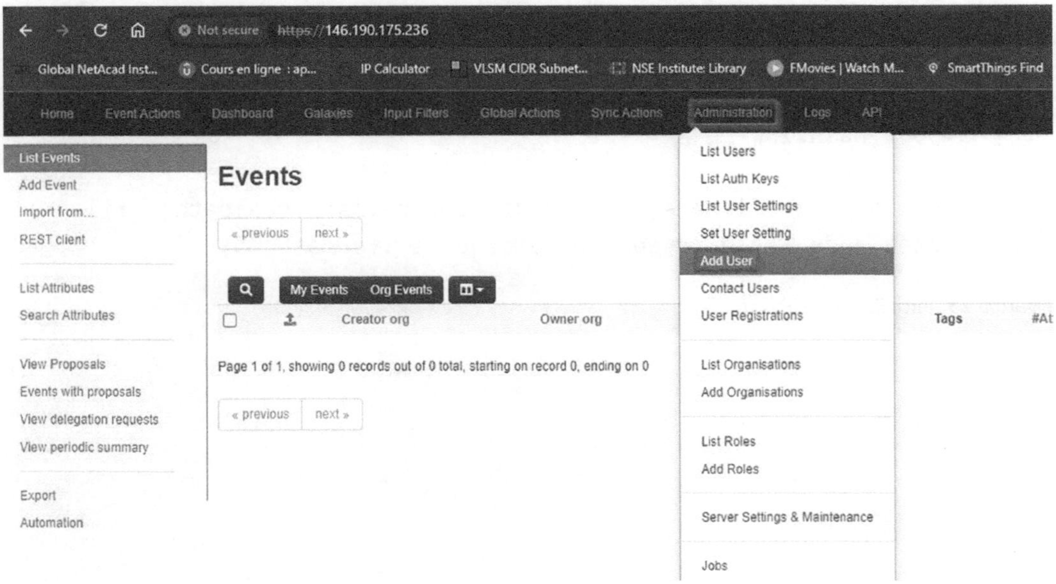

Figure 6.21 First step in adding a user.

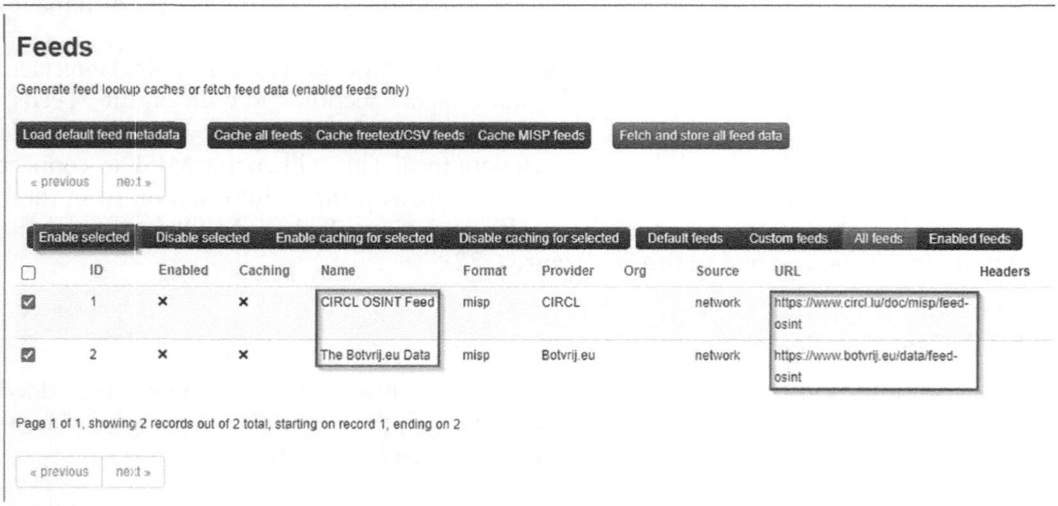

Figure 6.22 The second stage of adding a feed.

```
#!/var/ossec/framework/python/bin/python3
## MISP API Integration
#
import sys
import os
from socket import socket, AF_UNIX, SOCK_DGRAM
from datetime import date, datetime, timedelta
import time
import requests
from requests.exceptions import ConnectionError
import json
import ipaddress
import hashlib
import re
pwd = os.path.dirname(os.path.dirname(os.path.realpath(__file__)))
socket_addr = '{0}/queue/sockets/queue'.format(pwd)
```

Figure 6.23 Initialization section.

```
def send_event(msg, agent = None):
    if not agent or agent["id"] == "000":
        string = '1:misp:{0}'.format(json.dumps(msg))
    else:
        string = '1:[{0}] ({1}) {2}->misp:{3}'.format(agent["id"], agent["name"], agent["ip"] if "ip" in agent else "any", json.dumps(msg))
    sock = socket(AF_UNIX, SOCK_DGRAM)
    sock.connect(socket_addr)
    sock.send(string.encode())
    sock.close()
false = False
```

Figure 6.24 Send events function.

```python
# New Alert Output if MISP Alert or Error calling the API
alert_output = {}
# MISP Server Base URL
misp_base_url = "https://**your misp instance**/attributes/restSearch/"
# MISP Server API AUTH KEY
misp_api_auth_key = "*Your API Key"
# API - HTTP Headers
misp_apicall_headers = {"Content-Type":"application/json", "Authorization":f"{misp_api_auth_key}", "Accept":"application/json"}
```

Figure 6.25 Initializing MISP variables.

```python
## Extract Sysmon for Windows/Sysmon for Linux and Sysmon Event ID
event_source = alert["rule"]["groups"][0]
event_type = alert["rule"]["groups"][2]
## Regex Pattern used based on SHA256 lenght (64 characters)
regex_file_hash = re.compile('\w{64}')
```

Figure 6.26 Extraction of Sysmon parameters.

```python
if event_source == 'windows':
    if event_type == 'sysmon_event1':
        try:
            wazuh_event_param = regex_file_hash.search(alert["data"]["win"]["eventdata"]["hashes"]).group(0)
        except IndexError:
            sys.exit()
    elif event_type == 'sysmon_event3' and alert["data"]["win"]["eventdata"]["destinationIsIpv6"] == 'false':
        try:
            dst_ip = alert["data"]["win"]["eventdata"]["destinationIp"]
            if ipaddress.ip_address(dst_ip).is_global:
                wazuh_event_param = dst_ip
            else:
                sys.exit()
        except IndexError:
            sys.exit()
    elif event_type == 'sysmon_event3' and alert_output["data"]["win"]["eventdata"]["destinationIsIpv6"] == 'true':
        sys.exit()
    elif event_type == 'sysmon_event6':
        try:
            wazuh_event_param = regex_file_hash.search(alert["data"]["win"]["eventdata"]["hashes"]).group(0)
        except IndexError:
            sys.exit()
    elif event_type == 'sysmon_event7':
        try:
            wazuh_event_param = regex_file_hash.search(alert["data"]["win"]["eventdata"]["hashes"]).group(0)
        except IndexError:
            sys.exit()
```

Figure 6.27 Processing Windows Sysmon events.

Processing Windows Sysmon Events: This part manages the logic specific to each type of Sysmon event by constructing appropriate MISP search requests, querying the MISP API, and processing the response (Figure 6.27).

Sending Alerts to Wazuh after MISP Search: If the MISP response contains attributes (IoCs), the script generates an MISP alert object and sends it to Wazuh (Figure 6.28).

```
            if (misp_api_response["response"]["Attribute"]):
        # Generate Alert Output from MISP Response
                alert_output["misp"] = {}
                alert_output["misp"]["event_id"] = misp_api_response["response"]["Attribute"][0]["event_id"]
                alert_output["misp"]["category"] = misp_api_response["response"]["Attribute"][0]["category"]
                alert_output["misp"]["value"] = misp_api_response["response"]["Attribute"][0]["value"]
                alert_output["misp"]["type"] = misp_api_response["response"]["Attribute"][0]["type"]
                send_event(alert_output, alert["agent"])
    else:
        sys.exit()
```

Figure 6.28 Sending alerts to Wazuh after MISP search.

```
 GNU nano 4.8                                   ossec.conf
    <key>aa093264ef885029653eea20dfcf51ae</key>
    <port>1516</port>
    <bind_addr>0.0.0.0</bind_addr>
    <nodes>
        <node>wazuh.manager</node>
    </nodes>
    <hidden>no</hidden>
    <disabled>yes</disabled>
  </cluster>

</ossec_config>

<ossec_config>
  <localfile>
    <log_format>syslog</log_format>
    <location>/var/ossec/logs/active-responses.log</location>
  </localfile>

<integration>
    <name>custom-misp.py</name>
 <group>sysmon_event1,sysmon_event3,sysmon_event6,sysmon_event7,sysmon_event_15,sysmon_event_22,syscheck</group>
    <alert_format>json</alert_format>
  </integration>

</ossec_config>
```

Figure 6.29 Adding the "ossec.conf" integration block.

6.5.2.2.2 Adding the "ossec.conf" integration block

We will now integrate our custom-misp.py script into OSSEC's configuration by adding the following integration block to the ossec.conf file. This block specifies the parameters necessary for Wazuh to use our custom script when it detects events linked to Sysmon and other monitored events (Figure 6.29).

In this integration block, <name> specifies the name of our script, <group> defines the event groups this script is associated with (such as Sysmon and syscheck events), and <alert format> indicates the alert format, which is JSON in our case.

By adding this block to the ossec.conf file, we inform OSSEC of using our custom script for specific event handling. This customization allows Wazuh to trigger our script when it detects events corresponding to the specified groups, thus facilitating the integration and automation of threat response.

6.5.2.2.3 Definition of an alert rule on Wazuh

In this configuration for Wazuh, a set of rules specific to the MISP integration has been defined to detect, report, and act on Threat Intelligence events originating from MISP. The

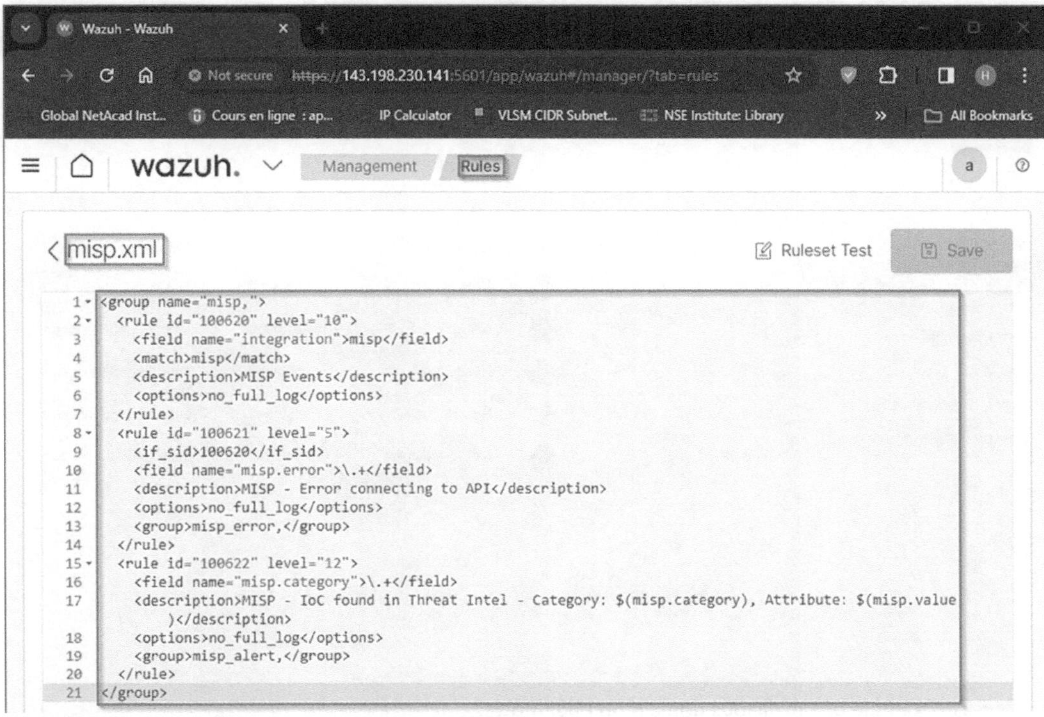

Figure 6.30 "misp.xml" file.

main rule (ID 100620) identifies MISP-related events with specific alert levels. In case of a connection error to the MISP API, rule 100621 is activated to generate a moderate-level alert (level 5). In addition, rule 100622 detects the presence of IoC in MISP's Threat Intelligence, generating high-level alerts (level 12). These rules are configured to exclude full logs in alerts ("no full log") and are grouped according to their context, in particular "misp error" and "misp alert". These rules aim to provide precise visibility of MISP-related events and to automate responses based on the nature of these events (Figure 6.30).

6.5.2.3 Integration tests

We followed an in-depth methodology to implement and test the integration between MISP and Wazuh to ensure the whole process worked correctly.

First, we scanned the MISP DB using the "list attributes" command to identify a malicious domain we wanted to use for testing purposes. We noted the ID and name of this domain for future reference (Figure 6.31).

Then, to simulate suspicious activity, we ran the "ping" command on the identified malicious domain. This generated network-related Sysmon events on our system, capturing traffic related to our test (Figure 6.32).

In the Wazuh Dashboard, we checked the alerts to ensure that the Sysmon events had been correctly captured. We successfully identified an alert corresponding to the Sysmon event triggered by the ping, validating the successful detection of the suspicious activity (Figure 6.33).

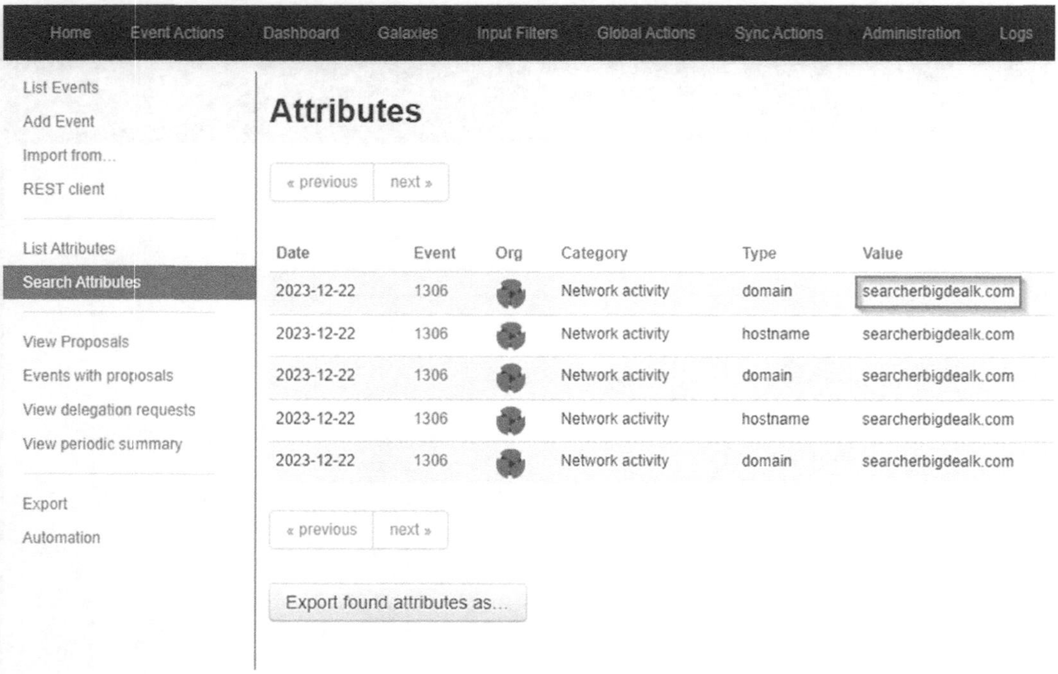

Figure 6.31 Example of a malicious domain on MISP flows.

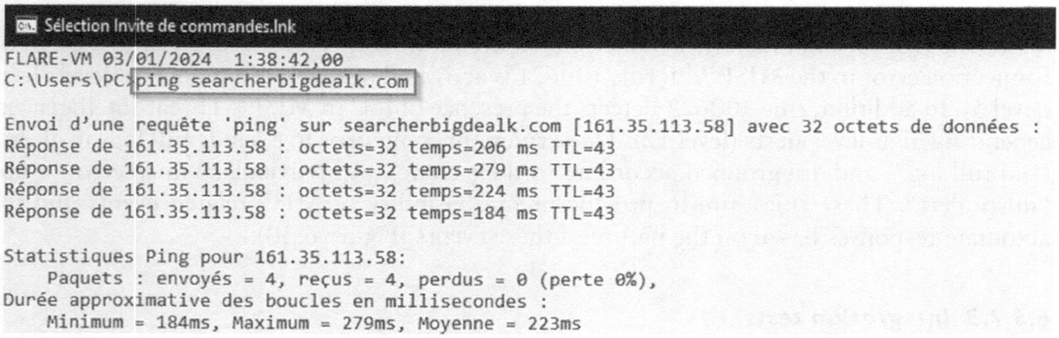

Figure 6.32 Pinging the malicious "bigsearcherdealk.com" domain.

For further analysis, we looked specifically at MISP-related Wazuh alerts. Using the rules previously defined in the configuration files, we searched for indications of a match between the Sysmon event and the IoCs stored in MISP. This confirmed the effectiveness of our integration, conclusively demonstrating the detection of a malicious IoC in network traffic through ping (Figure 6.34).

In summary, this systematic approach has validated and highlighted the operational effectiveness of the integration between MISP and Wazuh, underlining the system's ability to detect and respond to threats based on the IoCs stored in the MISP DB.

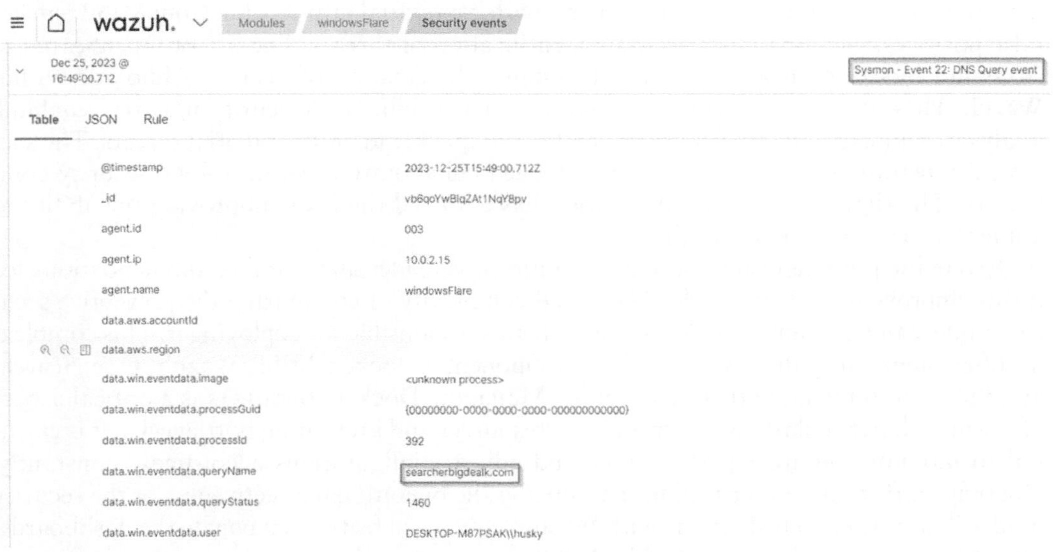

Figure 6.33 The Sysmon DNS alert on the Wazuh Dashboard.

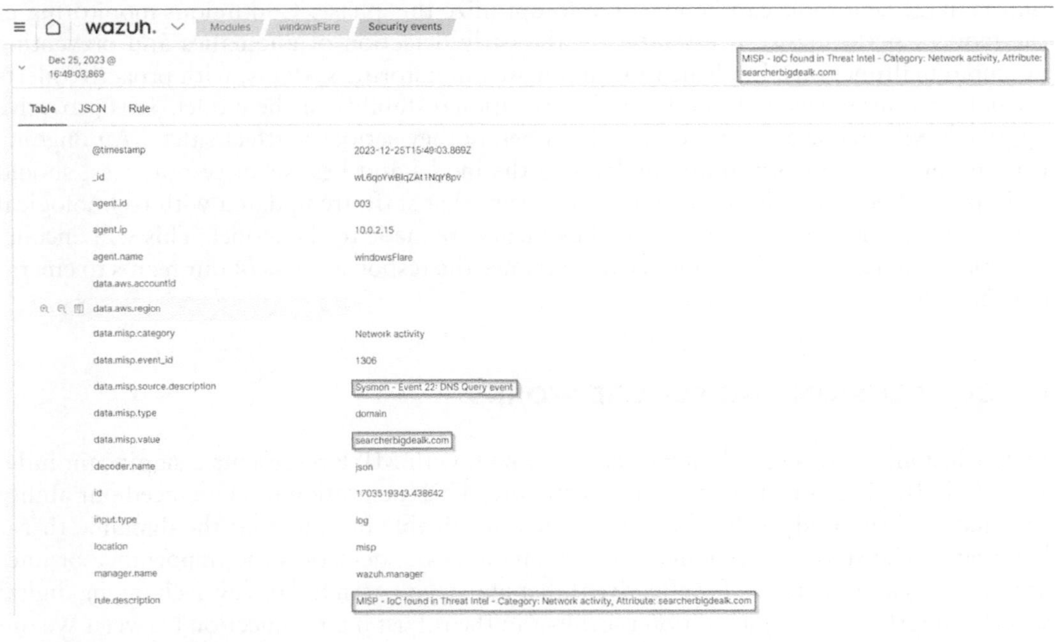

Figure 6.34 MISP IoC found an alert.

6.6 BENEFITS AND LIMITATIONS

The integration of the preventive security model, with its integrated architecture of Docker, MISP, Wazuh, and OpenSearch, has brought several significant benefits to the security environment. The integration of IoCs between MISP and Wazuh has considerably improved detection accuracy. Enriching Wazuh alerts with contextual information from MISP reduces false positives, enabling more accurate identification of threats. The use of OpenSearch to store and index aggregated data has facilitated the creation of visual dashboards within Wazuh. These dashboards provide deep contextual visibility into security activities, enabling analysts to understand attack patterns better and quickly identify potential threats. The successful integration with MISP has strengthened collaboration within the IT security community. The sharing of threat information has enriched the IoCs, improving the ability to anticipate new trends in cyber threats.

Despite the positive results, several limitations were identified, and recommendations for future improvements were made. The initial complexity of configuring the preventive security model can represent a challenge for the teams responsible for deploying it. This complexity often stems from the diversity of the components (Docker, MISP, Wazuh, OpenSearch) and the need to integrate them coherently. Managing Docker containers is a critical aspect of the model, particularly when optimizing resources and preventing bottlenecks. It is essential to monitor container performance and adjust configurations accordingly constantly. Ongoing staff training is crucial to maximizing the operational effectiveness of the security model. Teams need to be familiar with Wazuh's advanced features, navigate the dashboards, and understand the alerts enriched by MISP.

Configuration automation should be seen as a strategic investment. By adopting automation tools, teams can simplify the deployment process, reduce lead times, and minimize the potential errors associated with manual configuration. Automation scripts and configuration management tools can be explored to optimize this phase. Continuous monitoring of the model's performance is essential for the early detection of anomalies and preventing potential malfunctions. Implementing automated monitoring systems, with proactive alerts for potential problems, will reinforce the operational stability of the model. This proactive approach will enable anticipating problems before they seriously affect safety. An ongoing training program, incorporating updates to the model and best security practices, should be in place. Regular training sessions will ensure that staff are updated with technological developments, new threats, etc., and adjustments are made to the model. This will encourage optimum use of the safety model and enhance the responsiveness of our teams to emerging challenges.

6.7 CONCLUSION AND FUTURE WORKS

In conclusion, the successful integration of Wazuh with MISP represents a significant milestone in bolstering our IT security infrastructure. This integration has enhanced our ability to detect and respond to threats and provided invaluable insights into the dynamic threat landscape. The strategic combination of technologies underscores the importance of integrating tools and platforms for effective threat management in today's ever-changing digital world. The efficiency of integration is evident in the robust interconnection between Wazuh and MISP, facilitating seamless communication between threat detection systems and threat information-sharing platforms. The automated flow of IoCs from MISP to Wazuh ensures rapid availability of the latest threat information, empowering us to respond proactively to emerging threats.

Moreover, the integration has significantly improved our performance in threat management by increasing responsiveness and accuracy in threat detection. Leveraging MISP IoCs in Wazuh has minimized false positives, allowing us to focus our efforts on addressing real threats effectively. Operational benefits abound with the centralization of threat management, simplifying the monitoring, analysis, and response to security incidents from a unified platform. Real-time updates from MISP ensure that our system remains abreast of the latest attacker trends and tactics, enhancing our resilience against evolving threats.

We remain committed to continuous adaptation and innovation, positioning ourselves to navigate future challenges confidently. By harnessing the synergies between technologies and embracing a proactive approach to cybersecurity, we are poised to uphold the integrity and security of our IT infrastructure amidst the evolving threat landscape.

REFERENCES

Alaoui, E. A. A., Tekouabou, S. C. K., Maleh, Y., & Nayyar, A. (2022). Towards to intelligent routing for DTN protocols using machine learning techniques. *Simulation Modelling Practice and Theory*, *117*, 102475. https://doi.org/10.1016/j.simpat.2021.102475

Briliyant, O. C., Tirsa, N. P., & Hasditama, M. A. (2021). Towards an automated dissemination process of cyber threat intelligence data using stix. *2021 6th International Workshop on Big Data and Information Security (IWBIS)*, Depok, Indonesia, 109–114.

Cascavilla, G., Tamburri, D. A., & Van Den Heuvel, W.-J. (2021). Cybercrime threat intelligence: A systematic multi-vocal literature review. *Computers & Security*, *105*, 102258.

Chantzios, T., Koloveas, P., Skiadopoulos, S., Kolokotronis, N., Tryfonopoulos, C., Bilali, V.-G., & Kavallieros, D. (2019). The quest for the appropriate cyber-threat intelligence sharing platform. *DATA*, 369–376. https://doi.org/10.5220/0007978103690376

de Melo e Silva, A., Costa Gondim, J. J., de Oliveira Albuquerque, R., & García Villalba, L. J. (2020). A methodology to evaluate standards and platforms within cyber threat intelligence. *Future Internet*, *12*(6), 108.

Doerr, C. (2018). Cyber threat intelligences standards-a high level overview. *TU Delft CTI Labs*.

Faiella, M., Granadillo, G. G., Medeiros, I., Azevedo, R., & Zarzosa, S. G. (2019). Enriching Threat Intelligence Platforms Capabilities. *ICETE (2)*, 37–48. https://doi.org/10.5220/0007830400370048

Guarascio, M., Cassavia, N., Pisani, F. S., & Manco, G. (2022). Boosting cyber-threat intelligence via collaborative intrusion detection. *Future Generation Computer Systems*, *135*, 30–43.

Hadi, H. J., Riaz, M. A., Abbas, Z., & Nisa, K. U. (2023). Cyber threat intelligence model: An evaluation of taxonomies and sharing platforms. In Maleh, Yassine, Mamoun Alazab, Loai Tawalbeh, and Imed Romdhani, eds., *Big Data Analytics and Intelligent Systems for Cyber Threat Intelligence* (pp. 3–33). River Publishers.

Maleh, Y. (2023). Deep learning fusion for multimedia malware classification. In Ahmed A. Abd El-Latif, Mudasir Ahmad Wani, Yassine Maleh, Mohammed A. El-Affendi, eds., *Recent Advancements in Multimedia Data Processing and Security: Issues, Challenges, and Techniques* (pp. 46–73). IGI Global.

Maleh, Y., Fatani, I. F. E., & Gholami, K. El. (2022). A systematic review on software defined networks security: Threats and mitigations. In Y. Maleh, M. Alazab, N. Gherabi, L. Tawalbeh, & A. A. Abd El-Latif (Eds.), *Advances in Information, Communication and Cybersecurity* (pp. 591–606). Springer International Publishing.

Maleh, Y., Shojafar, M., Alazab, M., & Baddi, Y. (2021). *Machine Intelligence and Big Data Analytics for Cybersecurity Applications*. Springer International Publishing AG.

Möller, D. P. F. (2023). Threats and threat intelligence. In *Guide to Cybersecurity in Digital Transformation: Trends, Methods, Technologies, Applications and Best Practices* (pp. 71–129). Springer.

Preuveneers, D., & Joosen, W. (2021). Sharing machine learning models as indicators of compromise for cyber threat intelligence. *Journal of Cybersecurity and Privacy*, *1*(1), 140–163.

Safarzadeh, M., Gharaee, H., & Panahi, A. H. (2019). A novel and comprehensive evaluation methodology for SIEM. *Information Security Practice and Experience: 15th International Conference, ISPEC 2019*, Kuala Lumpur, Malaysia, November 26–28, 2019, Proceedings 15, 476–488.

Schlette, D., Caselli, M., & Pernul, G. (2021). A comparative study on cyber threat intelligence: The security incident response perspective. *IEEE Communications Surveys & Tutorials*, *23*(4), 2525–2556.

Stojkovski, B., Lenzini, G., Koenig, V., & Rivas, S. (2021). What's in a cyber threat intelligence sharing platform? A mixed-methods user experience investigation of MISP. *Annual Computer Security Applications Conference*, Virtual Event, USA, 385–398.

Tounsi, W. (2019). What is cyber threat intelligence and how is it evolving? *Cyber-Vigilance and Digital Trust: Cyber Security in the Era of Cloud Computing and IoT*, Wiem Tounsi, Wiley, 1–49.

Wagner, T. D., Mahbub, K., Palomar, E., & Abdallah, A. E. (2019). Cyber threat intelligence sharing: Survey and research directions. *Computers & Security*, *87*, 101589.

Chapter 7

Security study of Web applications through a white-box audit approach

A case study

Afef Jmal Maâlej and Mohamed Salem Eleze

7.1 INTRODUCTION

Websites are becoming increasingly effective communication tools. Nevertheless, Web applications are vulnerable to attack and can give attackers access to sensitive information or unauthorized access to accounts. The number of vulnerabilities in Web applications has increased dramatically over the past decade. Many are due to improper validation and sanitization of input. In particular, the field of IT security is increasingly important for companies because data and system security are critical to their success. Cyberattacks are increasingly frequent and sophisticated, and therefore, it is essential to put in place security measures to protect business applications and data. Thus, identifying these vulnerabilities is essential for developing high-quality and secure Web applications.

The company FRS in Tunisia specializes in developing IT solutions for businesses. Given the growing importance of IT security, this company wants to make sure its applications are secure, and its customers' data is protected. In this context, our role consists of performing a security audit of a Web application that is developed by the FRS company using Springboot and Angular technologies. Software testing is a key aspect of IT security auditing because it allows the detection of potential vulnerabilities in applications. Indeed, these tests can evaluate the robustness and reliability of an application by simulating different situations to highlight possible malfunctions or security errors.

In this chapter, we will focus on the application of software testing as part of the IT security audit. Our security audit consists of a white-box audit, which examines the code application source to identify potential vulnerabilities. Once the security audits are complete, we provide recommendations to improve the application security. These recommendations include both technical and organizational security measures to help enhance Web application security.

The rest of this chapter is organized as follows. Section 7.2 is dedicated to present the needs and concerns of Web application security. Section 7.3 provides an introduction to white-box security audit. Section 7.4 draws the development environment review. The audit of source code is outlined in Section 7.5. Afterward, our white-box audit methodology is highlighted in Section 6, and different results are described in Section 7.7. Finally, we conclude, in Section 7.8, with a summary of chapter contributions, and we identify potential areas of future research.

7.2 WEB APPLICATION SECURITY: NEEDS AND CONCERNS

Web applications are unique and different with respect to traditional desktop applications. This uniqueness leads to new challenges in the testing and quality assurance domains.

DOI: 10.1201/9781032714806-9

- The open operating environment of typical Web applications makes it widely visible and susceptible to various attacks, such as denial-of-service (DoS) and distributed denial-of-service (DDoS) attacks [1]. This creates difficulty in predicting and simulating a realistic workload. The differences in implementation and levels of standards compliance add further complexity on and across the browsers while delivering coherent user experiences. The proliferation of numerous popular browsers and inadequate compatibility testing create innumerable challenges [2].
- The multilingual features of the backend and frontend and numerous components under different programming languages create an additional challenge for fully automated continuous integration (CI) practices. The heterogeneous nature of application development frameworks and different encoding standards further enhance the trials [3].
- Real-time multiuser environments, along with multithreaded nature, create difficulty in detecting and reproducing resource contention issues. The effective management of resources such as HTTP, files, database connections, and threads is crucial to the security, scalability, usability, and functionality of Web applications and their associated challenges [3].

For a popular e-commerce Web application, for example, exploitation of vulnerabilities may result in a loss of company reputation. Hence, the detection of security bugs will improve the quality and reliability of Web applications, along with preventing economic losses [4]. The threat landscape for Web applications is constantly changing. Key factors in this evolution are the advancements made by attackers, the deployment of increasingly complex systems, and the release of new technologies [5]. All these proclamations need to identify the risks associated with Web applications and categorize the typical risk severity, risk consequences, and detection accuracy during testing [6].

7.3 WHITE-BOX SECURITY AUDIT

Pareek [7] explained the types of security testing used for Web applications, including white-box, black-box, and gray-box testing. In addition, the author described the different phases of security testing for Web applications. A review of OWASP's top 10 Web application security risks was also conducted.

Adopting the white-box security audit strategy, our team would have access to as much knowledge about the target environment as feasible, just like an actual employee would. This approach is intended to be ready for the worst-case situation in which an attacker has extensive knowledge of your infrastructure [8]. In particular, white-box testing permits the preparation of scenarios such as insider threats or an attacker that has obtained detailed internal information [9]. Since the audit team has visible access to the crucial data and specifics essential for attacking the organization, this procedure typically reveals more vulnerabilities and is significantly faster. In addition, it expands the scope of testing to include tasks that are typically not included in a traditional black-box audit, such as source code audits and application design reviews.

White-box auditing can be performed manually or by using automated tools that scan the application's source code for known vulnerabilities or common security vulnerabilities. Auditors can also use reverse engineering techniques to understand the application's internal workings and identify potential vulnerabilities that could be exploited by attackers.

The white-box audit can be performed at different development levels, including the initial development phase, the test phase, and the maintenance phase. The results of the white-box audit can help developers improve application security by correcting identified vulnerabilities and implementing additional security measures.

7.4 DEVELOPMENT ENVIRONMENT REVIEW

Auditing the development environment is an essential step in assessing the security of an application. The development environment can be a source of vulnerabilities if best safety practices are not followed. This audit ensures that the development used to create the application is consistent with standards and best security practices. It also allows you to identify any configuration or installation faults that could expose the application to security risks.

It should be noted that PostgreSQL 9.5, Spring Boot 2.6.7, and Java 8, used to develop the application, are not only widespread but also have known vulnerabilities and associated risks that can affect application security.

To assess their potential impact, we analyzed the risks associated with these tools. Table 7.1 presents some examples of known vulnerabilities associated with these tools, identified by their CVE (Common Vulnerabilities and Exposures) code [10]. In fact, CVEs are unique identifiers assigned to known security vulnerabilities in software and systems. They make it easier to communicate and understand vulnerabilities between different stakeholders, including developers, security auditors, and end users.

7.5 AUDIT OF SOURCE CODE

7.5.1 Study of our choices

In the following, we detail the choice of tools used for the audit process. For that, we provide a comparative analysis with alternative tools not chosen. This comparison could offer insights into the specific benefits and limitations of the selected tools over others, contributing to a more informed decision-making process for practitioners in the field.

7.5.1.1 Comparison between GitHub and SVN

We chose GitHub as our source code management platform in our audit because of its many advantages over SVN as depicted in Table 7.2.

Table 7.1 Software, versions, vulnerabilities, and examples of associated vulnerabilities

Software	*Version*	*Associated vulnerabilities and risks*	*Example of vulnerability*	*CVE-ID*
PostgreSQL	9.5	Known unpatched vulnerabilities, data loss, SQL injection, elevation of privileges	Remote arbitrary code execution via SQL injection attack	CVE-2019-9193
Spring Boot	2.6.7	Known unfixed vulnerabilities, security vulnerabilities in authorization and identity management, remote code execution, denial of service	Remote code execution through a JSON file deserialization attack	CVE-2022-26371
Java	8	Known unfixed vulnerabilities, security failures in authorization and identity management, remote code execution, denial-of-service	Remote arbitrary code execution via security failure in RMI protocol	CVE-2021-2163

Table 7.2 Comparison between GitHub and SVN

Criterion	*GitHub*	*SVN*
System type	Distributed version management system	Centralized version management system
Usage	Mainly used for open-source projects	Used in companies for proprietary projects
Accessibility	Accessible wherever there is an Internet connection	Accessible only within the internal corporate network
Integration	Easy integration with many tools and services	Limited integration with other tools and services
Collaboration	Promotes collaboration and teamwork	Less user-friendly collaboration and teamwork
Features	Many features for project management	Limited project management features
Performance	Fast performance for small and medium size	Slower performance for larger sizes

Table 7.3 Comparison between Jenkins, Travis CI, and CircleCI

Feature	*Jenkins*	*Travis CI*	*CircleCI*
Open source	Yes	No	No
Continuous integration	Yes	Yes	Yes
Integration with third-party tools	Very good	Good	Good
Personalization	Very good	Good	Good
Security features	Very good	Good	Good
Extensibility	Very good	Good	Good

7.5.1.2 Comparison between Jenkins, Travis CI, and CircleCI

Our choice for the main platform of our audit is Jenkins because of its great flexibility, ease of use, compatibility with many development tools, and its extended plugin library (Table 7.3).

7.5.1.3 Comparison between SonarQube, Codacy, and PMD

After comparing the functionalities of SonarQube, Codacy, and PMD (Table 7.4), we chose SonarQube for several reasons. First, it offers a full range of security and code quality features, and it supports a wide range of programming languages. In addition, SonarQube's intuitive user interface allows easy reading and understanding of analysis reports. Moreover, SonarQube is regularly updated with new features and security policies, ensuring continuous protection against security threats and vulnerabilities. Finally, SonarQube has an active community and responsive technical support, which is an important factor in choosing a reliable security tool.

7.5.2 Architecture of tools used for source code auditing

The architecture used to perform a white-box security audit of the source code with GitHub, Jenkins, SonarQube, and PostgreSQL is shown in Figure 7.1.

This architecture was used to detect vulnerabilities and security failures in the Web application source code before it went into production. It has also provided valuable information to correct detected security issues.

Table 7.4 Comparison between SonarQube, Codacy, and PMD

Features	*SonarQube*	*Codacy*	*PMD*
Supported programming languages	27+	30+	Java, C++, PL/SQL, etc.
Integration with development tools	Eclipse, IntelliJ, Visual Studio, Jenkins, Bamboo, etc.	GitHub, GitLab, Bit-bucket, Jenkins, etc.	Eclipse, IntelliJ, Ant, Maven, etc.
Security	Code security scanning, vulnerability detection, compliance monitoring	Code security scanning, vulnerability detection, compliance monitoring	Code security scanning, vulnerability detection
Code quality	Code quality analysis, code duplication detection, code coverage metrics	Code quality analysis, code duplication detection, code coverage metrics	Code quality analysis, code duplication detection
Price	Free and payable	Payable	Free

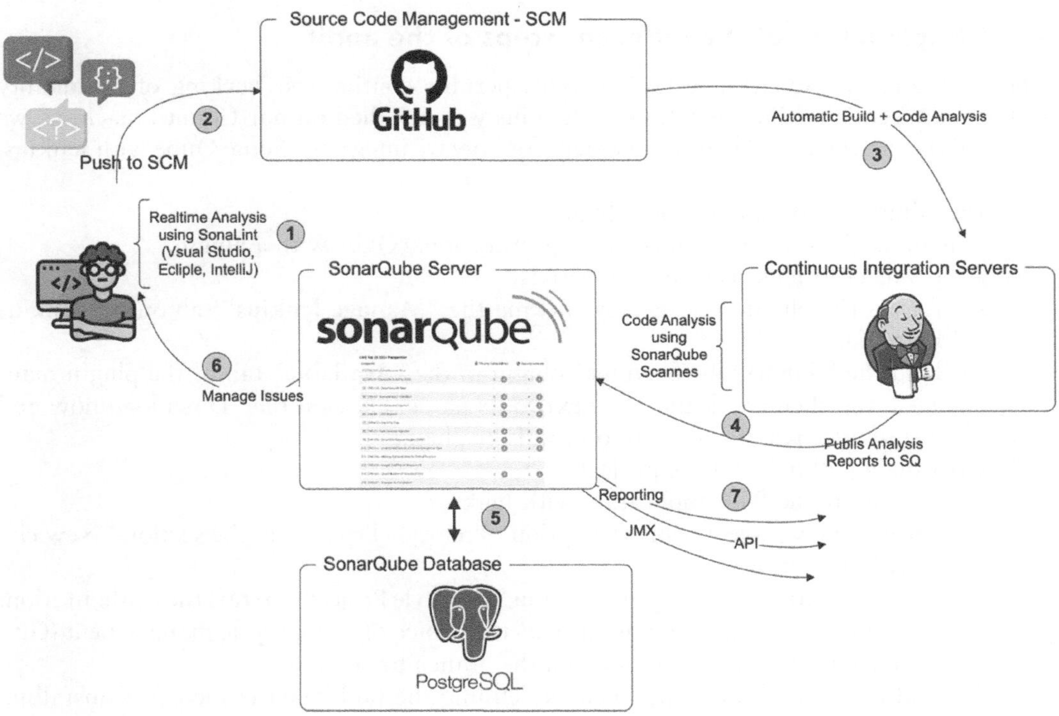

Figure 7.1 Architecture used for source code auditing.

7.6 WHITE-BOX AUDIT METHODOLOGY

7.6.1 Overview of audit methodology

CI with Jenkins and GitHub in combination with SonarQube for white-box software auditing is a common practice in software development.

Jenkins [11] is an open-source CI tool that allows applications to be tested, built, and deployed in an automated manner. GitHub, meanwhile, is a source-code management platform that facilitates collaboration among members of a development team. Together, these tools automate the testing and deployment processes of an application, saving time and minimizing human errors.

SonarQube [12] can be integrated with Jenkins to perform static code analysis whenever new code is submitted to the GitHub repository. This integration permits quick detection of vulnerabilities, programming errors, and bad coding practices in the application source code. SonarQube detailed reports can be viewed directly from the Jenkins interface, making it easy to detect and correct code problems.

In short, using CI with Jenkins and GitHub in combination with SonarQube for white-box software auditing is a recommended practice for companies that want to improve the quality and security of their applications. This practice allows for quick detection of code quality issues and security vulnerabilities while automating testing.

Besides, it is highly recommended to use up-to-date versions of these development tools to take advantage of the latest bug fixes and security enhancements. Also, specific safety recommendations for each tool, such as secure configurations and best practices, should be followed.

7.6.2 Explanation of the different steps of the audit

The integration of SonarQube with Jenkins permits continuous checking of the quality of the source code during the CI. Once Jenkins was installed on our Ubuntu machine, we accessed the interface, and here are the steps for how we integrated SonarQube with Jenkins.

a. **Installing SonarQube scanner plugin**

   Jenkins gives the possibility of integrating SonarQube with a plugin.

   - Connect to Jenkins using credentials.
   - Access the plugin manager by clicking the "**Manage Jenkins**" tab on the Jenkins home page.
   - Find the "**SonarQube Scanner**" plugin in the "**Available**" tab of the plugin manager, and then check the box next to its name, click on the "Download now and install after restart" button to install the plugin.

b. **Creating a Jenkins freestyle project**

   To integrate the Web application with Jenkins,

   - Create a new project using the option "Freestyle Project" in the section "New element" of Jenkins.
   - Name the project and select the option "Freestyle Project" to start the configuration.
   - Configure the project details, such as the choice of source code management (Git), the URL of the Git repository, and the branch to be built.
   - Add other configurations, such as defining the build environment and installing the necessary plugins for the project.

c. **Configuring the built environment for the SonarQube scanner**

   - Scroll down in project settings in Jenkins and go to "Build Environment."
   - Select the "**Prepare SonarQube scanner environment**" checkbox.
   - Provide the authentication token generated for the SonarQube project (or the one we created for our project) to connect to our SonarQube instance.

   To configure the scanner of SonarQube, it is necessary to create a file named "**sonar-scanner.properties**" and add it to the project. This file must contain the necessary properties to connect to SonarQube, such as the project key, the host URL, and the authentication token. It is also possible to consider more properties if needed. An example of the file "**sonar-scanner.properties**" is depicted in Figure 7.2.

```
-Dsonar.projectKey=Your Project key
-Dsonar.host.url= http://localhost:9000
-Dsonar.login=Your Token
```

Figure 7.2 Github- sonar-scanner.properties.

Figure 7.3 Jenkins- Invoquer les cibles Maven de haut niveau.

Figure 7.4 Jenkins- Jenkins location.

Then, in order to use SonarQube Scanner with Maven, it is necessary to add the construction objectives and specify the version of Maven to use (see Figure 7.3).

d. Configuring Jenkins location

- Go to the Jenkins dashboard, then open "Manage Jenkins" and click "Configure System."
- In the "Jenkins Location" section, retrieve the IP address of the machine and configure the Jenkins URL with the corresponding IP address, as shown in Figure 7.4.

e. Configuring the SonarQube environment

- Add a name to your SonarQube server and provide the URL for that server.
- In the **"SonarQube authentication"** section, add the authentication token generated on the Sonar server, click Apply, and **Save**. Now, the Jenkins freestyle job is configured to run and generate the SonarQube report.

7.7 WHITE-BOX AUDIT RESULTS

7.7.1 Overview of SonarQube results

The Sonar server is installed on an Ubuntu 22.04 machine at port 9000 with an IP address 192.168.152.131. We started the Sonar server on this machine and configured Jenkins to communicate with it. After building the project in Jenkins, we were able to observe the results in the console output and see the latest code analysis on the SonarQube dashboard, as shown in Figure 7.5.

7.7.2 Generating source code audit report

We chose to use the **Sonar CNES Report** to generate the white-box audit report. This plugin allows export code analysis from a SonarQube server in different file formats, such as docx, xlsx, csv, markdown, and text. The following steps have been performed to install and run this program.

- Installation:
  Copy the sonar-cnes-report.jar file to the SonarQube plugin folder (the path under Linux should look like/opt/sonarqube/extensions/plugins).
  Restart SonarQube (On Linux: sudo systemctl restart sonar).
- Execution: In the SonarQube dashboard, click on **"More"** then **"CNES Report."**
  After selecting the project, branch, and document types we want to generate, we click on the "Generate" button to start the process of generating the report.

7.7.3 Analysis and interpretation of results

Now that we have generated the white-box audit report using the Sonar CNES Report and reviewed the dashboard provided by SonarQube, it is time to move on to the next step: analyzing and interpreting the results. To better understand the overall quality of the code and identify potential vulnerabilities, we will look at various aspects of our report, such as code quality indicators, bugs, security vulnerabilities, duplicate codes, and badly formatted codes.

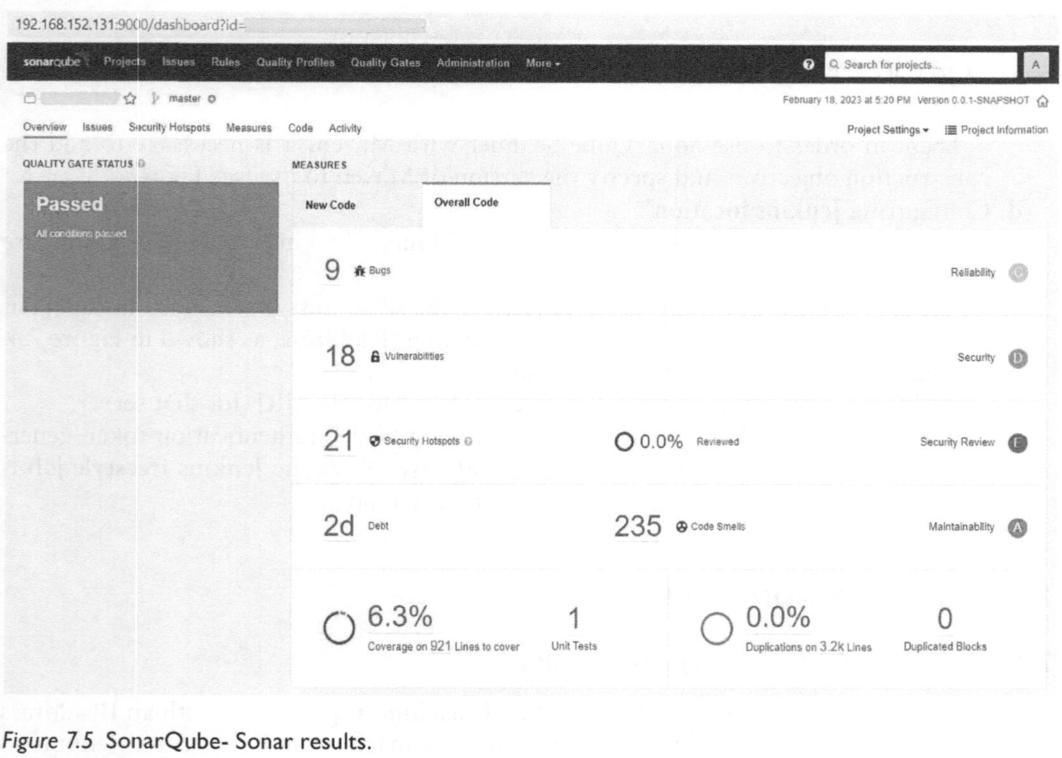

Figure 7.5 SonarQube- Sonar results.

Table 7.5 Problems per gravity and type

Type/severity	*Info*	*Minor*	*Major*	*Critical*	*Blocker*
Bug	0	2	7	0	0
Vulnerability	0	0	0	18	0
Code smell	2	192	35	4	2

7.7.4 Problems by severity and type

The following is a summary of the application code quality and safety statistics (Table 7.5). These statistics reveal the presence of different types of code quality problems, such as programming errors (BUG), potential security vulnerabilities (VULNERABILITY), and programming practices that may lead to future problems (CODE SMELL). The severity of these problems varies from minor to critical, with critical problems considered deadlocks. The data indicates the presence of 2 major bugs, 7 critical bugs, 18 critical vulnerabilities, and 192 code problems to be corrected to achieve high quality and safety standards.

7.7.5 List of problems

- **Vulnerability:**

  We have identified a vulnerability in the application source code that is classified in the security category **OWASP [13] Top 10: A01 - Broken Access Control 2021 and A05 - Broken Access Control**. This vulnerability is defined by the **Common Weakness Enumeration (CWE) [14]: CWE-915: Improperly Controlled Modification of Dynamically Determined Object Attributes**. In particular, we found that the application controllers were the source of **18** occurrences of this vulnerability.

  The specific vulnerability is **"Persistent entities should not be used as arguments of "@RequestMapping" methods."** This vulnerability can potentially expose sensitive information stored in the database by allowing an attacker to execute an **SQL injection**. This is caused by using persistent entities as arguments to **"@RequestMapping"** methods in the application controllers. Persistent entities are Java objects that are mapped to records in the database. **"@RequestMapping"** methods can allow an attacker to send malicious queries to the database, compromising application security. In addition, the vulnerability is considered **CRITICAL** because it allows an attacker to interact with the database directly through the Web interface bypassing security controls and compromising the **confidentiality** and **integrity** of stored data. Addressing this vulnerability is critical to ensure application security.

  The **"Persistent entities should not be used as arguments of "@RequestMapping" methods"** vulnerability detected by the SonarQube tool can be exploited to expose sensitive data or cause errors in the application. The recommendation is to use Data Transfer Objects (DTOs) instead of persistent entities as arguments to the "@RequestMapping" method.

  Figure 7.6 shows the occurrence number of this vulnerability in each controller of the application, indicating that they are indeed the source of the vulnerability.
- **BUG:**

  The results showed that there were 9 bugs in the source code of the application. This analysis provides detailed information about the different types of code quality problems, as well as their severity and their location in the source code.

	Lines of Code	Bugs	Vulnerabilities
controller	692	2	18
Planification	118	1	2
AssetController.java	47	0	1
AuthController.java	64	0	0
CampaignController.java	56	0	1
CampaignElementController.java	13	0	0
ChannelController.java	39	0	2
DeviceController.java	32	0	2
DeviceGroupController.java	43	0	2
FolderController.java	36	0	1
LongPoolingController.java	57	1	0
MicroProgramController.java	54	0	2
OrganizationController.java	32	0	2
SiteController.java	47	0	3
UserController.java	54	0	0

Figure 7.6 SonarQube- vulnerability statistics.

Table 7.6 shows a detailed summary of the nine bugs identified in the application. This provides a better understanding of the nature of each bug, its number of occurrences, and the level of risk associated with it.

The "Severity" column of Table 7.6 provides the SonarQube risk levels for each identified bug based on their potential impact on application security and stability. Risk levels are defined as follows:

- **Blocker:** A critical issue that blocks code execution or causes a major security violation.
- **Critical:** A critical issue that can lead to a major security breach or cause the application to fail.
- **Major:** Important issue that can potentially lead to errors or unwanted behavior in the application.
- **Minor:** A minor problem that does not directly affect the operation of the application but may affect its quality or performance.
- **Info:** An informational problem that does not pose a threat to the application but can provide useful information for improving code.

- **Code smell:**

Code smells are indicators of potential problems in software source code. They can be seen as warning signs that draw the attention of developers to sections of code that require special attention. Code smells can be caused by inappropriate programming

Table 7.6 Bugs

Bug type	*Description*	*Nombre of occ.*	*Severity*
"InterruptedException" should not be ignored	InterruptedException exceptions should not be ignored because they may indicate synchronization problems in code.	1	Major
Math operands should be cast before the assignment	Mathematical operands must be converted to their appropriate type before being assigned, otherwise, this can lead to unexpected results.	1	Minor
Strings and Boxed types should be compared using "equals()"	Character strings and boxed types must be compared using the "equals()" method because the "==" operator compares references, not values.	1	Major
Conditionally executed code should be reachable	Conditionally executed code should not be unreachable, otherwise it may indicate a logic problem in the code.	1	Major
Optional value should only be accessed after calling is- Present()	Optional values should only be accessed after checking for presence by calling the "isPresent()" method; otherwise, NullPointerException exceptions may occur.	4	Major
"equals(Object obj)" and "hashCode()" should be overridden in pairs	The methods "equals(Object obj)" and "hash- Code()" must be redefined together for the classes that will be used in collections; otherwise, this can lead to unexpected behaviors.	1	Minor

practices or bad development habits, which can lead to logic errors, bugs, or security breaches. In this context, identifying and correcting "code smells" is an important step to ensure both application quality and security.

We have identified a total of 235 "code smells" in the source code of the application. The SonarQube dashboard presents a detailed view of the rules and tags associated with these "code smells," as well as their severity and the number of occurrences. The "code smells" provide an overview of potential problems in the source code, identifying problem areas that need special attention.

To keep any code secure, it is important to identify and resolve smells at the development stage. This ensures that your application is less likely to be vulnerable to attacks and compromises of sensitive data. Here are some recommendations to ensure code security and avoid code smells:

1. Follow clear and consistent coding conventions to improve readability and facilitate vulnerability detection.
2. Avoid code duplication to avoid repetitive security errors.
3. Implement security tests to identify and remediate vulnerabilities as early as possible.
4. Use built-in security features or proven libraries for sensitive operations.
5. Apply security principles such as least privilege, validation and data escape, and protection from common attacks.
6. Conduct regular security-focused code reviews to identify and correct potential vulnerabilities.
7. Use static code analysis tools or vulnerability scanners to automatically detect security issues.
8. Update dependencies and libraries used to benefit from the security patches.
9. Train developers on security best practices and make them aware of common threats.

By following these recommendations, code security is improved, and code smell risks are reduced. This helps protect the application and the data it contains.

- **Security access points:**

  Security Hotspots are areas of code that have been identified as potentially vulnerable to security attacks. During the audit, 21 Security Hotspots were identified, including one for the **Cross-Site Request Forgery** (CSRF) vulnerability and another 20 for the **Insecure Configuration** vulnerability.

Table 7.7 shows a summary of the 21 identified security hotspots.

Table 7.8 provides specific recommendations for enhancing the security of source code access points identified during white-box auditing with the SonarQube tool. These measures will help improve the security of the application by addressing the detected vulnerabilities.

Table 7.7 Security hotspots

Security hotspot type	*Review priority*	*Description*	*Number of occurrences*
Cross-Site Request Forgery (CSRF)	HIGH	A **CSRF (Cross-Site Request Forgery)** attack is a deception of a trusted user of a Web application to perform sensitive actions that he did not intend to perform. The attacker can do this by clicking on a link corresponding to a privileged action or by having the user visit a malicious website that incorporates a hidden Web query. Web browsers automatically include cookies, allowing the attacker to authenticate sensitive actions. This security access point is linked to the application security configuration and has been categorized as **A05 – Security Misconfiguration** in the **OWASP Top 10 2021 list**.	1
Insecure Configuration	LOW	Misconfiguration vulnerability **CORS (Cross-Origin Resource Sharing)** is a security vulnerability identified in the **A05 – Security Mis-configuration** category of the **OWASP Top 10 2021** list. This vulnerability affects the same- origin policy, which by default prevents cross- origin HTTP requests from a JavaScript frontend to a resource with a different origin. However, additional HTTP headers, called CORS, can be added to the response to change the access control policy. If misconfigured, it can allow an attacker to access sensitive information or compromise the system by exposing API keys or using weak passwords.	20

Table 7.8 Security access points related to source code

Security access point	*Recommendations*	*References*
Cross-Site Request Forgery (CSRF)	Use anti-CSRF tokens to verify the origin of queries. Validate the HTTP Referer and Origin headers to verify the request origin.	OWASP: A05:2021 – Security misconfiguration
Insecure configuration	Avoid default configurations and sensitive values in the source code. Apply secure and appropriate configuration settings for the used servers and services.	OWASP: A05:2021 – Security misconfiguration

7.8 CONCLUSION

Security methods involve enhancing security procedures during the software development life cycle and throughout the application life cycle. All application security initiatives should aim to reduce the possibility of malicious actors gaining unauthorized access to systems, applications, or data. The ultimate purpose of application security is to prevent attackers from accessing, altering, or destroying sensitive or proprietary data.

In this chapter, we contribute to bridging the gap between academia and industrial practice. In fact, we studied the white-box audit of an industrial Web project by highlighting its crucial role in the assessment of application security. Emphasis was placed on the importance of conducting a thorough analysis of the development environment and source code to detect potential vulnerabilities and security vulnerabilities. Particularly, a white-box audit is a thorough security analysis that provides a better understanding of where security problems originate. It also uncovers vulnerabilities that are not visible during a pentest, but that may cause a security risk even so. In addition, we examined in detail key tools such as SonarQube, Jenkins and GitHub, which play an important role in the white-box audit. These tools provide automated capabilities to identify common security issues and generate detailed reports. This makes it easier to take corrective actions to improve application security.

Typical challenges in the field of security are emerging. Hence, industrial and academic collaboration is highly desirable to strengthen and improve security methods, develop new testing techniques, and address evolving and emerging security challenges. It is an obvious fact that organizations in today's scenario need more holistic approaches to tackle threats of security breaches.

Perspectives of our work concern a black-box audit, as a complement to a white one, which assesses security from the application as an external user without having access to the source code by using various attack techniques to detect potential vulnerabilities. In addition, we aim to provide more best practices and experience from applying novel approaches to large-scale industrial projects in the context of software architecture, considering emerging technologies.

REFERENCES

1. Raghavan, S. V., & Dawson, E.: *An Investigation into the Detection and Mitigation of Denial of Service (DoS) Attacks: Critical Information Infrastructure Protection*. Springer Science & Business Media (2011)
2. Robert S., Philip S.: Client-Side Attacks and Defense. Syngress. ISBN: 978-1-59749-590-5 (2012)
3. Mayhew D. J.: The usability engineering lifecycle, In: *98th Conference Summary on Human Factors in Computing Systems*, pp. 127–128, ACM, New York, NY (1998)
4. Li, X., & Xue, Y.: A Survey on Web Application Security. Technical report, Vanderbilt University (2011)
5. Dukes, L., Yuan, X., & Akowuah, F.: A case study on web application security testing with tools and manual testing, In: *Southeastcon Proceedings*, Jacksonville, FL, USA, pp. 1–6, IEEE Computer Society (2013)
6. Sampath S., Mihaylov V., Souter A., Pollock L.: A scalable approach to user-session based testing of Web applications through concept analysis, In: *19th International Conference on Automated Software Engineering*, Linz, Austria, pp. 132–141 (2004)
7. Pareek, K. A study of web application penetration testing. *IJITE* (2019), 7, 1776–2321.
8. White Box Security Audit. https://www.securitybrigade.com/research/types-of-security-audits/

9. Bourque P. and Dupuis R., editors.: *Guide to the Software Engineering Body of Knowledge, Version 3.0 SWEBOK*. IEEE (2014). https://www.computer.org/web/swebok.
10. CVE (Common Vulnerabilities and Exposures). https://www.cvedetails.com/
11. Jenkins. https://www.jenkins.io/
12. Sonarqube. https://docs.sonarqube.org/
13. OWASP (Open Web Application Security Project). https://owasp.org/
14. CWE (Common Weakness Enumeration). https://cwe.mitre.org/

Chapter 8

Case study method

A step-by-step black-box audit for security study of Web applications

Afef Jmal Maâlej and Mohamed Salem Eleze

8.1 INTRODUCTION

Because data and system security are essential to business performance, the field of IT security is becoming more and more significant. In this context, cyberattacks are becoming more common and complex, so it is crucial to implement security measures to protect company applications and data. The Tunisian company FRS specializes in creating IT solutions for corporations. As with other companies, the FRS firm needs to ensure that both its apps and the data of its clients are secure, given the growing significance of IT security. Our job in this situation entails conducting a security audit of a Web application created by the FRS company utilizing Springboot and Angular technologies. Because it enables the identification of potential application vulnerabilities, software testing is a crucial component of IT security audits. In fact, these tests can assess an application's resilience and dependability by simulating various scenarios and highlighting any possible bugs or security vulnerabilities.

In this chapter, our focus is on how software testing is used as a component of an IT security audit. Our security audit is a black-box audit, which evaluates security from the application's perspective as an external user without access to the source code by finding possible vulnerabilities through a variety of attack techniques. One of the advantages of this approach is that testers get a fresh look at the target, and thus, a new assessment of potential entry points from an attacker's point of view. This avoids, for example, focusing tests only on what is perceived as important to secure, while the risks of other elements may be underestimated. It is possible to conduct a black-box security test without notifying the teams in charge of detecting attacks to see the company's ability to detect an attack and react appropriately. Following the completion of the security assessments, we propose tips for enhancing application security. To help improve the security of Web applications, these recommendations include both technological and organizational security measures.

The rest of this chapter is organized as follows. Section 8.2 provides an introduction to black-box security audit. Section 8.3 describes the development environment used for our auditing approach. Afterward, our black-box audit methodology is highlighted in Section 8.4, and different results and recommendations are exposed in Section 8.5. In Section 8.6, we present a literature review of some works dealing with security testing of Web applications. Finally, we conclude in Section 8.7 with a summary of article contributions and identify prospective study areas for the future.

DOI: 10.1201/9781032714806-10

8.2 BLACK-BOX SECURITY AUDIT

Black-box auditing [1] is a type of audit that tests the security of an application or system without any prior knowledge of its internal structure. This is similar to an outside attacker looking for security vulnerabilities. Testers only know the name of the target organization and often an IP address or a URL. The attack surface is, therefore, broad. Time is first spent exploring the various elements included in the target before prioritizing the attacks according to the elements discovered during this phase.

The primary purpose of a black-box audit is to identify security vulnerabilities that could be exploited by attackers. These can include attacks such as SQL[1] injections, XSS[2] attacks, CSRF[3] attacks, brute force attacks, and leaks of sensitive information.

Black-box audits are particularly useful for assessing the security of applications that have been developed by third parties or for which there is no complete documentation available. They also identify vulnerabilities that may have been introduced over time as a result of updates or changes to the operating environment. Black-box audits can also be used to test APIs[4], which are increasingly used in modern applications for communicating with remote systems or Web services. Overall, black-box audits are an essential part of any overall security strategy and should be performed regularly to ensure application or system security.

8.3 DEVELOPMENT ENVIRONMENT USED FOR OUR APPROACH

- **Swagger** [2] is an open-source framework for designing, building, documenting, and deploying RESTful APIs. It provides a user interface for generating API documentation, as well as a range of tools to facilitate API design and testing. Swagger is compatible with many programming languages and has a wide range of third-party tools for code generation and API testing. It also offers advanced features for security and version management.
- **OWASP ZAP**[5] [3] is an open-source security audit tool used by security testers to detect vulnerabilities in Web applications. This is an automated penetration testing tool that can be used to test [4] vulnerabilities, including SQL injections, XSS vulnerabilities, and CSRF attacks. ZAP is also compatible with most common platforms and programming languages. It has a simple and lightweight user interface, making it faster and easier to use for fast security audits [5].

8.4 BLACK-BOX AUDIT METHODOLOGY

8.4.1 Overview of audit methodology

In our black-box audit, we used two tools: OWASP ZAP and Swagger. OWASP ZAP was used for vulnerability scanning of APIs, while Swagger was used to determine the full documentation of the application. We chose these two tools because of their advanced functionalities and compatibility with different programming languages.

In combination, Swagger and OWASP ZAP can provide a complete solution for API security auditing. Swagger allows to generate the complete API documentation, which is essential for a black-box audit. In addition, this documentation can be easily integrated into OWASP ZAP to facilitate the security audit. By using these two tools together, we can identify potential vulnerabilities in the API and take the necessary steps to correct them [6].

8.4.2 Explanation of the different steps of the audit

Integrating Swagger with SpringBoot application: With the aim to integrate Swagger into the SpringBoot application, we created a dedicated configuration class called **"SwaggerConfig."** This class uses the **"@EnableSwagger2"** and **"@Configuration"** annotations to enable Swagger support and set the necessary configuration settings.

Once the configuration was complete, we were able to access the Swagger documentation by launching the application on IntelliJ IDEA and accessing the following URL: **http://localhost:8080/swagger-ui.html**. This interactive documentation allowed us to view all the exposed operations by the APIs, as well as the input and output parameters for each operation. In summary, the integration of Swagger into the SpringBoot application allowed us to automatically generate a complete interactive API documentation.

Integrating Swagger documentation with OWASP ZAP: After installing and configuring OWASP ZAP on Ubuntu 22.04, we started the OWASP ZAP tool and the application on IntelliJ IDEA in parallel, as the application was deployed on our Windows machine.

To perform analysis on the application, we imported the Swagger documentation from the API by clicking the **"Import"** tab and then **"Import OpenAPI from Definition URL."** We then pasted the URL of the Swagger documentation into the field "URL pointing to OpenAPI def" and clicked the **"Import"** button using the Windows machine IP address instead of "localhost" in the URL.

Configuring Targeted Attacks and Services in OWASP ZAP: Once we have imported the Swagger API documentation, here is an overview of the imported APIs on OWASP ZAP:

- We have a total of 78 APIs grouped into 10 different categories.
- For each API, we have access to detailed information such as URL, parameters, supported HTTP methods, accepted and returned data types, etc.
- We can also view requests and responses for each API, allowing us to better understand how they work and detect potential vulnerabilities.

Figure 8.1 summarizes the presentation of imported APIs on OWASP ZAP.

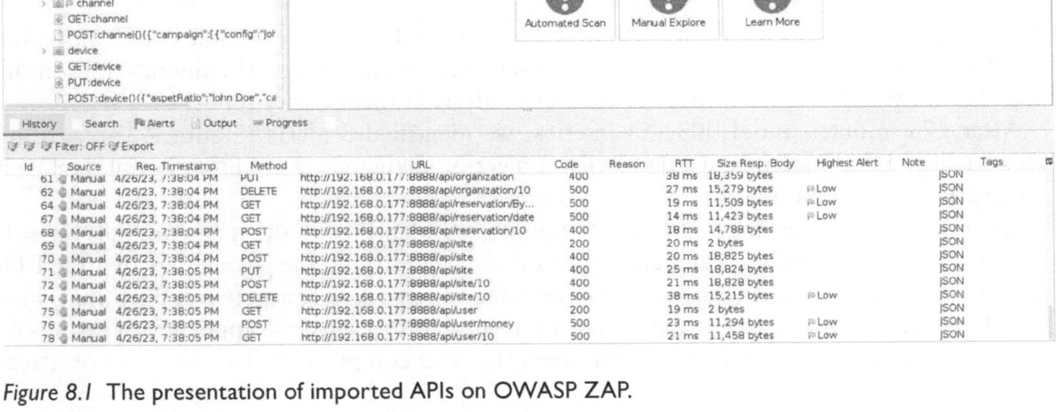

Figure 8.1 The presentation of imported APIs on OWASP ZAP.

Figure 8.2 OWASP ZAP- configuration of an active scan attack.

Next, we configured the active scan attack by selecting the attacks to start and the services to target. We have deselected the services that are not needed or used by the application to be audited to avoid overloading the application and the OWASP ZAP tool with unnecessary queries and to speed up the vulnerability scanning process (Figure 8.2). This step is important to ensure that the OWASP ZAP tool performs an accurate and complete analysis of the application.

After configuring the active scan attack, we launched the attack by clicking the **"Start Scan"** button in the OWASP ZAP user interface toolbar. We then observed how the attack unfolded, generating a large number of requests and responses between the application and OWASP ZAP [7].

8.5 BLACK-BOX AUDIT RESULTS

8.5.1 Obtained results by OWASP ZAP

We were also able to see in real time the detected vulnerabilities by the tool, as well as details on their impact and severity level. This step identified several potential vulnerabilities in the application, which were recorded for further analysis in the security report.

After 275 minutes and 1,003,154 queries, we identified **3 alerts** (Figure 8.3) regarding an attack by **SQL injection**, which involves injecting malicious SQL into a query to a Web application using a database.

After 1 minute 40 seconds and 9,577 requests, we identified **4 alerts** (Figure 8.4) regarding an attack by **Parameter Tampering** that involves modifying the parameters of an HTTP request to try to cause unexpected behaviors or errors in the targeted application. This attack can allow an attacker to gain unauthorized access to sensitive application features or data or cause errors in application operations that can compromise data integrity or affect application availability.

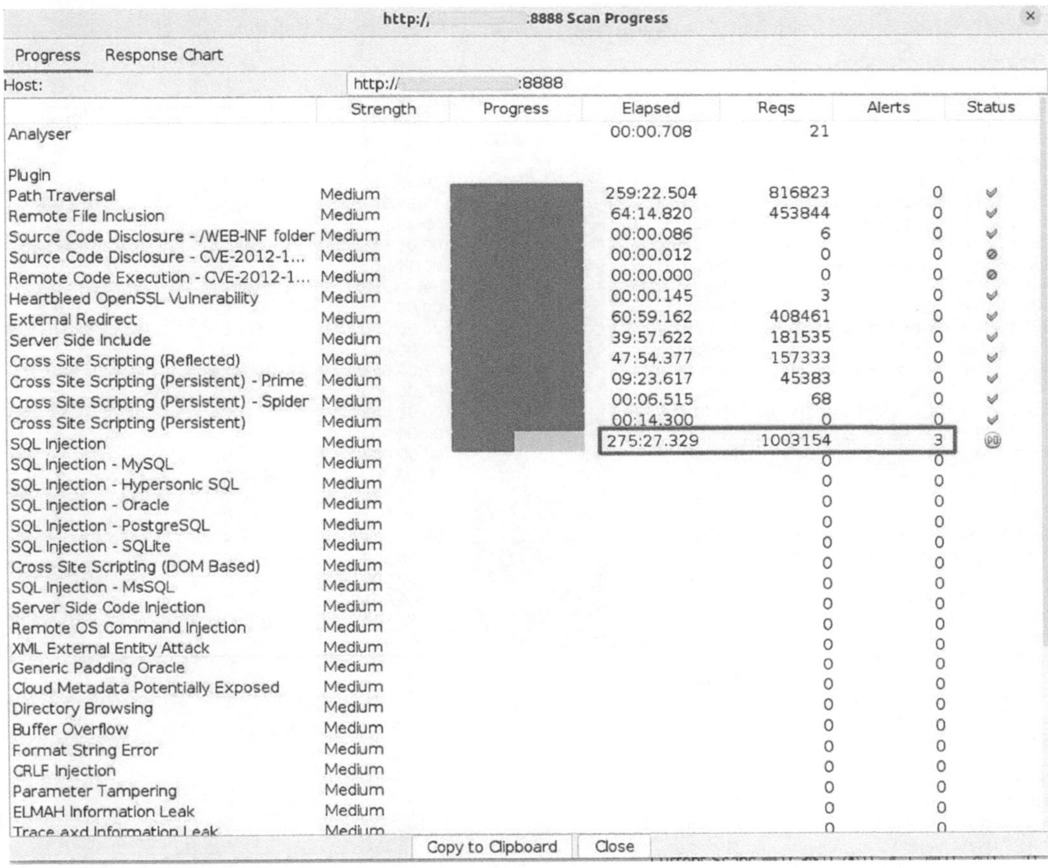

	Strength	Progress	Elapsed	Reqs	Alerts	Status
Analyser			00:00.708	21		
Plugin						
Path Traversal	Medium		259:22.504	816823	0	
Remote File Inclusion	Medium		64:14.820	453844	0	
Source Code Disclosure - /WEB-INF folder	Medium		00:00.086	6	0	
Source Code Disclosure - CVE-2012-1...	Medium		00:00.012	0	0	
Remote Code Execution - CVE-2012-1...	Medium		00:00.000	0	0	
Heartbleed OpenSSL Vulnerability	Medium		00:00.145	3	0	
External Redirect	Medium		60:59.162	408461	0	
Server Side Include	Medium		39:57.622	181535	0	
Cross Site Scripting (Reflected)	Medium		47:54.377	157333	0	
Cross Site Scripting (Persistent) - Prime	Medium		09:23.617	45383	0	
Cross Site Scripting (Persistent) - Spider	Medium		00:06.515	68	0	
Cross Site Scripting (Persistent)	Medium		00:14.300	0	0	
SQL Injection	Medium		275:27.329	1003154	3	
SQL Injection - MySQL	Medium			0	0	
SQL Injection - Hypersonic SQL	Medium			0	0	
SQL Injection - Oracle	Medium			0	0	
SQL Injection - PostgreSQL	Medium			0	0	
SQL Injection - SQLite	Medium			0	0	
Cross Site Scripting (DOM Based)	Medium			0	0	
SQL Injection - MsSQL	Medium			0	0	
Server Side Code Injection	Medium			0	0	
Remote OS Command Injection	Medium			0	0	
XML External Entity Attack	Medium			0	0	
Generic Padding Oracle	Medium			0	0	
Cloud Metadata Potentially Exposed	Medium			0	0	
Directory Browsing	Medium			0	0	
Buffer Overflow	Medium			0	0	
Format String Error	Medium			0	0	
CRLF Injection	Medium			0	0	
Parameter Tampering	Medium			0	0	
ELMAH Information Leak	Medium			0	0	
Trace.axd Information Leak	Medium			0	0	

Figure 8.3 OWASP ZAP- SQL injection alerts.

After 13 seconds and 384 requests, we identified **22 alerts** (Figure 8.5) regarding an attack by **XSTL Injection** that injected malicious XSL statements into a request to cause unintended application behavior or to access sensitive data.

After 23 seconds and 816 requests, we identified **684 alerts** (Figure 8.6), regarding an attack by **User Agent Fuzzer** that involves sending HTTP requests using different user agents, which are request headers identifying the client used to make the request. By sending requests with malicious or modified user agents, the attacker can attempt to cause unexpected behaviors in the application, such as processing errors or leaks of sensitive information.

Note: It is important to note that these results are only an indication of the potential vulnerabilities of the application and do not guarantee that all the vulnerabilities will be absent. Therefore, it is recommended that one conduct a comprehensive security scan of the application using several tools and techniques to detect as many vulnerabilities as possible and enhance application security.

8.5.2 Black-box audit report generation

After the analysis was completed, we generated an audit report by selecting the **"Generate Report"** option in the OWASP ZAP user interface. It is possible to choose the desired report format (HTML, XML, JSON, PDF, and so on) and to set the location where the report

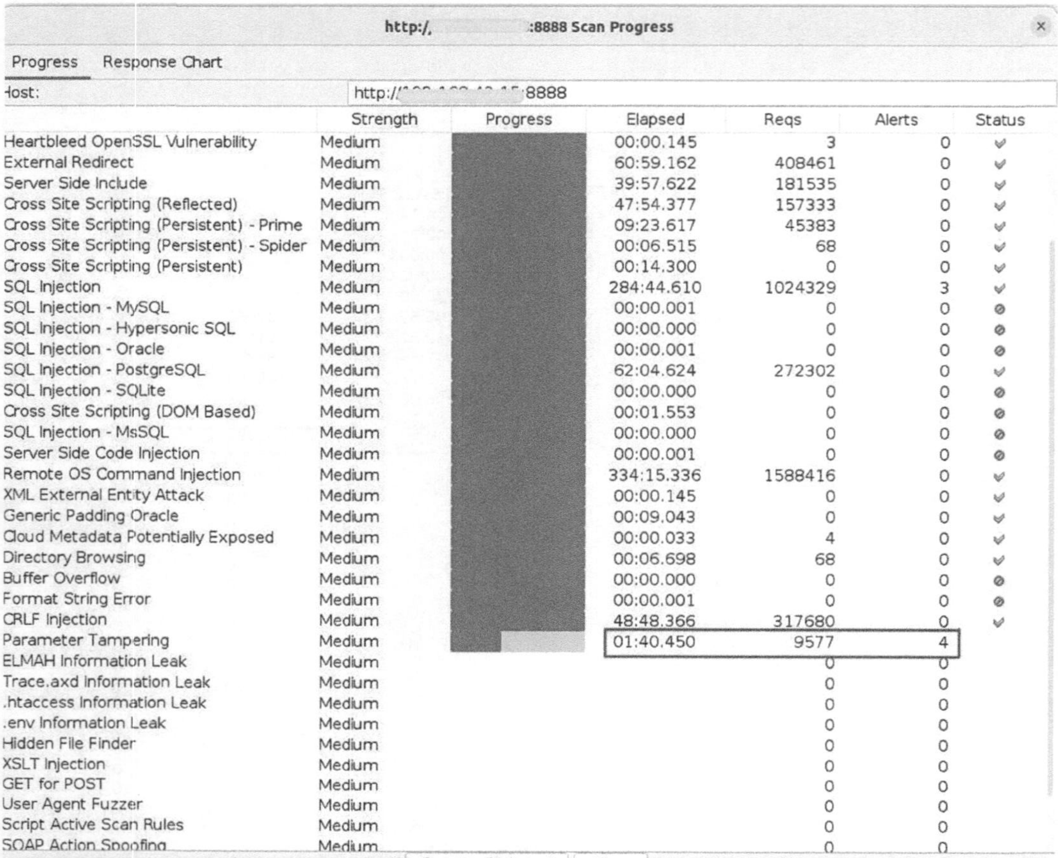

http:/[redacted]:8888 Scan Progress

Progress Response Chart

Host: http://[redacted]:8888

	Strength	Progress	Elapsed	Reqs	Alerts	Status
Heartbleed OpenSSL Vulnerability	Medium		00:00.145	3	0	✓
External Redirect	Medium		60:59.162	408461	0	✓
Server Side Include	Medium		39:57.622	181535	0	✓
Cross Site Scripting (Reflected)	Medium		47:54.377	157333	0	✓
Cross Site Scripting (Persistent) - Prime	Medium		09:23.617	45383	0	✓
Cross Site Scripting (Persistent) - Spider	Medium		00:06.515	68	0	✓
Cross Site Scripting (Persistent)	Medium		00:14.300	0	0	✓
SQL Injection	Medium		284:44.610	1024329	3	✓
SQL Injection - MySQL	Medium		00:00.001	0	0	⊘
SQL Injection - Hypersonic SQL	Medium		00:00.000	0	0	⊘
SQL Injection - Oracle	Medium		00:00.001	0	0	⊘
SQL Injection - PostgreSQL	Medium		62:04.624	272302	0	✓
SQL Injection - SQLite	Medium		00:00.000	0	0	⊘
Cross Site Scripting (DOM Based)	Medium		00:01.553	0	0	⊘
SQL Injection - MsSQL	Medium		00:00.000	0	0	⊘
Server Side Code Injection	Medium		00:00.001	0	0	⊘
Remote OS Command Injection	Medium		334:15.336	1588416	0	✓
XML External Entity Attack	Medium		00:00.145	0	0	✓
Generic Padding Oracle	Medium		00:09.043	0	0	✓
Cloud Metadata Potentially Exposed	Medium		00:00.033	4	0	✓
Directory Browsing	Medium		00:06.698	68	0	✓
Buffer Overflow	Medium		00:00.000	0	0	⊘
Format String Error	Medium		00:00.001	0	0	⊘
CRLF Injection	Medium		48:48.366	317680	0	✓
Parameter Tampering	Medium		01:40.450	9577	4	
ELMAH Information Leak	Medium			0	0	
Trace.axd Information Leak	Medium			0	0	
.htaccess Information Leak	Medium			0	0	
.env Information Leak	Medium			0	0	
Hidden File Finder	Medium			0	0	
XSLT Injection	Medium			0	0	
GET for POST	Medium			0	0	
User Agent Fuzzer	Medium			0	0	
Script Active Scan Rules	Medium			0	0	
SOAP Action Spoofing	Medium			0	0	

Copy to Clipboard Close

Figure 8.4 OWASP ZAP- parameter tampering.

should be saved. Then, OWASP ZAP generated a security audit report that lists all detected vulnerabilities and provides recommendations for remediation. The report provides a high-level overview of the vulnerabilities detected in the application, showing the number of vulnerabilities by category. Vulnerabilities are classified according to severity using a risk rating scale from 0 (low) to 5 (high). The main categories of vulnerabilities detected by OWASP ZAP include:

- **Code injection:** Code injection vulnerabilities can allow an attacker to inject malicious code into the application. Examples include SQL injections, client-side script injections (XSS), server-side script injections (SSRF), and so on.
- **Authentication vulnerabilities:** Authentication vulnerabilities can allow an attacker to compromise authentication and take control of a user account. Examples include weak passwords, lack of user account control, and so on.
- **Session management vulnerabilities:** Session management vulnerabilities can allow an attacker to steal or manipulate user sessions. Examples include nonsecure cookies, the absence of CSRF (Cross-Site Request Forgery) tokens, etc.
- **Data security vulnerabilities:** Data security vulnerabilities can allow an attacker to access sensitive data, such as credit card information or personally identifiable information. Examples include data leakage, encryption vulnerabilities, and so on.

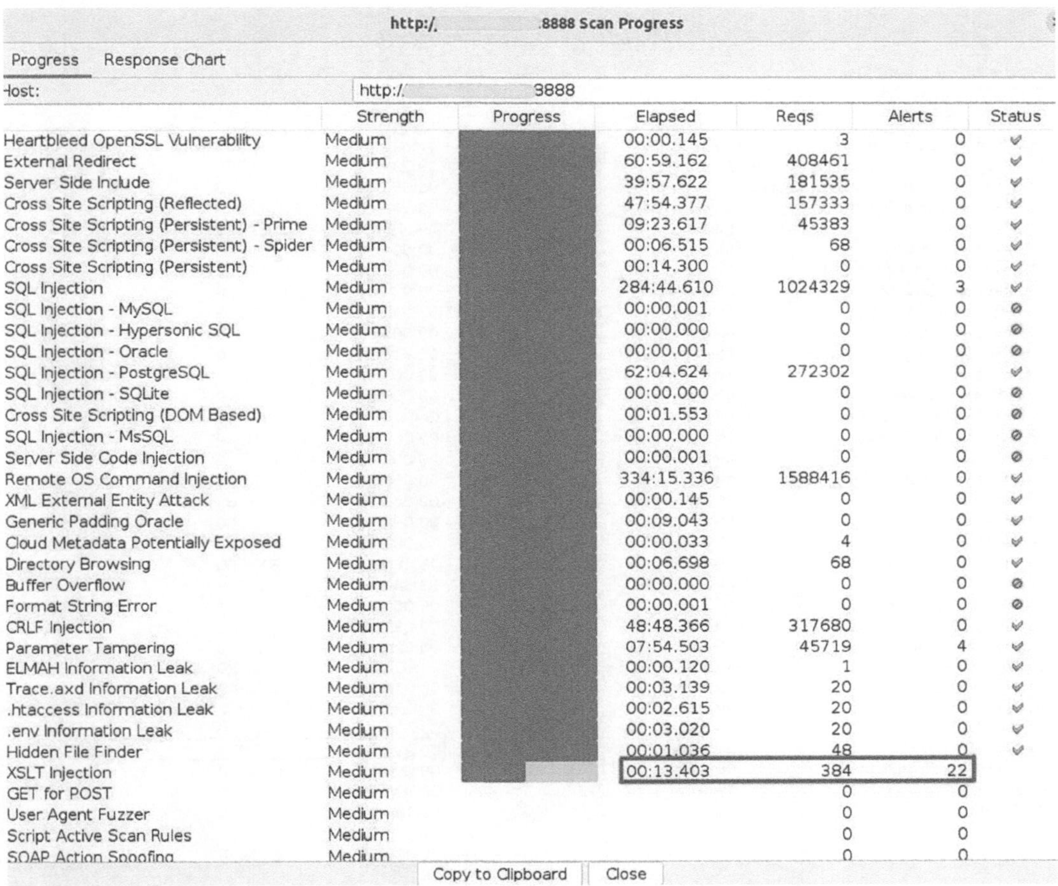

http:/. .8888 Scan Progress

Progress Response Chart

Host: http:/. 3888

	Strength	Progress	Elapsed	Reqs	Alerts	Status
Heartbleed OpenSSL Vulnerability	Medium		00:00.145	3	0	✓
External Redirect	Medium		60:59.162	408461	0	✓
Server Side Include	Medium		39:57.622	181535	0	✓
Cross Site Scripting (Reflected)	Medium		47:54.377	157333	0	✓
Cross Site Scripting (Persistent) - Prime	Medium		09:23.617	45383	0	✓
Cross Site Scripting (Persistent) - Spider	Medium		00:06.515	68	0	✓
Cross Site Scripting (Persistent)	Medium		00:14.300	0	0	✓
SQL Injection	Medium		284:44.610	1024329	3	✓
SQL Injection - MySQL	Medium		00:00.001	0	0	⊘
SQL Injection - Hypersonic SQL	Medium		00:00.000	0	0	⊘
SQL Injection - Oracle	Medium		00:00.001	0	0	⊘
SQL Injection - PostgreSQL	Medium		62:04.624	272302	0	✓
SQL Injection - SQLite	Medium		00:00.000	0	0	⊘
Cross Site Scripting (DOM Based)	Medium		00:01.553	0	0	⊘
SQL Injection - MsSQL	Medium		00:00.000	0	0	⊘
Server Side Code Injection	Medium		00:00.001	0	0	⊘
Remote OS Command Injection	Medium		334:15.336	1588416	0	✓
XML External Entity Attack	Medium		00:00.145	0	0	✓
Generic Padding Oracle	Medium		00:09.043	0	0	✓
Cloud Metadata Potentially Exposed	Medium		00:00.033	4	0	✓
Directory Browsing	Medium		00:06.698	68	0	✓
Buffer Overflow	Medium		00:00.000	0	0	⊘
Format String Error	Medium		00:00.001	0	0	⊘
CRLF Injection	Medium		48:48.366	317680	0	✓
Parameter Tampering	Medium		07:54.503	45719	4	✓
ELMAH Information Leak	Medium		00:00.120	1	0	✓
Trace.axd Information Leak	Medium		00:03.139	20	0	✓
.htaccess Information Leak	Medium		00:02.615	20	0	✓
.env Information Leak	Medium		00:03.020	20	0	✓
Hidden File Finder	Medium		00:01.036	48	0	✓
XSLT Injection	Medium		00:13.403	384	22	
GET for POST	Medium			0	0	
User Agent Fuzzer	Medium			0	0	
Script Active Scan Rules	Medium			0	0	
SOAP Action Spoofing	Medium			0	0	

Copy to Clipboard Close

Figure 8.5 OWASP ZAP- XSTL injection.

- **Server security vulnerabilities:** Server security vulnerabilities can allow an attacker to take control of the server or execute malicious code remotely. Examples include operating system vulnerabilities, Web server vulnerabilities, and so on.

8.5.3 Analysis and interpretation of results

The following is an analysis and interpretation of the results obtained with the OWASP ZAP tool. To illustrate this issue, we have created several summary tables that provide an overview of various important data. The following tables provide a summary of the results of the performed security analysis [8].

8.5.3.1 Alert summary

Table 8.1 summarizes detected alerts based on their risk level.

8.5.3.2 Alerts

Table 8.2 depicts the various vulnerabilities detected during the security scan performed on the website. Vulnerabilities are categorized by name, risk level, and the number of detected instances.

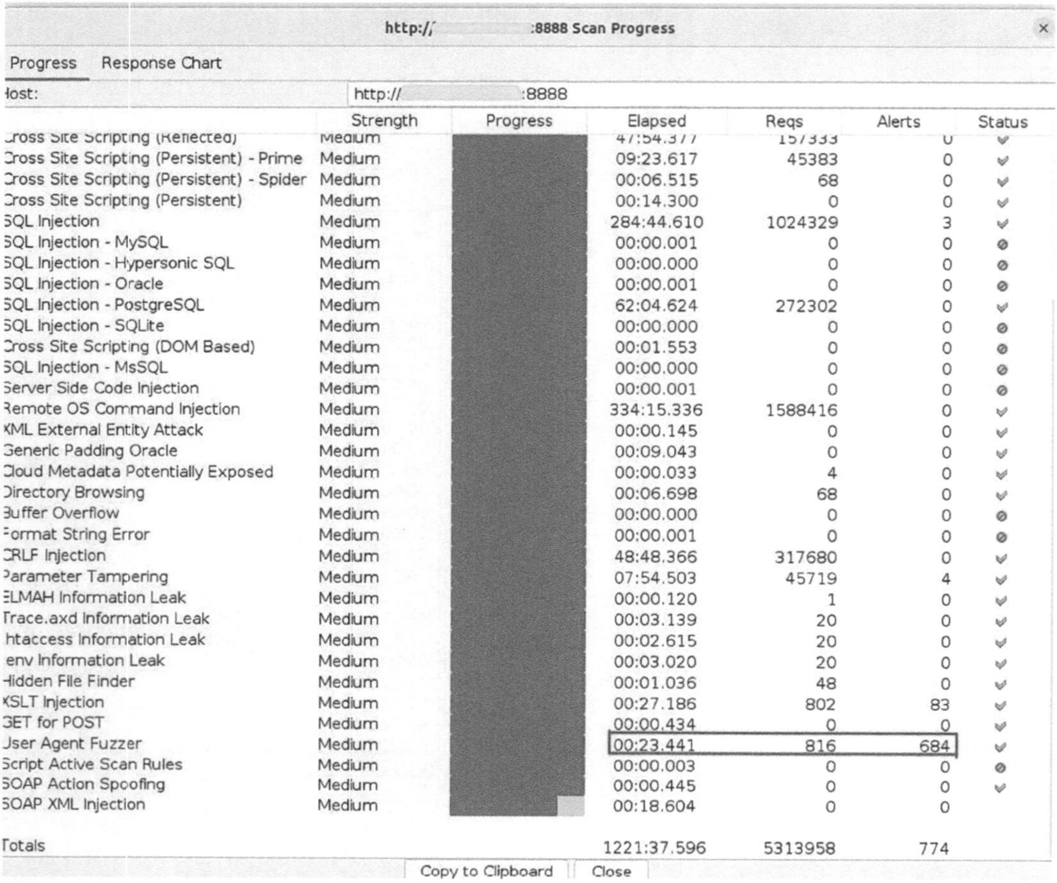

http:// :8888 Scan Progress

Progress | Response Chart

Host: http:// :8888

	Strength	Progress	Elapsed	Reqs	Alerts	Status
Cross Site Scripting (Reflected)	Medium		47:54.377	157333	0	✓
Cross Site Scripting (Persistent) - Prime	Medium		09:23.617	45383	0	✓
Cross Site Scripting (Persistent) - Spider	Medium		00:06.515	68	0	✓
Cross Site Scripting (Persistent)	Medium		00:14.300	0	0	✓
SQL Injection	Medium		284:44.610	1024329	3	✓
SQL Injection - MySQL	Medium		00:00.001	0	0	⊘
SQL Injection - Hypersonic SQL	Medium		00:00.000	0	0	⊘
SQL Injection - Oracle	Medium		00:00.001	0	0	⊘
SQL Injection - PostgreSQL	Medium		62:04.624	272302	0	✓
SQL Injection - SQLite	Medium		00:00.000	0	0	⊘
Cross Site Scripting (DOM Based)	Medium		00:01.553	0	0	⊘
SQL Injection - MsSQL	Medium		00:00.000	0	0	⊘
Server Side Code Injection	Medium		00:00.001	0	0	⊘
Remote OS Command Injection	Medium		334:15.336	1588416	0	✓
XML External Entity Attack	Medium		00:00.145	0	0	✓
Generic Padding Oracle	Medium		00:09.043	0	0	✓
Cloud Metadata Potentially Exposed	Medium		00:00.033	4	0	✓
Directory Browsing	Medium		00:06.698	68	0	✓
Buffer Overflow	Medium		00:00.000	0	0	⊘
Format String Error	Medium		00:00.001	0	0	⊘
CRLF Injection	Medium		48:48.366	317680	0	✓
Parameter Tampering	Medium		07:54.503	45719	4	✓
ELMAH Information Leak	Medium		00:00.120	1	0	✓
Trace.axd Information Leak	Medium		00:03.139	20	0	✓
.htaccess Information Leak	Medium		00:02.615	20	0	✓
.env Information Leak	Medium		00:03.020	20	0	✓
Hidden File Finder	Medium		00:01.036	48	0	✓
XSLT Injection	Medium		00:27.186	802	83	✓
GET for POST	Medium		00:00.434	0	0	✓
User Agent Fuzzer	Medium		00:23.441	816	684	✓
Script Active Scan Rules	Medium		00:00.003	0	0	⊘
SOAP Action Spoofing	Medium		00:00.445	0	0	✓
SOAP XML Injection	Medium		00:18.604	0	0	
Totals			1221:37.596	5313958	774	

Copy to Clipboard | Close

Figure 8.6 OWASP ZAP- user agent fuzzer.

8.5.3.3 Statistics of HTTP response codes

Table 8.3 shows statistics of HTTP response codes for a set of executed requests. HTTP response codes are three-digit codes returned by a Web server in response to a request from a client.

8.5.3.4 Detected security risks by OWASP ZAP

Table 8.4 shows a list of security risks detected by OWASP ZAP. Risks are classified according to severity, from low to high. Each risk is accompanied by a description that explains the potential implications of the vulnerability.

8.5.3.5 API vulnerabilities

Vulnerabilities detected at the API level by the OWASP ZAP tool can expose sensitive information, such as debug error messages or confidential data present in the URL. These vulnerabilities can be exploited by attackers to access private information or perform more advanced attacks on the application. Table 8.5 summarizes the actions that should be taken to address these vulnerabilities.

Table 8.1 Risks

Risk level	*Number of alerts*
High	0
Medium	0
Low	2
Informational	1
False positive	0

Table 8.2 Alerts

Vulnerability name	*Risk level*	*Number of instances*
Application error disclosure	Low	25
Disclosure of information – Debug error messages	Low	25
Disclosure of information – Sensitive information in URL	Information	1

Table 8.3 Statistics of HTTP response codes

HTTP response code	*Number of responses*
403 Forbidden	55
404 Not found	124
405 Method not allowed	14
200 OK	123
400 Bad request	563253
500 Internal server error	278

Table 8.4 Security risks detected by OWASP ZAP

Severity	*Title*	*Description*
Low	Application error disclosure	This page contains an error/warning message that may disclose sensitive information, such as the location of the file that produced the unhandled exception. This information can be used to launch further attacks against the Web application.
Low	Information disclosure – Debug error messages	The response appears to contain common error messages returned by platforms such as ASP.NET and Web servers such as IIS and Apache. The list of common debug messages can be configured.
Info	Information disclosure – Sensitive information in URL	The request seems to contain sensitive information disclosed in the URL. This can violate PCI (Payment Card Industry) compliance policies for most organizations. The string list for this check can be configured to add or remove values that are specific to its environment.

8.5.3.6 API-related security access points

Table 8.6 provides specific recommendations for enhancing security access point security for APIs which are identified during the black-box audit with the OWASP ZAP tool. These measures will help improve the security of the application APIs by addressing the detected vulnerabilities.

Table 8.5 Recommendations for API vulnerabilities

Type of vulnerability	*Recommendation*	*References*
Application error disclosure	Errors must be managed appropriately to prevent sensitive information from being exposed to users. Error messages must be clear and informative but must not contain sensitive information.	- CWE-200 [9]: Exposure of Sensitive Information to an Unauthorized Actor - OWASP [10] 2017 A06 (Security Misconfiguration) - OWASP 2021 A05 (Security Misconfiguration)
Information disclosure – Debug error messages	Error messages should not contain sensitive debug information. Debug information should only be exposed to those who have permission to access it.	CWE-200: Exposure of Sensitive Information to an Unauthorized Actor OWASP 2017 A03 (Sensitive Data Exposure) - OWASP 2021 A01 (Broken Access Control)
Information disclosure – Sensitive information in URL	Sensitive information should not be passed in the application URL. It must be transmitted securely, for example, using an encryption protocol such as HTTPS.	CWE-200: Exposure of Sensitive Information to an Unauthorized Actor OWASP 2017 A03 (Sensitive Data Exposure) - OWASP 2021 A01 (Broken Access Control)

Table 8.6 API-related security access points

Type of attack	*Recommendations*	*References*
SQL injection	Use prepared queries or ORMs (object-relational mapping) to avoid SQL injections. Validate and filter user entries.	OWASP Top 10 – A03 : 2021 (Injection)
Parameter tampering	Validate and verify settings and data sent by the user. Use encryption techniques to protect sensitive data.	OWASP Top 10 – A08 : 2021 (Software and Data Integrity Failures)
XSTL injection	Apply appropriate validation to the XML data. Use secure libraries or frameworks for processing XML data.	OWASP Top 10 – A05 : 2021 (Security Misconfiguration)
User agent fuzzier	Validate and filter user inputs to prevent fuzzing attacks. Implement mechanisms to limit requests or IPs when abnormal activity is detected.	OWASP Top 10 – A03 : 2021 (Injection)

8.5.4 Organizational security measures

Organizational security measures are a set of practices and policies that enhance the security of a company's information systems. These include adopting security policies, training and raising employee awareness, developing incident response plans, implementing access and confidentiality controls, and conducting regular security audits. For FRS company, we recommend implementing the organizational security measures shown in Table 8.7.

Table 8.7 Organizational security measures

Organizational security measures	*Description*
Using SSL/TLS certificates	Provides a secure connection between the user's browser and the Web server by encrypting the data that is exchanged.
Strong authentication	Requires users to provide two authentication factors, such as a password and code generated by a mobile application, to access the application.
Password management	Implements strong password policies, such as minimum length, use of special characters, and so on, and encourages users to change their passwords regularly.
Validate entries	Verifies that user input does not contain malicious code or commands that could compromise application or server security.
Error handling	Displays generic error messages to avoid giving sensitive information to malicious users.
Protection against injection attacks	Uses prepared queries to avoid SQL injections or other types of injections that can compromise application or server security.
Regular software updates	Ensures that software and libraries used in the application are updated regularly to avoid known security vulnerabilities.
Securing APIs	Uses authentication and validation mechanisms to protect the application's APIs from DDoS (distributed denial-of-service) attacks or identity fraud.
Monitoring activity logs	Monitor application activity logs for unauthorized access attempts or suspicious activities.
Security incident response plan	Develop a security incident response plan to manage incidents in an efficient and coordinated manner in case of an attack or security incident.

8.6 RELATED WORKS

Ablahd [11] proposed a system that detects Web application vulnerabilities before they can be exploited by attackers. To detect these vulnerabilities, a special scanner was built using Python 3.7's built-in tools, including AST, CFG, Flask, and Django. This proposed system solves two types of risks that can infect a Web application due to vulnerability. The proposed scanner detects injection flaws such as command execution and XSS. A flexible set of tools was designed for the scanner, called SCANSCX, and several vulnerable applications were designed for testing and evaluating SCANSCX's abilities.

Albahar et al. [12] compared security testing tools for detecting vulnerabilities in Web applications based on approved standards and methods to facilitate testers' selection of the most appropriate tools. To enhance the effectiveness of Web security testers in real life, the authors proposed a benchmarking framework that incorporates the latest research into benchmarking and evaluation criteria, in addition to new criteria that provide more coverage with benchmarking metrics. Moreover, a score-based comparative analysis was used to evaluate the tool's abilities. In their study, Burp Suite Professional was rated the highest out of the commercial tools, while OWASP ZAP was rated the highest out of the noncommercial tools.

Wardana et al. [13] declared that vulnerability assessment and security testing on published websites following specific standards are critical for information security. Testing was done considering the four primary stages of preparation, discovery, attack, and reporting.

One high-level, two medium-level, and four low-level vulnerabilities were identified. During the penetration testing, it was found that while specific security measures had been implemented, including encryption and authentication techniques, they were not utilized to their full potential. Furthermore, other measures could have been implemented to improve the security of the websites, such as using firewalls and intrusion detection systems. These security measures could have provided more robust protection against potential attacks and alerted the organization to any malicious activity occurring on their websites. This assessment and testing showed that while specific security measures were in place, there was still a need to strengthen security by taking advantage of all the available resources. The testing demonstrated the need for improving security measures to prevent potential malicious actors from exploiting vulnerabilities.

Wibowo et al. [14] stated that Web applications are required to respond to the ease of use of Internet technology. Cross-site scripting (XSS) is one of the most widespread security threats or attacks. Numerous solutions may be utilized to avoid cyberattacks, such as OWASP Security Shepherd, a secure platform with various tools and techniques, including XSS, to protect Web applications from cyberattacks. Using a combination of secure coding practices, automated tools, and manual code reviews, OWASP Security Shepherd provides an effective solution to protect Web applications from XSS attacks. With OWASP Security Shepherd, developers, and organizations can mitigate the risks associated with XSS attacks by ensuring that their Web applications are properly tested and patched. In addition, OWASP Security Shepherd provides an intuitive interface and features such as easy-to-use reports, real-time monitoring, and support for multiple programming languages. This makes it possible for developers to quickly identify and address potential vulnerabilities in their Web applications, thereby improving the overall security of their online assets.

Nagendran et al. [15] described client-side and server-side attacks in classifying Web attacks. In addition, they provided an in-depth explanation of how to perform a manual test on Web applications to ensure their integrity and security, as well as a guide to test the OWASP's top 10 security vulnerabilities. Besides, the authors discussed manual Web application security testing methodologies, which they classified into five phases: recognition, scanning, exploitation, maintaining access and privilege escalation, and clearing tracks and reporting.

Mirjalili et al. [16] proposed the design and development of a distributed framework to automate Web security testing, the major components of which are an operational unit called an executor that conducts attacks and a control unit called an orchestrator that orchestrates them across consecutive stages. The authors defined the general activities carried out during a penetration test and presented a method for integrating the attackers that execute such jobs. Moreover, they defined a flexible method for integrating external tools to achieve the desired hacking goals by mapping vulnerabilities to the framework's integrated tools. To realize the full potential of this distributed framework, they presented a suite of tools with a Web-based user interface that integrates seamlessly into their system. Despite the benefits of distributed hacking, it suffers from process synchronization, although global knowledge is required; resource management, since global security knowledge is required; fault tolerance; and error recovery.

Based on the reviewed studies, Table 8.8 presents the key findings on some security test types and the suggested techniques, advantages, and limitations of each study.

Table 8.8 Summary of related works

Author	*Publication year*	*Security test type*	*Suggested technique*	*Advantages*	*Limitations*
Ablahd [11]	2023	Automated	A system was proposed that detects Web application vulnerabilities using Python 3.7 to identify injection flaws such as command execution and cross-site scripting.	The proposed scanner is easy to use (in each Web application) and flexible when it comes to updating.	The authors needed to adapt the scanner to detect other types of Web vulnerabilities.
Albahar et al. [12]	2022	Automated and manual	To enhance the effectiveness of Web security testers in real life, a benchmarking framework was proposed incorporating the latest benchmarking and evaluation research.	A comprehensive framework is offered with all the necessary features for security testers.	Benchmarking should be applied to other tools, and the framework should be extended to include more new metrics.
Wardana et al. [13]	2022	Manual	Manual security testing on published websites follows certain standards with four primary stages: 1. Preparation 2. Discovery 3. Attack 4. Report	To improve the security of the websites, other measures could have been implemented, such as using firewalls and intrusion detection systems. These security measures could have provided more robust protection against potential attacks and alerted the organization to any malicious activity occurring on their websites.	The testing demonstrated the need for an improvement in security measures to prevent potential malicious actors from exploiting vulnerabilities. In conclusion, the security testing revealed that while security measures were already in place, they needed to be strengthened and utilized more effectively.

(*Continued*)

Table 8.8 (Continued) Summary of related works

Author	*Publication year*	*Security test type*	*Suggested technique*	*Advantages*	*Limitations*
Wibowo et al. [14]	2021	Automated and manual	An integrated approach for OWASP Security Shepherd based on using a combination of secure coding practices, automated tools, and manual code reviews.	OWASP Security Shepherd provides the following: 1. An effective solution for protecting Web applications from XSS attacks; 2. An intuitive interface and features such as easy-to-use reports, real-time monitoring, and support for multiple programming languages.	The present Web application firewalls only offer basic protection rules that do not consider advancements in the sector. The authors wanted to build and create a lightweight and adaptable Web application firewall in the future as part of their ongoing development.
Nagendran et al. [15]	2019	Manual	Web application security testing with the following five phases: 1. Recognition 2. Scanning 3. Exploitation 4. Maintaining access and privilege escalation 5. Clearing tracks and reporting	An in-depth explanation was provided for how to perform a manual security test on Web applications.	Performing manual security tests requires a great deal of expertise in working with HTTP requests and responses.
Mirjalili et al. [16]	2014	Automated	Automated penetration testing framework with the following two major components: 1. An operational unit called an executor that conducts attacks; 2. A control unit called an orchestrator that orchestrates attacks across consecutive stages.	The distributed hacking framework provides scalability, a distributed nature, and ease of use.	Suffers from process synchronization, resource management, fault tolerance, and error recovery.

8.7 CONCLUSION

Security methods comprise enhancing security procedures during the software development life cycle and throughout the application life cycle. All application security initiatives should aim to reduce the possibility of malicious actors gaining unauthorized access to systems, applications, or data. The ultimate purpose of application security is to prevent attackers from accessing, altering, or destroying sensitive or proprietary data.

In this chapter, we contribute to bridging the gap between academia and industrial practice. In fact, we studied the black-box audit of an industrial Web project by highlighting its crucial role in the assessment of application security. First, our approach identified potential security vulnerabilities, providing valuable information to further protect the application. Using Swagger has helped to understand the structure of the application and facilitate its documentation, while OWASP ZAP has provided automated tests to detect common vulnerabilities. By combining these tools, it was possible to obtain complete coverage during the audit, thus making it possible to remedy the identified vulnerabilities and strengthen the overall security of the application. We also proposed some recommendations to help the FRS company further enhance the security of their application.

Academic efforts are needed in parallel to clearly understand, define, and deliver more reliable testing models, methods, tools, and techniques that can syndicate traditional testing approaches. On the contrary, industry should open its doors to cooperate and collaborate with academics, and vice versa. It should be promoted at least through panel discussions, lecture series of experts, and keynote addresses from both academia and industry experts.

Perspectives of our work concern a white box audit as a complement to a black one, which allows us to prepare scenarios such as insider threats or an attacker that has obtained detailed internal information. Since the audit team has visible access to the crucial data and specifics essential for attacking the organization, this procedure typically reveals more vulnerabilities and is significantly faster. In addition, we aim to provide more best practices and experience from applying novel approaches to large-scale industrial projects in the context of software architecture.

NOTES

1 Structured Query Language
2 Cross-Site Scripting
3 Cross-Site Request Forgery
4 Application Programming Interface
5 Zed Attack Proxy

REFERENCES

1. Black Box Security Audit. https://www.securitybrigade.com/types-of-security-audits/
2. Swagger. https://swagger.io/
3. OWASP ZAP. https://www.zaproxy.org/docs/
4. CVE (Common Vulnerabilities and Exposures). https://www.cvedetails.com/
5. CVSS (Common Vulnerability Scoring System). https://www.techtarget.com/searchsecurity/definition/CVSS-Common-Vulnerability-Scoring-System
6. White Box Security Audit. https://www.securitybrigade.com/types-of-security-audits/
7. Jenkins. https://www.jenkins.io/
8. Sonarqube. https://docs.sonarqube.org/

9. CWE (Common Weakness Enumeration). https://cwe.mitre.org/
10. OWASP (Open Web Application Security Project). https://owasp.org/
11. Ablahd, A.Z. Using python to detect web application vulnerability. *Res Militaris* 2023, 13, 1045–1058.
12. Albahar, M.; Alansari, D.; Jurcut, A. An empirical comparison of pen-testing tools for detecting web app vulnerabilities. *Electronics* 2022, 11, 2991.
13. Wardana, W.; Almaarif, A.; Widjajarto, A. Vulnerability assessment and penetration testing on the xyz website using NIST 800-115 standard. *J. Ilm. Indones.* 2022, 7, 520–529.
14. Wibowo, R.M.; Sulaksono, A. Web vulnerability through cross site scripting (XSS) detection with OWASP security shepherd. *Indones. J. Inf. Syst.* 2021, 3, 149–159.
15. Nagendran, K.; Adithyan, A.; Chethana, R.; Camillus, P.; Varshini, K.B.S. Web application penetration testing. *IJITEE* 2019, 8, 1029–1035.
16. Mirjalili, M.; Nowroozi, A.; Alidoosti, M. A survey on a web penetration test. *Adv. Comput. Sci. Int. J.* 2014, 3, 117–121.

Chapter 9

Security in cloud-based IoT

A survey

Abdelhalim Hnini, Anas Anouar, Ayoub Khadrani, Chaima Dhiba, Salmaa Naffah, and Imane Chlioui

9.1 INTRODUCTION

The Internet of Things (IoT) prototype consists of intelligent, self-configuring sensors (things) linked to a live, global network architecture. On the IoT, "things" refers to any object, including communicative devices and inanimate objects. IoT is a technology that is expanding quickly. IoT, which has problems with performance, reliability, security, and privacy, is typically identified by physical and compact sensors with limited capacity and storage. It is a network of real sensors that is controlled and monitored online. The IoT provides smart sensors using its foundational technologies, such as apps, Internet protocols, vast and universal computing, sensor networks, and communication technologies. The IoT prototype consists of intelligent, self-configuring sensors (things) linked to a live, global network architecture.

Cloud computing is a significantly more developed technology with practically infinite computing and storage capabilities. The creation and fusion of customer, adaptable, and worldwide processing resources is known as cloud computing. Memory, computation, storage, network bandwidth, and virtual machines are different cloud resources. Access to remote configurable computer resources is made more accessible with cloud computing, and the amount of lab required to maintain services is reduced. The different virtualized assets makeup cloud computing are dispersed across numerous systems and locations. IoT services frequently employ it because of its flexibility and scalability. Consequently, considerable IoT and cloud integration development is common. Therefore, a novel IT prototype where cloud and IoT are two complementary integrated technologies is needed to disrupt the present and future of the Internet.

IoT technology integrates cutting-edge communication, networking, cloud computing, sensing, and actuation technologies. It is projected to open the door for ground-breaking applications in various fields, including healthcare, security and surveillance, transportation, and industry. The sheer quantity of potentially susceptible connected devices causes genuine security, privacy, and governance dangers, casting doubt on the IoT's entire future. IoT applications are anticipated to have a wide range of effects on people's lives and bring about a lot of conveniences. Still, if security and privacy cannot be guaranteed, this can have several unfavorable effects. This survey focuses on the security aspects of cloud-based IoT and discusses up-to-date IoT security solutions based on cloud.

The organization of this study is shown in the following. We shall introduce the IoT, cloud computing, and its relation to sheep. Next, we will discuss the security concerns and difficulties associated with IoT. Then, we will present cloud-based IoT security challenges, with some solutions to security issues in the final section.

DOI: 10.1201/9781032714806-11

9.2 INTERNET OF THINGS

The IoT can be defined as a collection of two terms: one is the Internet. It is described as a network of networks that connects billions of users using standard Internet protocols [1]. The Internet uses different technologies to connect different industries and sectors. Multiple devices such as mobiles, personal systems, and enterprise organizations are connected to the Internet.

The second term is things. The term refers to a device or object that becomes a smart object [2]. Furthermore, it is also a part of all objects in this real world. If we were to define the IoT, it would not be precise and concise, but Agrawal et al. [3] defined it as the interaction between the physical and digital worlds. The digital world interacts with the physical world through various sensors and actuators.

IoT can also be defined as "An open and comprehensive network of intelligent objects that can auto-organize, share information, data, and resources, reacting and acting in the face of situations and changes in the environment" [4–6].

In this section, we recall the basics of IoT, overview its characteristics, and take an overview of its general architecture.

9.2.1 IoT generic architecture

While the traditional Internet connects people to the network, IoT takes a different approach, providing machine-to-machine (M2M) and human-to-machine (H2M) connectivity for heterogeneous types of machines to serve multiple application purposes (e.g., identification, location, tracking, monitoring, and control) [7].

Connecting many heterogeneous machines [8] generates much traffic, so big data stores must be handled [9,10]. Therefore, the long-established TCP/IP architecture for network connectivity cannot meet the requirements of IoT in many aspects, including privacy and security (e.g., data protection, machine security, data confidentiality, data encryption, network security) [11], scalability, reliability, interoperability, and quality of service [12].

Although many architectures have been proposed for IoT, there is still a need for reference architectures [13,14]. The basic architectural model proposed in the literature is a three-layer architecture [12,15–17] (Figure 9.1). It includes the Perception layer, Network layer, and Application layer.

- **Perception layer:** The perception layer, also known as the physical layer, collects data/information and perceives the physical world. In this layer, all actors act according to the information collected by sensors on different objects and perform specific actions through their corresponding objects [18].
- **Network layer:** This layer is the middle layer, providing the interface connection between the application and perception layers. It is also responsible for initial data processing, transfer, and device connection [3].
- **Application layer:** This layer is the realization of the IoT. It implements the functions of sensors and actuators. We can think of it as software that works with the sensors of other virtual smart objects.

Another proposed layered architecture is the five-layer architecture (Figure 9.1) [12,15–17]. The five layers from top to bottom are: Business, application, processing, transport, and

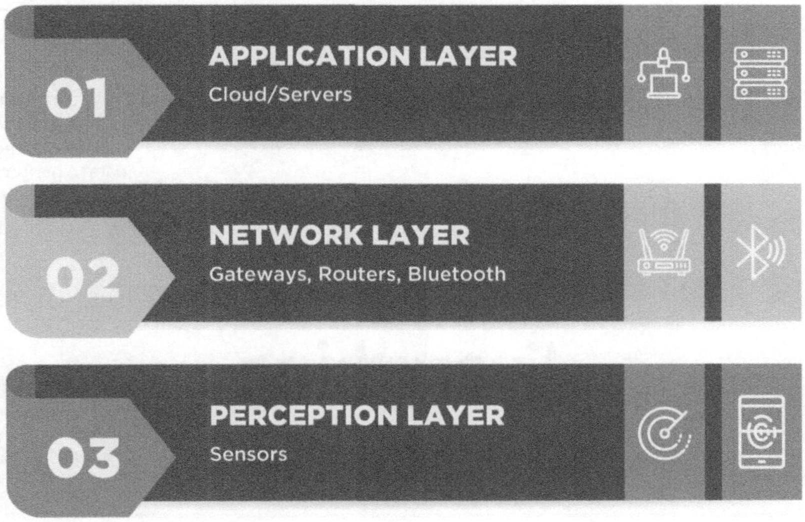

Figure 9.1 IoT three-layer architecture.

perception layers. The functions of the perception layer, transport layer (network layer), and application layer are the same as the three-layer architecture. The remaining architectural layers are:

- **Perception layer:** It works similarly to the three-layer architecture described earlier.
- **Transport layer:** It takes data from the perception layer and transmits it to the next layer and vice versa. This is done with the help of networks such as LAN, wireless technologies, 3G, 4G, LTE, and RFID [3].
- **Processing layer:** The third layer should do the main work as it processes all the information gathered by the perception layer. Store large amounts of data using technologies such as cloud computing and DBMS. We then analyze how the data is retrieved when required to perform the desired task [18].
- **Application layer:** This layer requires applications with appropriate equipment to perform their intended tasks.
- **Business layer:** This layer is the last layer of the architecture and manages all system functions and many others. One of them is data protection [5].

9.2.2 The IoT to the Internet of Everything (IoE)

According to Cisco, due to the convergence between networks of people, processes, data, and objects, IoT is moving toward the IoE. It is a multidimensional Internet that combines the IoT and big data (Figure 9.2).

- **People:** Connecting people more relevantly and with more value.
- **Processes:** Deliver the correct information to the right person (or machine) at the right time.

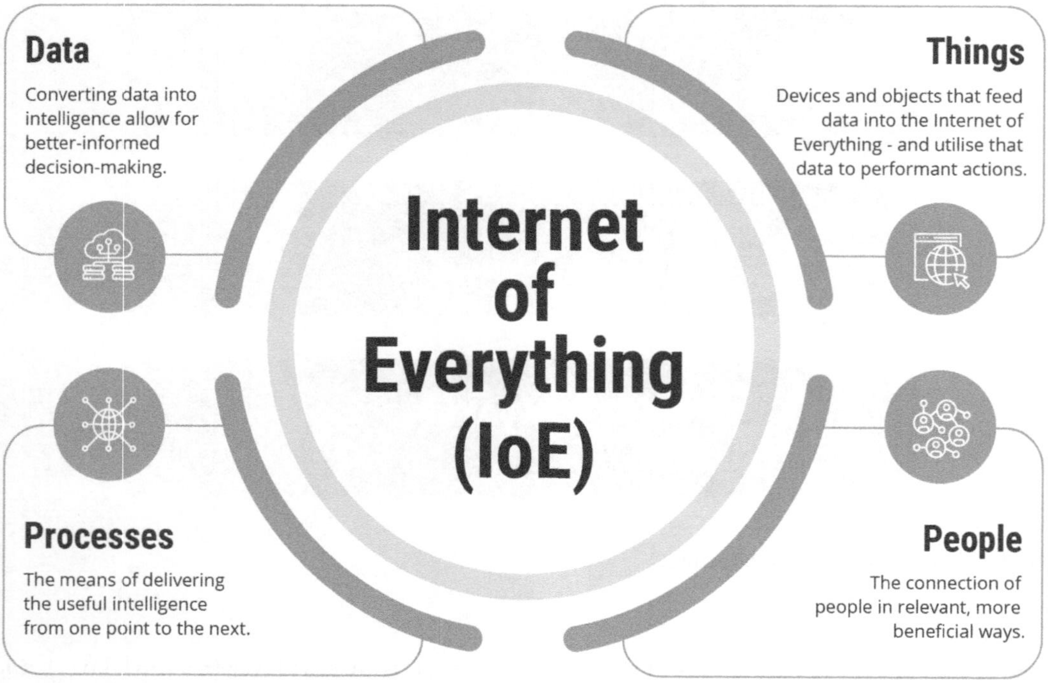

Figure 9.2 Internet of everything.

- **Data:** Use data to bring out the most useful information for decision-making.
- **Things:** Physical devices and objects connected to the Internet for intelligent decision-making.

IoE presents a broader vision of the IoT. The network is distributed and decentralized. It is equipped with artificial intelligence at all levels to protect the networks better, allow the users to have personalized data, and help make decisions. This is a vision of IoE marketing.

9.3 CLOUD COMPUTING

Imagine your typical workday spent on your computer. How many computer resources are you using during peak hours? Averages for most users are about 10% for the processor, 60% for memory, and 20% for network bandwidth.

Assuming your workplace has hundreds or even thousands of computers, wouldn't it be great to pool and utilize the idle computing resources of all your company's computers? This way, your company gets the most out of it.

Now, let's apply the same idea to a data center, where many servers (web servers, application servers, and database servers) are used similarly with minimal utilization. Likewise, they may pool and distribute hardware resources across servers to utilize them more efficiently; otherwise, you'll have to invest upfront in computing resources you use only occasionally. Others can then use additional pool resources you do not use in your company over your company's network. Or, if your company hires a third party to provide resource pools, you can access your computer resources over the Internet.

But what if your company or department only pays for the computing resources it uses? Instead of investing upfront capital when purchasing a computer, your business can simply sign up with a service provider on a pay-as-you-go billing model. This effectively means that your company is shifting from a CapEx model to an OpEx model for its computing needs. Cloud computing is helpful in this situation.

Cloud computing is described as a "model for enabling convenient, on-demand network access to a shared pool of configurable computing resources that can be rapidly provisioned and released with minimal management effort or service provider interaction" by the National Institute of Standards and Technology (NIST). Cloud computing is now a standard component of company transformation initiatives rather than an anomaly.

Even so, many businesses find it challenging to overcome the obstacles to fully integrating cloud computing into their infrastructure.

9.3.1 Characteristics of cloud computing

Businesses will find it simpler to leverage the potential benefits of cloud computing services as they become more technologically and commercially developed. However, understanding what cloud computing is and how it works is just as crucial. Figure 9.3 shows the eight features of cloud computing.

The properties of cloud computing are as follows:

- **Resources pooling:** This indicates that the cloud provider employed a multileaner architecture to distribute computing resources to different clients. Depending on client needs, various physical and virtual resources have been assigned and reassigned. The client generally has little control over or access to information regarding the resources offered, although they can select the location more abstractly.
- **On-demand self-service:** This is one of the most significant and practical benefits of cloud computing since it allows users to monitor server uptimes, capacity, and network storage continuously. With this capability, the user may also monitor computer functionality.

Figure 9.3 Characteristics of cloud computing.

- **Simple upkeep:** The servers are simple to maintain, have minimal downtime, and are always up unless in rare circumstances. Every time it releases an upgrade, cloud computing gets better and better. The upgrades run faster with fixed issues than before and are more system-friendly.
- **Large network access:** The user can access the cloud data using a device and an Internet connection or upload data to the cloud from any location. Through the network and the Internet, one can access these features.
- **Availability:** Depending on usage, the cloud's capabilities can be modified and increased. This evaluation enables the customer to, if necessary, purchase more cloud storage for a very low cost.
- **Automatic system:** Cloud computing supports a certain service level of measurement capabilities and automatically analyses the necessary data. The utilization can be monitored, controlled, and reported. Accountability is given to both the host and the client.
- **Economical:** Since the firm (host) must purchase the storage, which can be made available to other businesses, saving the host from monthly or yearly charges, it is a one-time expenditure. Only a few additional expenses and the cost of routine maintenance are significantly less.
- **Security:** One of the best aspects of cloud computing is cloud security. To prevent data loss, even if a server is damaged, it offers a snapshot of the data that is being stored. The data is kept on storage devices that nobody else can access or hack. Storage services are prompt and dependable.
- **Pay as you go:** Customers only pay for the space or the service in cloud computing. There is no need to pay any additional fees or hidden costs. Space is frequently provided without charge, and the service is affordable.
- **Measured service:** Resources the company uses for monitoring and recording in the cloud. Charge-per-use capabilities are used to evaluate this resource utilization. As a result, the service provider can assess and report resource usage on the virtual server instances running in the cloud or on a general basis. You will be paid as a model, depending on how often the manufacturing company uses you.

9.3.2 Virtualization in cloud computing

Cloud computing is based on virtualization, a method that improves the utilization of actual computer hardware. Using software, virtualization may divide the hardware components of a single computer, such as processors, memory, storage, and more, into several virtual computers, commonly referred to as VMs. Despite only using a piece of the underlying computer hardware, each VM runs its OS and functions such as a standalone machine.

As a result, virtualization enables considerably more efficient use of physical computer hardware, enabling an organization to get a higher return on its hardware investment.

Today, virtualization is a standard approach in business IT architecture. Technology also powers the cloud computing industry. Because of virtualization, cloud service providers can service customers using their physical computing hardware, and cloud customers can buy only the computer resources they require at the time they require them and expand them affordably as their workloads grow.

Creating a virtual platform, including virtual computer networks, virtual storage devices, and virtual computer hardware, is known as virtualization.

Hardware virtualization is accomplished using a program known as a hypervisor. Software is integrated into the server hardware component using a virtual machine hypervisor. The physical hardware that the hypervisor controls, the client shares, and the provider. The Virtual Machine Monitor (VVM) can eliminate actual hardware and implement hardware

Figure 9.4 Virtualization in cloud computing.

virtualization. Several process extensions aid in accelerating virtualization processes and improving hypervisor performance. Server socializing refers to the process of virtualizing the server platform. Using a hypervisor adds an abstract layer between the program and the hardware. Virtual representations, like virtual processors, happen once a hypervisor is implemented. We are unable to use actual processors after installation. Popular hypervisors include VMWare vSphere and Hyper-V, which are based on ESXi. Virtualization involves transforming physical servers into virtual representations within cloud computing environments. Essentially, virtual software mimics hardware resources, facilitating the smooth execution of cloud operations. This emulation extends to operating systems (OSs), network resources, and servers, offering significant benefits for businesses. Moreover, cloud computing serves as an efficient method for hosting various web-based applications, contributing to the growing interest in hardware, OSs, servers, and storage virtualization, as shown in Figure 9.4.

Virtualization technology is a prerequisite for cloud computing. Virtualization can be divided into two categories: server virtualization and application virtualization. Through application virtualization, several users can access an application hosted on a single machine. The application may be housed on top-notch virtual machines in the cloud, but because so many people access it, its expenditures are split among them. This reduces the cost of delivering the application to the user. The application can be executed on low-cost hardware, like a low-end workstation or a thin-client terminal, without the end user having high-end hardware.

In addition, the user is not restricted to a single device or location to utilize the virtual program or access its data if the data used by it is kept in the cloud. In these situations, the end user typically uses a mobile app or an Internet browser to access the virtual application.

Server virtualization hosts virtual computers on regular physical hardware, such as networks, storage devices, or computer hardware. Any number of virtual machines could run on a real host machine, allowing several machines to share a single set of hardware. Virtual machines don't have to use the same OS or set of programs; each can be installed with its OS and unique set of applications. A significant financial advantage of server virtualization is the ability to condense many physical machines into a small number of physical servers that host the virtual machines. Aside from the obvious decrease in equipment acquisition costs, the rise in computational efficiency results in decreased space, maintenance, cooling, and electricity expenses. Another advantage is that using fewer physical devices and spending less on electricity is better for the environment.

You can expand your resources to meet any change—either an increase or a decrease—in demand when you pool the virtual machines together so they can be quickly activated and switched on in a way that allows them to join or leave the pool. Elasticity, made feasible by server virtualization cost-effectively, is the instantaneous change in the number of virtual machines inside a pool.

What distinguishes cloud computing from virtualization at this point? Let's review the attributes listed in the NIST definition of cloud computing: on-demand self-service, quick elasticity, and measurable service provision.

None of them are automatically offered by virtualization. Although demand management, reporting, billing, and other business procedures and tools are necessary, virtualization can operate as an enabling technology to make these features possible. You must think about how to standardize your service offerings, make them accessible through straightforward portals, track usage and cost data, measure their availability, orchestrate them to meet demand, provide a security framework, provide instantaneous reporting, and have a billing or charging mechanism based on usage if you want to deploy a cloud effectively. Virtualization, in and of itself, is not a service, which is another way to look at this. It can be used to build an infrastructure-as-a-service offering with other devices and procedures [19].

9.3.3 Cloud computing architecture

Instead of using a storage device or hard drive, a user can save files to the cloud, making it possible to retrieve the data from anywhere with an Internet connection. Infrastructure-as-a-service (IaaS), platform-as-a-service (PaaS), and software-as-a-service (SaaS) are three broad categories of cloud services. Depending on the deployment model, public, private, and hybrid clouds can all be categorized under the cloud umbrella.

In addition, the cloud can be separated into front-end and back-end layers. The front-end layer is the one with which users interact. Through cloud computing software, this layer enables users to access data that has been saved on the cloud.

- **User interface:** The interface you use daily and are familiar with is called the user interface. Thanks to the seamless environment it generates, end users can execute work in the cloud without ever having to launch any local software. For instance, you use your tablet, desktop computer, or smartphone—all of which depend on interfaces to function properly. Google Docs, Gmail, and Evernote are a few popular user interfaces.
- **Software:** The front end's software architecture is what the user sees and uses. Client-side software applications or browsers presenting data to users comprise the bulk of front-end software architecture.
- **Client device or network:** The client-side device is the hardware on the end user's end. It can be any input device, such as your mouse, keyboard, or sound card, since it is an essential component of the front-end architecture. The client-side device doesn't need to be exceptionally powerful in cloud computing to handle the enormous load. That task will be handled and processed by your cloud.

The back-end layer is composed of computers, servers, central servers, and databases, as well as software. This layer, which is the foundation of the cloud, is only in charge of safely storing data.

Given that it is responsible for supporting the framework of a cloud-based system, an ideal back-end cloud architecture is created to be as durable and resilient as possible. The following are the crucial elements of a solid back-end cloud architecture:

- **Application:** The user interface clients employ makes up a sizable portion of the application architecture. It provides back-end services so that users can access client data. This layer handles their demands and requirements.
- **Service:** The back-end of cloud computing is a wonderful area of development that enhances the usefulness of any cloud hosting system. Every cloud-based task can be carried out thanks to services. Popular services include management and application development, storage, and web services, enabling various activities to be completed quickly and effectively in a cloud runtime environment.
- **Cloud runtime:** The phrase "Cloud Runtime" is a catchphrase that describes the idea of services being easily accessible. This is comparable to a cloud OS, which utilizes virtualization technology and enables users to instantly access many networked servers. These servers function independently from one another and together make up the primary server known as hypervisors, even though they are supported by virtualization infrastructure. Leading hypervisors include, among others, VMWare Fusion, Oracle VM for x86, and Oracle Virtual Box.
- **Storage:** A cloud application's data is kept in cloud storage. It is typical to employ a specific area of the cloud for this. Examples are hard drives (HDDs), solid-state drives (SSDs), Intel Optane DC Persistent Memory, and others. Most of the system's storage is HDDs in server bays, with solid-state storage or flash devices making up a smaller portion. The capacity from these two resources is split up into portions in a cloud computing system using partitioning software suited for running numerous services on top of an OS.
- **Infrastructure and architecture:** Infrastructure is the driving force behind all cloud software services. It's a broad term encompassing various technologies, including network cards, accelerator cards, the CPU, motherboard, and graphics processing unit (GPU). A model for your infrastructure is based on how busy your clients are.
- **Management:** The "middleware" of a cloud computing system is the management software. This software ensures that each task receives a fair amount of attention and that different resources are allocated to each activity if it relates to several business fronts.
- **Security:** Safety is essential. One thing must be kept in mind when creating the security structure of the architecture: debugging processes must be tracked. Bugs are handled daily, if not hourly. Regular system backups are essential because nothing is useful if data can't be recovered when needed. Firewall software is necessary for the security infrastructure since the cloud is not immune to virtual threats [20]. Figure 9.5 shows the components of cloud computing architecture.

9.3.4 Types of cloud computing

You may categorize cloud computing either depending on the deployment model or the kind of service. Depending on the exact deployment model, we can categorize the cloud as public, private, or hybrid. Depending on the service the cloud model provides, it can be categorized as IaaS, PaaS, and SaaS, as shown in Figure 9.6.

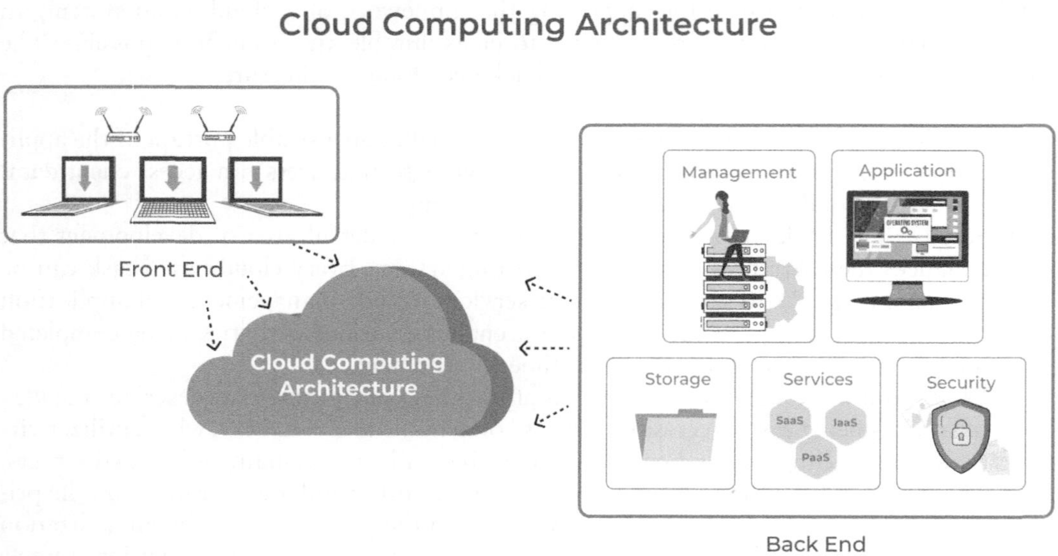

Figure 9.5 Cloud computing architecture.

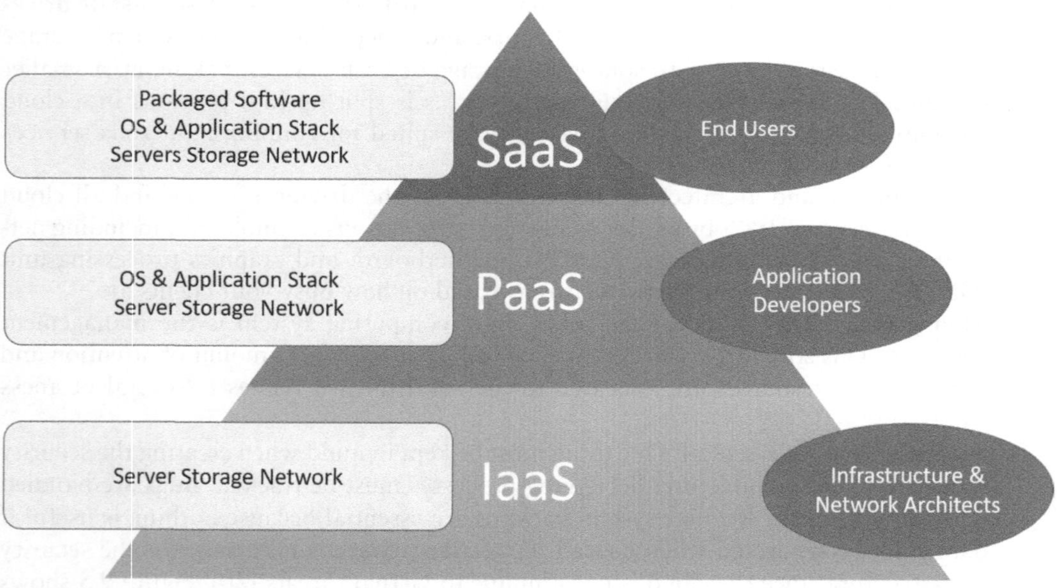

Figure 9.6 Types of cloud computing.

Three cloud deployment models are described as public, private, and hybrid cloud.

- **Public cloud:** As the name suggests, a public cloud is accessible to the general public. In this context, the public can use cloud services or an enterprise. Instead of a private or limited network, the Internet is often used to access public cloud services. The deployment paradigm that most people are familiar with is the public cloud. The public cloud is applicable at all levels of abstraction. As a result, public clouds are those that offer infrastructure, platforms, software, data, or business processes as a service. The public cloud model is nearly always exclusively accessible over the Internet with a monthly fee for use, in contrast to cloud services of the other deployment models, which may be accessed via networks that might be local, wide, or international (the Internet). Google Print, Google Docs, Microsoft Office 365, Amazon EC2, and Amazon Cloud Player are examples of public cloud services. These share a monthly operating expense pricing structure and incur little to no upfront capital costs that the consumer bears.

- **Private cloud:** A corporation, business unit, or even a single person might be included in the scope of a private cloud. A business may have a private cloud that provides services over a wide area network (WAN). A WAN can be compared to a corporate Internet protected from outsiders by firewalls and other security measures. Similar to a WAN, a local area network (LAN) is typically limited to a single location, such as a residence or a place of business. To use services across a LAN, an individual could create his or her private cloud. As a result, a private cloud only allows a small number of customers to use its services through a LAN or WAN. Private cloud services may, under some conditions, be made available via the Internet, but only with access restrictions that restrict access to private companies. Data integrity and security concerns can make private cloud service delivery via the Internet challenging. Private clouds typically have both an operating expense component and a nonrecurring component.
- **Hybrid cloud:** A hybrid cloud combines two or more cloud deployment types, each with distinct advantages. A single deployment model, a hybrid cloud of three private clouds, can be used to create it, although public clouds can also be used in their place. The deployment model doesn't need to be the same. A hybrid cloud with both a private and public cloud can be created using many models. A hybrid cloud can even exist inside another hybrid cloud, allowing for replicating the other component clouds for load balancing or business continuity needs, as shown in Figure 9.7.

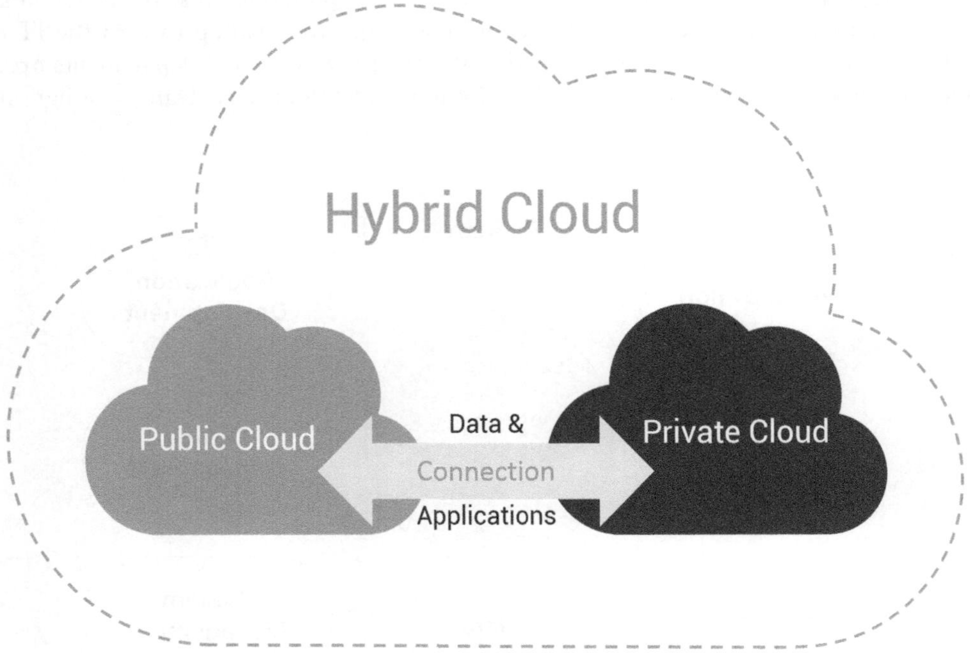

Figure 9.7 Hybrid cloud.

There are three cloud service models described as follows:

- **The SaaS** paradigm is a method for distributing software in which programmers upload their creations to a cloud-based delivery platform. Customers often use a browser to access these programs online. SaaS subscriptions can be affordable because the consumer doesn't need to invest in additional IT equipment. Enabling teams to operate from any location as needed also contributes to greater internal efficiency. Google Workspace, Dropbox, Salesforce, Cisco WebEx, and others are examples.
- **The PaaS model:** This architecture is frequently administered by an organization, allowing customers to successfully maintain their cloud-based apps and develop and run them. A third-party company will provide technical resources, such as hosting facilities or development tools, in the PaaS paradigm. Google App Engine, Apache Stratos, Heroku, Force.com, and other platforms are a few examples.
- **The IaaS concept:** This approach gives businesses access to the infrastructure they need to function. It all comes together in one spot, including virtual and non-virtual servers, storage, and data center space. Cloud computing, a type of utility computing in which the service provider provides access to distributed systems and resources, is another service offered by IaaS. Linode, Rackspace, Amazon Web Services (AWS), Microsoft Azure, Google, Compute Engine (GCE), and other providers are a few examples [21].

9.4 APPLICATION DOMAINS OF IOT CLOUD PLATFORMS

IoT cloud platforms are designed for specific application domains, such as application development, device management, system management, heterogeneity management, data management, analysis, deployment, monitoring, visualization, and, ultimately, research purposes, as shown in Figure 9.8. There are many more platforms on the market now, 26 of which are the most popular. Furthermore, the IoT Cloud Platform has been revised based on applicability and suitability preferences across several domains. Ten different areas were selected based on where most IoT cloud platforms are heading toward the IT market. Regarding management, it is assumed that only a few technical departments are best suited for these platforms, such as Devices, Systems, Heterogeneity, Data, Configuration,

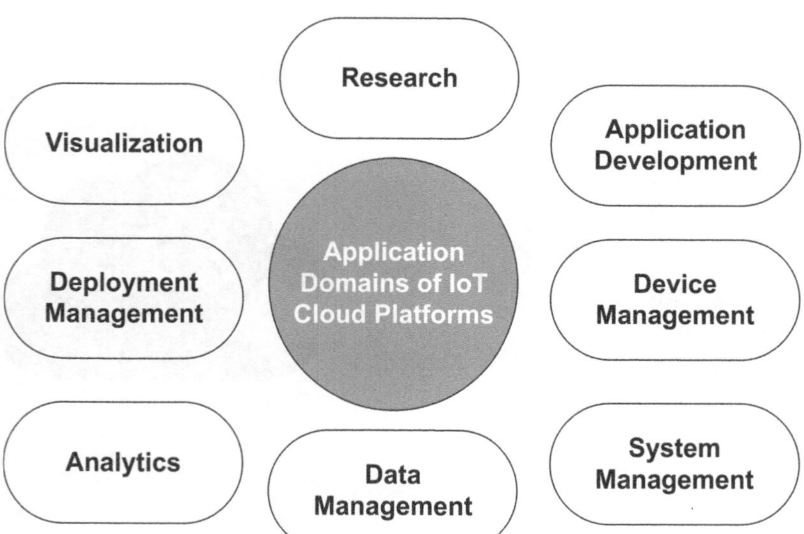

Figure 9.8 Application domains of IoT cloud platforms.

and Monitoring. Likewise, select the Analytics, Research, and Visualization domains to accommodate the rest of the platforms.

9.5 SECURITY ISSUES AND CHALLENGES OF IOT CLOUD

In the previous section, we discussed the structure of a three-layer IoT system and explained the purpose of each layer. It's important to note that each layer in an IoT system can be vulnerable to various attacks or risks.

9.5.1 IoT attacks

This section will delve into the specific threats and attacks commonly seen in each layer of the three-layer IoT architecture [22].

9.5.1.1 Perception layer

The perception layer, also known as the sensor layer, acts like a person's senses. It is responsible for detecting and gathering information from various sources. This layer often includes various sensors, such as RFID, 2D barcodes, and others, chosen based on the application's needs. These sensors collect information about location, environmental changes, motion, vibration, etc. However, the perception layer is also a primary target for attackers who may try to replace the sensors with their own. As a result, most threats to this layer are related to sensors [22]. Some common security threats in the perception layer include:

- **Node capture attack:** A physical attack involves an attacker's capture and control of a node. In such an attack, the attacker may manipulate the hardware components of the node or IoT device or even replace the captured node entirely. This attack can allow the attacker to access important and sensitive data from the captured node, such as routing tables, radio keys, and group communication keys. Additionally, the attacker may inject false data into the captured node, transmitting incorrect data within the system and disrupting the services provided by the IoT application [23].
- **Denial-of-service (DoS) attack:** A DoS attack aims to prevent legitimate users from accessing devices or network resources. This is typically done by overwhelming the targeted devices or resources with a large volume of requests, making it difficult or impossible for some or all legitimate users to access them. In the following sections, we will discuss this attack in more detail [23].
- **Sleep deprivation attack:** The sleep deprivation attack targets battery-powered computing devices, such as sensor nodes, which aim to conserve energy by remaining in a low-power sleep mode for as long as possible without negatively affecting the node's applications. The attacker launches this attack by legitimately interacting with the victim's device. Still, the purpose of these interactions is to prevent the victim node from entering its energy-saving sleep mode. This can significantly reduce the lifetime of the victim's device. One of the challenges of this attack is that it can be difficult to detect, as it is carried out through seemingly innocent interactions. Stajano first proposed the concept of the sleep deprivation attack in.
- **Side-Channel Attack (SCA):** SCAs are a type of attack that exploits information leakages in a system. These leakages can be related to timing, power, electromagnetic signals, sound, light, and other factors. SCA is a noninvasive and passive attack, meaning that the attacker does not need to remove the chips or actively tamper with the device's internal components to perform the attack. Instead, the attacker observes, collects, and analyses information leakages during processing to retrieve sensitive information from

the device. These attacks are mostly used to target cryptographic devices, as they can be used to break systems that are thought to be secure by exploiting their physical implementation rather than directly targeting the standard cryptographic algorithms. By measuring and analyzing leaked information, such as power consumption or timing data, attackers can recover secret parameters and potentially compromise the device's security.

- **Malicious code injection attack:** Physical attacks are a type of attack where the adversary physically interacts with the target device and inserts malicious code into it. This can be done by connecting a malicious device to the target or reprogramming the target system. This attack type is also called "malicious code injection." It can give the adversary full control over the target system if successful. These attacks can be challenging to detect and prevent, as they involve physical access to the device. It is important to protect against physical attacks, such as implementing physical security measures and regularly updating software and security measures [23].
- **Replay attack:** A replay attack involves an attacker capturing and reproducing a signal used to control a device, such as a remote control for a lock on a door. This type of attack can be particularly effective for IoT devices, which often use radio signals operating in the 433 MHz range to communicate with receivers on the devices. In this case, different modulation patterns are used to control specific devices. While this example may seem simple, it demonstrates how replay attacks can be used to compromise the security of certain devices.

9.5.1.2 Network layer

The network layer, the transmission layer, acts as a bridge between the perception and application layers in a smart system. It is responsible for transmitting information sensors collected from physical objects to other devices or networks. The medium for this transmission can be either wireless or wire-based. The network layer also connects smart devices, network devices, and networks, making it a sensitive target for attacks by hackers. It is important to ensure the integrity and authenticity of the information being transported in the network, as these are common security concerns at the network layer [24]. Some examples of security threats and problems that can affect the network layer include:

- **Man-in-The-Middle (MiTM) attack:** An MiTM attack is a type of cyberattack in which an attacker intercepts and manipulates the communication between a sender and a receiver, who believes they are communicating directly. The attacker can change the messages being exchanged as needed, which poses a serious threat to online security because it allows the attacker to capture and manipulate information in real time. This attack can have serious consequences, such as stealing sensitive information or injecting false data into the communication. It is important to protect against MiTM attacks by using secure communication protocols and regularly updating software and security measures.
- **Sinkhole attack:** In a sinkhole attack, a cyber attacker redirects traffic from a genuine network to a malicious one they have created and controlled. This fake network, also known as a "sinkhole," appears legitimate to the victim, but allows the attacker to intercept and potentially modify the victim's data when they connect. Sinkhole attacks can disrupt communication within a network, gather sensitive information, or disable entire networks. They can be executed using various methods, such as faking IP addresses or altering routing tables. To defend against sinkhole attacks, individuals and organizations must set up secure network configurations and monitoring systems.
- **A Sybil attack:** A Sybil attack is a type of attack in which a small number of entities create multiple fake identities to gain disproportionate influence in a system. In this attack, the adversary attempts to create many nodes, which may or may not be

randomly generated, to appear and behave as distinct nodes. With multiple identities, the adversary can access a specific object or multiple objects in a peer-to-peer (P2P) network. This increases the attacker's ability to intercept message routing and disrupt the operation of the overlay network. Sybil attacks can be difficult to detect and have serious consequences for the security and integrity of a P2P network.

9.5.1.3 Application layer

The application layer of an IoT system refers to all the applications that use or are deployed with IoT technology. Such applications include smart homes, smart cities, smart health, and animal tracking. The application layer is responsible for providing services to these applications, which can vary depending on the information collected by sensors. One of the main security concerns in the application layer is the vulnerability of IoT-based smart homes to threats and vulnerabilities from both inside and outside the home. To ensure strong security in an IoT-based smart home, it is important to address the issue of devices with weak computational power and limited storage, such as ZigBee devices. Some common security threats and problems that can affect the application layer include:

- **Cross-site scripting (XSS):** XSS is a vulnerability that allows an attacker to inject malicious code into a link on a legitimate website. When a user clicks the link, the harmless request and the malicious script are sent to the website. The website responds not only to the original request but also includes the attacker's script, which is executed by the user's web browser because it appears to come from a trusted source. A successful XSS attack can go unnoticed by the user. It can result in various consequences, including hijacked web accounts, stolen web sessions, remote control of the browser, or redirection to malicious locations. The impact of an XSS attack can be significant and not limited to these examples.
- **Malicious viruses:** Malware is malicious code designed to cause harm or damage to a system. It can be present in any software part and may not be detected or blocked by antivirus tools. Malware can either be self-executing or require the user to act to activate it. There are many different types of malware, including viruses, worms, Trojan horses, and ransomware, each with its characteristics and potential impacts on a system. It is important for individuals and organizations to be aware of the risks associated with malware and to implement appropriate security measures to protect against it.

9.5.2 Architectural model of cloud-based IoT

The architectural model depicted in Figure 9.9 illustrates the structure of a cloud-based IoT system. This model consists of four key zones, each playing a distinct role in the functioning of the IoT ecosystem.

- **Device zone:** This is the zone closest to the physical devices and sensors. It includes physical devices that capture and measure data, such as smartphones, RFID readers, and sensors.
- **Data transmission zone:** This zone is responsible for transporting data securely and reliably. It includes wireless networks, cellular networks, and satellite networks.
- **Cloud gateway zone:** This zone connects the devices to the cloud. It acts as an intermediary between the devices and the cloud, providing a secure connection and allowing for efficient data transfer.
- **Cloud services zone:** This zone is responsible for storing, managing, and analyzing the data. It includes cloud-based data storage, cloud-based analytics, and cloud-based machine learning services [25,26].

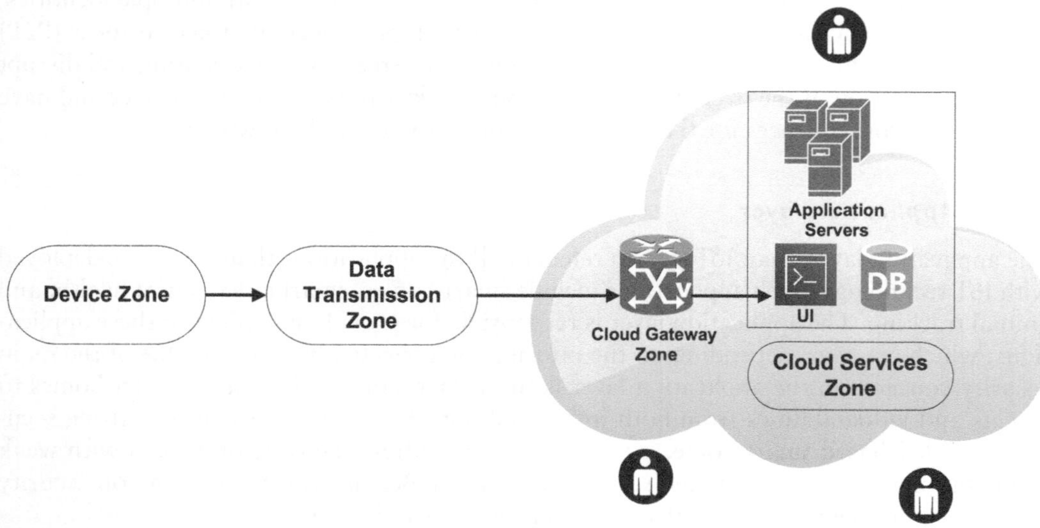

Figure 9.9 Architectural model of cloud-based IoT.

Table 9.1 Description of attacks and risks in cloud

Attacks and threats	*Description*
Information breaches	Security breaches and the use of protected data
Information loss	Data loss as a result of poor handling
Service or account hijacking	Attacks on the system aimed at stealing information
Applications and API attacks	Attacks to expose software interfaces or APIs
DOS	Attack on a machine or network that makes it inaccessible to the user
Malicious insider	Any insider can utilize the system for malicious purposes
Abuse and nefarious use of cloud services	Using cloud services for nefarious purposes or misuse of cloud services
Insufficient diligence	Risk due to insufficient and shortage of cloud knowledge
Shared technology	Due to shared resources, there have been several attacks

9.5.3 Cloud-based IoT attacks

Cloud security is typically the responsibility of cloud providers. Still, as more organizations have begun using cloud services to store and process their data and applications, attackers have also started targeting these services. Cloud computing in the IoT is used to store and manage IoT data, and can be vulnerable to various types of attacks and vulnerabilities. A cloud is a centralized server with computer resources that can be accessed remotely, making it a convenient way to handle large amounts of data generated by IoT devices. In traditional Internet systems, connections are made through physical links between web pages, but in the IoT, a combination of data is often needed for situation detection. Some common characteristics of IoT-based cloud attacks are listed in Table 9.1. It is important for individuals and organizations to be aware of the potential risks associated with cloud computing in the IoT and to implement appropriate security measures to protect against these threats.

This section provides some of the most powerful IoT-based cloud attacks in each zone. Their prevention techniques on the cloud experienced in the last few years are given as follows.

9.5.3.1 Device layer

The device layer of an IoT framework is primarily responsible for gathering data, recognizing objects, and controlling them. These layers are vulnerable to attacks that target the hardware components of the framework. The device layer can be divided into two parts: device nodes, such as sensors or controllers, and the device network, which connects to the transportation network.

Device nodes are responsible for acquiring and controlling data, while the device network transmits this data to the gateway or sends control instructions to the controller. Each device in the IoT can be detected and analyzed through the Internet using technologies such as Radio Frequency Identification (RFID), Wireless Sensor Network (WSN), and other methods. RFID and WSN can be used to ensure the reliability and secrecy of information in the IoT through password encryption technology. The main difference is that RFID is primarily used for object identification, while WSN is primarily for measuring real-world physical parameters related to the surrounding environment. Some common security threats in the device layer include:

- **Physical attacks/tampering:** These attacks are possible when an attacker has complete physical access to a tag. These attacks can be carried out in a laboratory setting and involve physically manipulating the tags. Examples of physical attacks against RFIDs include probe attacks, removing material using shaped charges or water etching, manipulating circuits, and clock glitching. These attacks often aim to extract data from the tag or alter it to create a copy.
- **Tag cloning:** Tag cloning is an attack in which a malicious actor creates a copy of an RFID tag. This small wireless device uses radio frequency technology to transmit information to a reader. These tags are commonly used in IoT systems to identify and track objects. In an IoT system, tag cloning can allow an attacker to impersonate a legitimate tag and gain unauthorized access to restricted areas or resources. For instance, if an attacker could clone an RFID tag that is used to access a secure facility, they could potentially enter the facility without being detected. An attacker typically needs physical access to the target tag and specialized equipment and knowledge to successfully clone a tag. They may use a device known as a proxmark or RFID cloner to capture the data from the tag and create a copy. They may also employ clock glitching or circuit manipulation techniques to bypass security measures and access the tag's data.
- **Node tempering:** An attacker can damage a sensor node by physically replacing the entire node or part of its equipment or by electronically probing the node to gain access and alter sensitive data such as shared cryptographic keys or routing tables.
- **Impersonation:** Impersonation involves temporarily altering one's identity for collusion attacks, which makes authentication challenging in a distributed environment.
- **Jamming adversaries:** Jamming attacks on remote devices in the IoT aim to weaken the system by emitting radio frequency signals without following a protocol. Radio interference can significantly impact system performance and disrupt the transmission and reception of data by legitimate nodes, causing the system to break down or behave erratically.
- **DoS attacks:** Attackers can exploit the processing power of nodes, making them unavailable. There are various types of DoS attacks, including battery draining, sleep deprivation, outage attacks, and battery draining.
- **Insecure initialization:** Ensuring the secure installation and configuration of the IoT at the physical layer ensures that the entire system functions properly without compromising security or disrupting system services. The physical layer communication should also be secured to prevent unauthorized access.

9.5.3.2 Data transmission layer

The data from the sensors in the device layer comes in an analog form and generates large volumes of data quickly. This data needs to be aggregated and converted into digital streams to be further processed. Data acquisition systems (DAS) perform these tasks by connecting to the sensor network, combining the outputs, and converting them to digital form. The data is transmitted to cloud gateways through Wi-Fi, wired LANs, or the Internet.

It is essential to establish a secure channel between the sensors and servers to ensure the integrity of the collected data. If the data is altered in any way, the analysis results will be significantly affected, and this could potentially cause serious harm. Therefore, as mentioned earlier, cryptographic solutions transmit and receive messages, protect privacy, and authenticate data.

The communication layer is a vulnerable attack surface because it relies on the Internet as the communication medium. It is known to be susceptible to various types of attacks on protocols, including well-known and unknown zero-day attacks. While attacks on edge computing, data accumulation, and data abstraction layers are less common, these layers can still be impacted by software bugs in third-party code and devices. Applications can also be targeted by attacks, such as SQL injections if they are poorly designed or tested. It's important to remember that user error can also be a security weakness, and system design should not rely on human interaction for security. Instead, it's best to design secure systems regardless of user behavior.

9.5.3.3 Gateway layer

The gateway layer in a cloud-based IoT system plays a crucial role in connecting devices to the Internet and facilitating communication between these devices and the cloud. However, the gateway layer is also a potential target for cyberattacks due to its central role in the system. In this context, it is important to ensure that the gateway layer is properly secured to prevent unauthorized access and protect against potential attacks. In this section, as for the various security challenges and vulnerabilities associated with the gateway layer in a cloud-based IoT system, we can mention:

- **Sniffer attack:** A sniffer attack, also known as a packet sniffing attack, is a type of cyberattack in which an attacker captures and analyses network traffic to extract sensitive information. In a cloud-based IoT system, a sniffer attack on the gateway layer could allow an attacker to intercept and read the communications between devices and the cloud, potentially exposing sensitive data and compromising the system's security. Sniffer attacks can be carried out using specialized software or hardware and can be difficult to detect, making them a significant threat to the security of a cloud-based IoT system.
- **Domain Name System (DNS) poisoning:** DNS poisoning, also known as DNS spoofing, is a type of cyberattack in which an attacker manipulates the DNS to redirect Internet traffic from legitimate websites to malicious ones. In a cloud-based IoT system, a DNS poisoning attack could potentially allow an attacker to redirect traffic from devices to a fake website that appears to be legitimate, potentially exposing sensitive data and compromising the security of the system. DNS poisoning attacks can be carried out using various techniques, such as modifying the DNS cache or redirecting traffic to a rogue DNS server. These attacks can be difficult to detect and can have serious consequences for the security of a cloud-based IoT system.

9.5.3.4 Cloud services layer

The cloud services layer in a cloud-based IoT system refers to the infrastructure and platforms that support the operation and management of IoT devices and applications. However, the cloud services layer is also a potential target for cyberattacks due to the sensitive information and critical functions it supports. Attackers may attempt to compromise the security of the cloud services layer to gain unauthorized access to data, disrupt service, or gain control over devices. In this context, ensuring the cloud services layer is properly secured to prevent these attacks and protect against potential threats is important. In this introduction, we will explore the various security challenges and vulnerabilities associated with the cloud services layer in a cloud-based IoT system, such as:

- **Trojan:** A Trojan, also known as a Trojan horse, is malicious software that disguises itself as legitimate software to gain access to a device or system. In a cloud-based IoT system, an attacker could use a Trojan to gain unauthorized access to devices or the cloud infrastructure, potentially exposing sensitive data and compromising the system's security. Trojans can be delivered to devices through various means, such as email attachments, downloads, or links, and can be difficult to detect due to their covert nature. It is important to take precautions to protect against Trojans and other malware to secure a cloud-based IoT system.
- **Mirai:** Mirai is a type of malware that targets IoT devices, such as routers, cameras, and smart appliances, to turn them into bots that can be used to launch distributed DoS (DDoS) attacks. In a cloud-based IoT system, an attacker could use Mirai malware to compromise the security of devices and the cloud infrastructure, potentially disrupting service and exposing sensitive data. Mirai was first discovered in 2016 and has been used in a number of high-profile DDoS attacks. It is important to protect against Mirai and other types of malware to secure a cloud-based IoT system.

9.6 SECURITY ISSUES SOLUTIONS

Security issues in IoT and cloud are a major concern today. With the rise of connected devices and cloud services, cybercriminals have found new ways to exploit these technologies. The security issues can range from simple data breaches to more complex attacks, such as DDoS attacks and malware. Security measures must be taken to protect the data and services being utilized. This can include encryption, authentication, and access control measures. In addition, organizations must keep their systems up-to-date to ensure security vulnerabilities are patched as soon as possible. At last, organizations must be aware of the risks associated with IoT and cloud and the need to develop a comprehensive security strategy to protect their networks and data.

This section provides some of the most powerful IoT-based cloud solutions in each zone. Their prevention techniques utilize approaches or procedures to avoid such threats while utilizing or building IoT systems with the cloud [25].

9.6.1 Device zone

The data integrity layer ensures that the data is not tampered with while in transit. This layer employs message authentication codes and digital signature algorithms to ensure data integrity.

The secure storage layer securely stores sensitive data on the cloud. This layer uses secure storage techniques such as file encryption, cloud storage encryption, and secure databases to ensure secure data storage.

Cryptographic solutions for the device zone model also uses several other technologies, such as secure boot, secure bootloaders, and secure firmware updates, to ensure secure deployment and operation of devices [27,28].

9.6.2 Data transmission zone

The data from the seniors of the device zone comes in an analog form, which must be converted into a digital format before being transmitted to the cloud platform. The data paquet needs to be encrypted with strong encryption algorithms such as AES-256 or RSA encryption to ensure the security of the data in transit.

The data should also be securely transmitted over secure protocols such as TLS (Transport Layer Security) or IPSec (Internet Protocol Security) [27,28].

9.6.3 Cloud gateway zone

A cloud gateway zone's security depends on the specific security measures implemented and enforced. Generally, the security of a cloud gateway zone is achieved through a combination of various security controls, such as encryption, authentication, authorization, and access control.

Encryption protects data in transit and at rest. Authentication verifies the user's identity and the validity of the data they are accessing. Authorization controls who has access to what data and resources. Access control restricts access to sensitive data and resources.

In addition to these security controls, organizations may also implement additional measures, such as identity and access management (IAM) solutions, firewalls, intrusion detection systems (IDS), and virtual private networks (VPNs). These additional measures help to ensure that only authorized users have access to the cloud gateway zone and that all data is properly protected [27,28].

9.6.4 Cloud services zone

Service layers assist as and when requested by the clients. For example, the service layer can impart required information like temperature and humidity estimations to the customer accessing this service. The significance of this layer for the IoT is that it provides the ability to provide high-superiority resources to meet client requirements. Various IoT contexts (i.e., acute city, healthcare and factory) are executed inside this layer; in addition, an Application Support Sub-layer (ASS), to maintain a wide range of merchandise facilities and to acknowledge smart estimation and resource distribution can be actualized all through particular middleware and distributed computing platforms.

The common issues of cloud computation services are service interruptions like data backup, system shutdown, and inability to reach data center. Also, DDoS attacks can be possible on cloud services, which prevent legitimate users from accessing the services, by making important cloud services use more resources like memory, storage space, and network bandwidth. As a result, the response of the cloud services turns out to be very slow, or they may even become unresponsive [27,28].

Figure 9.10 summarizes the security problems and possible solutions by layer for cloud-IoT.

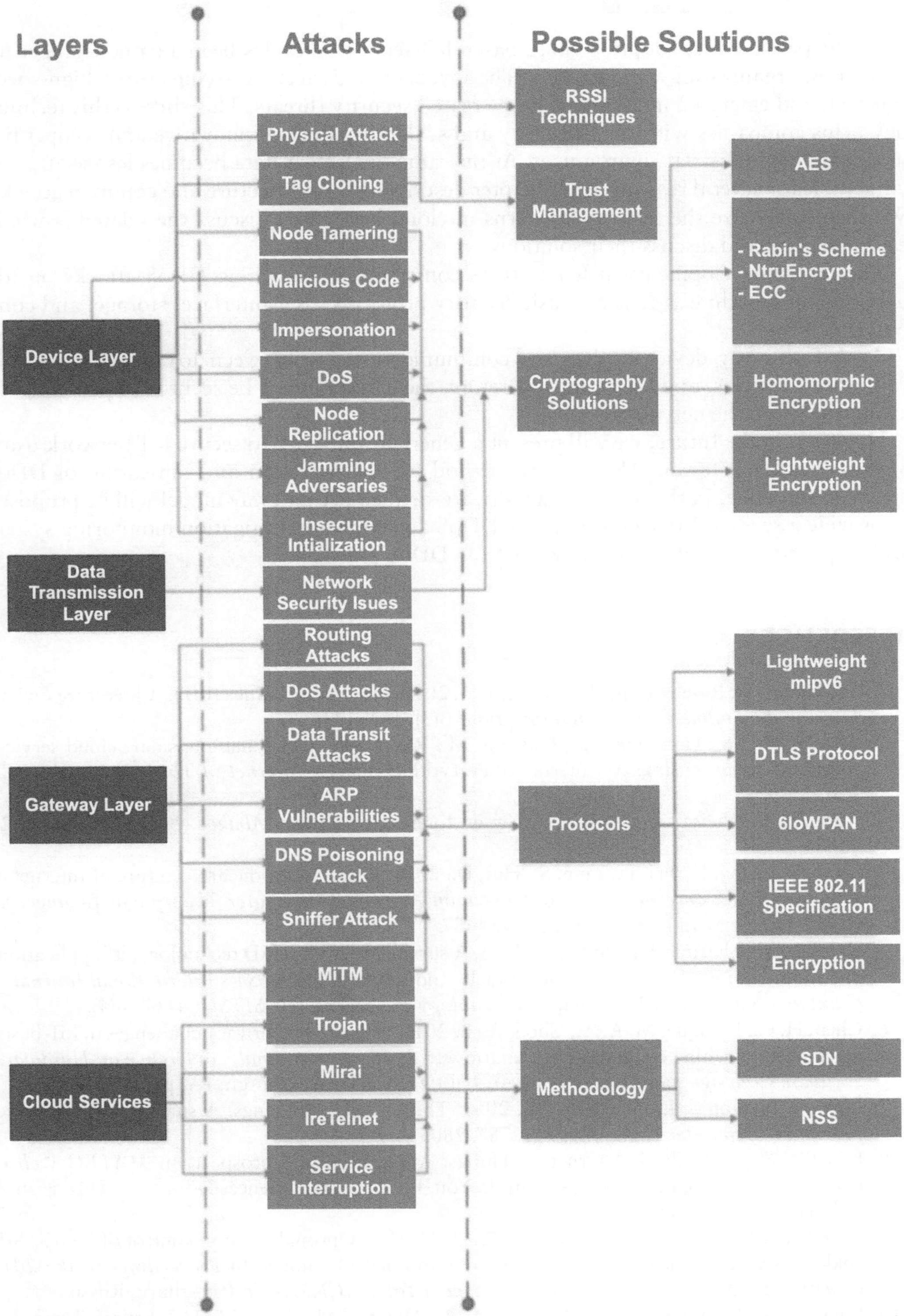

Figure 9.10 Layer-wise security issues and possible solutions for cloud-IoT.

9.7 CONCLUSION AND PERSPECTIVES

Over the past decade, adopting cloud-based IoT technologies has been a game-changer for industries, organizations, and hackers. The advent of modern cloud architectures, high-speed Internet, and emerging innovations have created security threats. This shift to this technology helps companies with the flexibility and scalability to stay innovative and competitive in a changing industrial environment. At the same time, their data becomes less secure and vulnerable for several reasons. This chapter describes IoT architecture and common attacks. We then categorize the security concerns in cloud-based IoT, discuss the related issues in each category, and discuss their solutions.

The recent developments in IoT botnets contribute to launching DDoS attacks on IoT networks. IoT architecture lacks basic security protocols, user interface, storage, and computation power.

An IoT gateway device bridges the communication gap between IoT devices, sensors, equipment, systems, and the cloud. Therefore, the gateway must be secured to prevent such cyberattacks on the network.

However, in the future, we will present a generic framework to secure IoT network using SDN as an IoT gateway. Also, a new method for the detection and mitigation of DDoS attack in the same. In this use case, a software-defined IoT gateway model will be proposed to provide a secured IoT gateway then a DDoS detection and mitigation monitoring system will be proposed to defend the network from DDoS attack.

REFERENCES

1. Madakam, S., Ramaswamy, R., Tripathi, S., 2015. Internet of Things (IoT): A literature review. *Journal of Computer and Communications* 3(3), 164–173.
2. Abdelwahab, S., Hamdaoui, B., Guizani, M., Rayes, A., 2014. Enabling smart cloud services through remote sensing: An internet of everything enabler. *Internet of Things Journal, IEEE* 1(3), 276–288.
3. Agrawal, S., Vieira, S., 2013. A survey on Internet of Things. *Abakós, Belo Horizonte* 1(2), 78–95.
4. Miao, W., Ting, L., Fei, L., Ling, S., Hui, D., 2010. Research on the architecture of Internet of things. In *IEEE International Conference on Advanced Computer Theory and Engineering (ICACTE)*, Sichuan Province, China, 484487.
5. Kamdar, N., Sharma, V., Nayak, S., 2016. A survey paper on RFID technology, its applications and classification of security/privacy attacks and solutions. *IRACST - International Journal of Computer Science and Information Technology & Security (IJCSITS)*, 64–68, 6(4).
6. Chen, H. C., Faruque, M. A. A., Chou, P. H., 2016. Security and privacy challenges in IoT-based machine-to-machine collaborative scenarios. In *2016 International Conference on Hardware/Software Codesign and System Synthesis (CODES+ISSS)*, Pittsburgh, PA, pp. 1–2.
7. Luigi, A., Antonio, I., Giacomo, M., 2010. The Internet of Things: A survey. *Science Direct Journal of Computer Networks*, 54, 2787–2805.
8. Liu, R., Wang, J., 2017. Internet of Things: Application and prospect. In *MATEC Web of Conferences;* Zhao, L., Xavior, A., Cai, J., You, L., Eds.; EDP Sciences France: Les Ulis, France, Volume 100, p. 02034
9. Taynitskiy, V., Gubar, E., Zhu, Q., 2017, 22–27 May. Optimal impulse control of bivirus SIR epidemics with application to heterogeneous Internet of Things. In *Proceedings of the 2017 Constructive Nonsmooth Analysis and Related Topics (CNSA), St.* Petersburg, Russia.
10. Rajakumari, S., Azhagumeena, S., Devi, A. B., Ananthi, M., 2017, 23–24 February. Upgraded living think-IoT and big data. In *Proceedings of the 2017 2nd International Conference on Computing and Communications Technologies (ICCCT)*, Chennai, India.

11. Dineshkumar, P., SenthilKumar, R., Sujatha, K., Ponmagal, R., Rajavarman, V., 2016, 9–11 December. Big data analytics of IoT based Health care monitoring system. In *Proceedings of the 2016 IEEE Uttar Pradesh Section International Conference on Electrical, Computer and Electronics Engineering (UPCON)*, Varanasi, India.
12. Shang, W., Yu, Y., Droms, R., Zhang, L. Challenges in IoT networking via TCP/IP architecture. Technical Report 04, NDN, Technical Report NDN-0038; Named Data Networking. https://named-data.net/techreports.html (accessed on 10 Augest 2018).
13. Khan, R., Khan, S.U., Zaheer, R., Khan, S., 2012, 17–19 December. Future Internet: The Internet of Things architecture, possible applications and key challenges. In *Proceedings of the 2012 10th International Conference on Frontiers of Information Technology*, Islamabad, India.
14. Weyrich, M., Ebert, C., 2016. Reference architectures for the Internet of Things. *IEEE Software*, 33, 112–116.
15. Bauer, M., Boussard, M., Bui, N., Loof, J. D., Magerkurth, C., Meissner, S., Nettsträter, A., Stefa, J., Thoma, M., Walewski, J.W., 2013. IoT reference architecture. In Alessandro Bassi, Martin Bauer, Martin Fiedler, Thorsten Kramp, Rob Kranenburg, Sebastian Lange, Stefan Meissner (eds.), *Enabling Things to Talk*; Springer: Berlin/Heidelberg, Germany, pp. 163–211.
16. Mashal, I., Alsaryrah, O., Chung, T. Y., Yang, C. Z., Kuo, W. H., Agrawal, D. P., 2015. Choices for interaction with things on Internet and underlying issues. *Ad Hoc Network*, 28, 68–90.
17. Said, O., Masud, M., 2013. Towards Internet of Things: Survey and future vision. *International Journal of Computer Networks*, 5, 1–17.
18. Wu, M., Lu, T. J., Ling, F. Y., Sun, J., Du, H. Y. 2010, 20–22 August. Research on the architecture of Internet of Things. In *Proceedings of the 2010 3rd International Conference on Advanced Computer Theory and Engineering (ICACTE)*, Chengdu, China.
19. Ruparelia, N. B., 2016. *Cloud Computing*, Cambridge, MA: Mit Press, 278 pages.
20. Vaezi, M., Zhang, Y., Vaezi, M., & Zhang, Y. (2017). Virtualization and cloud computing. *Cloud Mobile Networks: From RAN to EPC*, 11-31. Germany.
21. Sethi, P., Sarangi, S. R., 2017. Internet of Things: Architectures, protocols, and applications. *Journal of Electrical and Computer Engineering*, 2017, Article ID 9324035, 25 pages
22. Kavis, M., Architecting the Cloud K. Zhao and L. Ge, A survey on the Internet of Things security. In *2013 Ninth International Conference on Computational Intelligence and Security*, Leshan, 2013, pp. 663–667.
23. Maleh, Y., Sahid, A., Ezzati, A., & Belaissaoui, M. (2018). Key management protocols for smart sensor networks. In Yassine Maleh, Abdellah Ezzati, Mustapha Belaissaoui (eds.), *Security and Privacy in Smart Sensor Networks* (pp. 1-23). IGI Global. USA
24. Farooq, M. U., Waseem, M., Khairi, A., & Mazhar, S., 2015. A critical analysis on the security concerns of Internet of Things (IoT). *International Journal of Computer Applications*, 111(7), pp. 1–6.
25. Lin, J., Yu, W., Zhang, N., Yang, X., Zhang, H., Zhao, W., 2017. A survey on Internet of Things: Architecture, enabling technologies, security and privacy, and applications. *IEEE Internet of Things Journal*, 4(5), 1125–1142.
26. Chen, L., et al., 2017. Robustness, security and privacy in location-based services for future IoT: A survey. *IEEE Access*, 5, 8956–8977.
27. Asplund, M., Nadjm-Tehrani, S., 2016. Attitudes and perceptions of IoT security in critical societal services. *IEEE Access*, 4, 2130–2138.
28. Zhou, J., Cao, Z., Dong, X., Vasilakos, A., 2017. Security and privacy for cloud-based IoT: Challenges. *IEEE Communications Magazine*, 26–33.

Chapter 10

Exploring IoT penetration testing

From fundamentals to practical setup

Yassine Maleh and Youssef Baddi

10.1 INTRODUCTION

The term "Internet of Things" (IoT) can describe a network of interconnected computing devices that are able to sense their environment, gather data, and transmit that information through the use of built-in processors, communications gear, and sensors (Maleh et al., 2018). An "object" in the context of the IoT is any network-enabled device that may be embedded in any kind of material, whether it be natural, artificial, or machine-made (Maleh et al., 2018). The IoT leverages current and future sensing, networking, and robotics technologies to let users automate and integrate more deeply into systems, as well as conduct deeper analyses. Everyday machines and devices used in various sectors, including homes, offices, industry, transportation, buildings, and wearable technology, are becoming more and more capable of networking (Alaoui et al., 2022). This opens up a plethora of opportunities for businesses to enhance their operations and provide better customer satisfaction. A few of the most important characteristics of the loT include connectivity, sensors, AI, tiny devices, and active involvement. Figure 10.1 shows the IoT architecture.

10.1.1 IoT applications

There is a vast array of uses for loT devices. Almost every part of society makes use of them, as they raise living standards by making mundane jobs and personal duties easier in a number of ways (Maleh et al., 2022). Many different areas make use of loT technology, including healthcare, transportation, retail, security, smart homes and buildings, and industrial equipment (Gabriel et al., 2022). A few examples of when loT devices might be useful are:

- Thermostats, lights, security systems, and many other building systems are examples of Internet-connected smart devices that offer a variety of services to end users.
- Wearables, electrocardiograms, implanted pacemakers, electrocardiograms, electrocardiograms, surgical instruments, telemedicine, and other similar devices are used in the healthcare and life sciences industries.
- Three strategies are driving expansion in the Industrial IoT: using smart technology to revolutionize manufacturing processes, expanding production to increase profits, and developing innovative hybrid business models.
- The transportation industry's use of IoT technology is based on the same principles as those of other industries: the improvement of traffic conditions, navigation systems, and parking arrangements and communication between vehicles, roads, and pedestrians.

 DOI: 10.1201/9781032714806-12

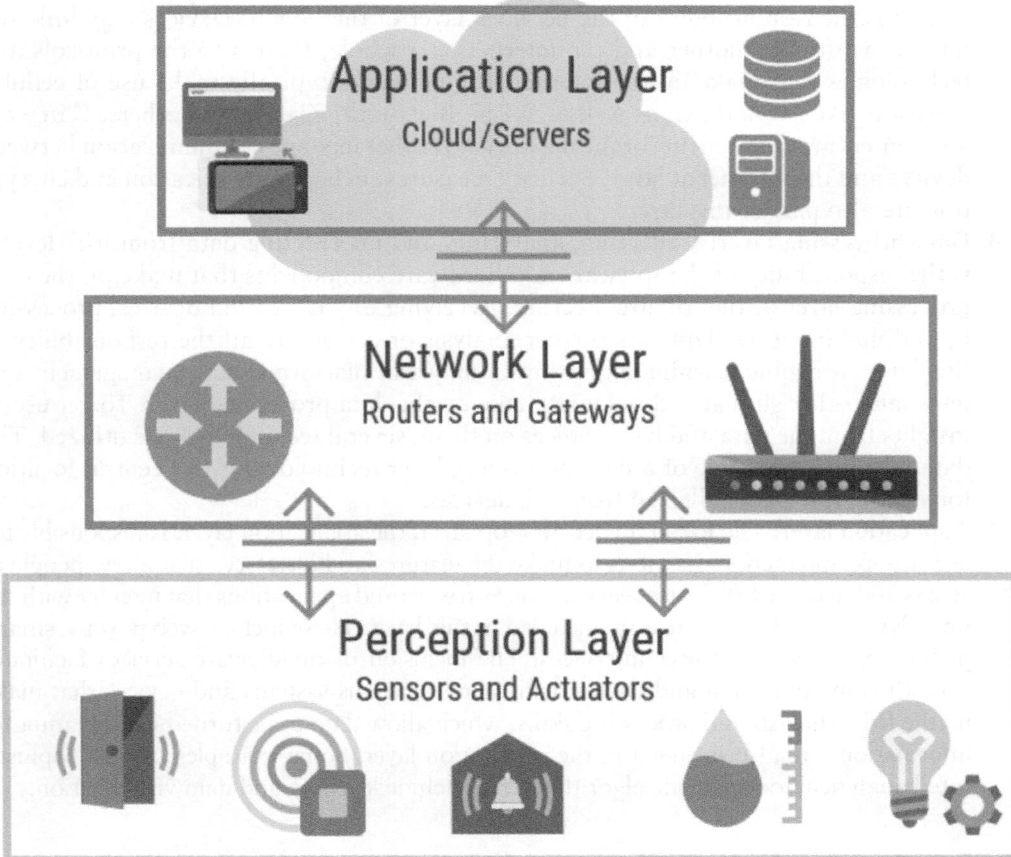

Figure 10.1 IoT architecture.

- The most common applications of IoT technology in retail include accepting payments, displaying advertisements, and keeping tabs on things to prevent theft or loss, all of which increase income.
- Devices commonly encountered in workplaces, such as photocopiers, printers, fax machines, and PBX monitoring systems, are referred to as IoT devices. These devices facilitate communication between endpoints and streamline long-distance data transmission.

10.1.2 IoT architecture

The architecture of the IoT depends on its functionality and implementation in different sectors. Nevertheless, there is a basic process on which the IoT is built. The IoT is made up of four layers that can be divided as follows: the sensing layer, the network layer, the data processing layer, and the application layer (El-Latif et al., 2022).

1. **Sensing layer:** The initial layer of the IoT architecture, the sensing layer, is in charge of gathering data from various sources. Environmental sensors and actuators that record data on factors like light, sound, humidity, and temperature make up this layer. These nodes communicate with the network layer using wireless or cable connections (Maleh, 2020).

2. **Network layer:** Ensuring communication and connectivity between devices in an IoT system is the responsibility of the network layer of the design. Devices may link and interact with one another and the internet as a whole, thanks to the protocols and technologies that make up the internet. IoT networks typically make use of cellular networks like 4G and 5G, as well as Wi-Fi, Bluetooth, Zigbee, and others. Gateways and routers are components of the network layer that mediate communication between devices and the internet at large. Security measures such as authentication and encryption are also part of this layer.
3. **Data processing layer:** Collecting, analyzing, and interpreting data from IoT devices is the responsibility of the software and hardware components that make up the data processing layer of the IoT architecture. Receiving raw data from devices, processing it, and making it available for further analysis or action are all the responsibility of this layer. Machine learning algorithms, analytical platforms, data management systems, and other similar technologies make up the data processing layer. To get useful insights from the data and base choices on them, several technologies are utilized. The data lake is an example of a data processing layer technology; it is a central location for storing raw data collected from IoT devices.
4. **Application layer:** The IoT architecture's top layer, the application layer, is responsible for user-to-app interaction. Its job is to make the features and interfaces that allow people to access and manage IoT devices easy to use. Software and applications that interact with the underlying IoT infrastructure are included in this layer. These include web portals, smartphone apps, and other user interfaces. The inclusion of middleware services facilitates smooth communication and data sharing across various systems and devices that make up the IoT. Analysis and processing skills, which allow data to be turned into meaningful information, are also included in the application layer. Some examples of such sophisticated analytical tools include algorithms for machine learning and data visualization.

10.1.3 IoT technologies and protocols

IoT encompasses a wide range of new technologies and skills. The challenge in the IoT space is the immaturity of the technologies and associated services, as well as that of the suppliers offering them. This represents a major challenge for organizations operating IoT. For successful communication between two endpoints, IoT mainly implements standard and network protocols. The main communication protocols and technologies with regard to the range between a source and a destination are as follows.

10.1.3.1 Short-range wireless communication

- **Bluetooth Low Energy (BLE):** Bluetooth Smart, or BLE for short, is a wireless LAN. Many fields, including medicine, safety, the arts, and physical fitness, will benefit from this technology.
- **Light-Fidelity (Li-Fi):** The two main distinctions between Wi-Fi and Li-Fi are the communication method and the speed. Using common incandescent light bulbs, the Li-Fi visible light communication (VLC) device can transmit data at a staggering 224 Gbps.
- **Near-Field Communication (NFC):** It is a method of establishing a connection between two electronic devices over a short distance by means of magnetic field induction. Contactless mobile payments, social networking, and document and product identification are its primary uses.
- **QR codes and barcodes:** These codes are labels with information about the product or item attached to them that can be read by machines. There are two types of barcodes:

one with one dimension (1D) and another with two dimensions (2D). QR codes, which stand for rapid response codes, are two-dimensional codes that hold product information and can be scanned using smartphones.

- **Radio frequency identification (RFID):** RFID tags may be read by electromagnetic fields to store data. It has several applications in various fields, including manufacturing, commerce, transportation, medicine, animals, and even household pets.
- **Thread:** Threads are a kind of IoT device networking protocol that is based on IPv6. Home automation is its primary use case, allowing for inter-device communication using wireless local area networks.
- Wi-Fi Wireless LANs rely heavily on Wi-Fi, a technology (LANs). With a range of about 50 meters and a maximum speed of 600 Mbps, 802.11n is now the most used Wi-Fi standard in households and businesses.
- **Wi-Fi direct:** Without a wireless access point, this protocol allows for peer-to-peer communication. Once you've selected the device to serve as an access point, only then can Wi-Fi Direct devices begin interacting.
- **Z-wave:** Developed mainly for use in home automation, Z-wave is a low-power, short-range wireless standard. It provides a dependable and easy method of remotely monitoring and operating HVAC systems, thermostats, garage doors, home theater systems, and more.
- **Zigbee:** Similar to the IEEE 203.15.4 standard, this one also allows for short-range communication. Zigbee is a wireless networking standard used by devices that seldom transmit data at low speeds within a small range (around 10–100 m).
- **ANT:** A multicast wireless sensor network technique, Adaptive Network Topology (ANT) is mostly utilized for short-range communication between devices connected to fitness and sports sensors.

10.1.3.2 Medium-range wireless communication

- **Halow:** This is an additional version of the Wi-Fi standard; it has a longer range, which is great for communicating in remote places. Low data rates mean less electricity and money spent on transmission.
- **Long-Term Evaluation advanced:** The LTE standard for mobile communications expands upon the capabilities of its predecessor, LTE, by enhancing data speed, range, efficiency, and performance.
- 6LoWPAN IPv6 over Low-Power Wireless Personal Area Networks (6LoWPAN) is an Internet protocol used for communication between small, low-power devices with limited processing capacity, such as various IoT devices (Maleh et al., 2021).
- **QUIC:** Quick UDP Internet Connections (QUIC) are multiplexed connections between IoT devices via the User Datagram Protocol (UDP); they offer security equivalent to SSL/TLS.

10.1.3.3 Long-range wireless communication

- **LPWAN:** One kind of wireless communication network that can carry data over great distances is known as low-power wide-area networking (LPWAN). There are a number of LPWAN protocols and technologies that can be used:
  - **LoRaWAN:** A long-range wide area network (LoRaWAN) is used to support applications such as mobile, industrial machine-to-machine (M2M) communications and secure two-way communications for loT devices, smart cities, and healthcare applications.

   - **Sigfox:** Utilized in devices that have restricted data transmission needs and short battery lives.
   - **Neul:** Used in a small fraction of the TV white space spectrum, it enables low-cost networks with great power, coverage, and quality.
- **Very Small Aperture Terminal (VSAT):** Broadband and narrowband data may be sent via VSAT, a communications technique that makes use of tiny parabolic antennas.
- **Cellular:** When you need to talk over a longer distance, cellular is the way to go. Although it allows for the transmission of high-quality data, it is expensive and energy-intensive.
- **Message Queuing Telemetry Transport (MQTT):** It is a lightweight ISO-standardized protocol used to transmit messages over long-range wireless communication. It can be used to establish connections with remote sites, for example via satellite links.
- **NB-IoT:** Narrowband IoT (NB-IoT) is a variant of LoRaWAN and Sigfox that uses higher-performance physical layer technology and spectrum for M2M communications.

10.1.3.4 Wired communication

- **Ethernet:** In modern times, Ethernet has become the de facto standard for network protocols. It's a LAN style that uses wires to link computers in a single location, such as a home, small business, or university.
- **Multimedia over Coax Alliance (MoCA):** Using preexisting coaxial cables, MoCA is possible to transmit high-definition video and associated content into households.
- **Power-Line Communication (PLC):** Data and energy can be sent from one endpoint to another using this protocol type using electrical lines. Broadband via power lines, industrial equipment, home automation, and other areas all rely on PLC (BPL).

10.1.4 IoT operating systems

IoT devices are made up of hardware and software components. Hardware components include terminals and gateways, while software components include operating systems. As a result of increased production of hardware components (gateways, sensor nodes, etc.), traditional IoT devices that previously operated without operating systems have begun to adopt new operating system implementations specifically programmed for IoT devices. These operating systems ensure device connectivity, user-friendliness, and interoperability.

Here are some of the operating systems used by IoT devices:

- **Windows 10 IoT:** This is a family of operating systems developed by Microsoft for embedded systems.
- **Amazon FreeRTOS:** this is an open-source operating system used in IoT micro-controllers that makes it easy to deploy, secure, connect, and manage low-energy, battery-powered edge devices.
- **Contiki:** This system is used in low-energy wireless devices such as street lighting, sound monitoring systems, and so on.
- **Fuchsia:** This is an open-source operating system developed by Google for various platforms, such as embedded systems, smartphones, tablets, and so on.
- **RIOT:** This operating system requires fewer resources and uses energy efficiently. As an example, it is compatible with sensors, actuator boards, embedded systems, and more.
- **Ubuntu core:** Robots, drones, edge gateways, and other devices use it under the Snappy name.

- **ARM Mbed OS:** This system is mainly used for low-power devices such as handhelds.
- **Zephyr:** This system is used in devices with low energy consumption and limited resources.
- **Embedded Linux:** Used for all small, medium, and large embedded systems.
- **NuttX RTOS:** This is an open-source operating system primarily developed to support 8- and 32-bit microcontrollers in embedded systems.
- **Integrity RTOS:** Most often found in the medical, aerospace/defense, industrial, and automotive fields.
- **Apache Mynewt:** This system supports devices running the BLE protocol.

10.1.5 IoT application protocols

- CoAP Constrained Application Protocol (CoAP) is a web transfer protocol used to transfer messages between constrained nodes and loT networks. This protocol is mainly used for M2M applications such as building automation and smart energy.
- **Edge:** edge computing helps the loT environment move computing processing to the edge of the network, enabling smart devices and gateways to run tasks and services from the edge of the cloud. Moving compute services to the network edge improves IoT caching, delivery, storage, and content management.
- **LWM2M:** Lightweight M2M (LWM2M) is an application-layer communication protocol used for application-level communication between IoT devices; it is used for IoT device management.
- **Physical Web:** Physical Web is a technology used to enable faster, more transparent interaction with nearby loT devices. It reveals the list of URLs broadcast by nearby devices using BLE tags.
- **XMPP:** eXtensible Messaging and Presence Protocol (XMPP) is an open real-time communication technology used for loT devices. This technology is used to develop interoperable devices, applications, and services for the loT environment.
- **MihiniM3DA:** Mihini/M3DA is software used for communication between an M2M server and applications running on an integrated gateway. It enables loT applications to exchange data and commands with an M2M server.

10.2 IOT SECURITY

The implementation of IoT devices is not without its costs, as the product selection continues to expand, putting a steady demand on overall security. Concerning the security of IoT devices, no globally binding laws have been passed as of yet (Maleh & Ezzati, 2016). Nevertheless, there are national laws and institutions that are advocating for a stricter security standard for IoT devices. Currently, suppliers are not particularly motivated by security concerns since it is difficult to hold them accountable in the event of a vulnerability. Consequently, the onus is mostly on the buyer rather than the manufacturer. As a result, protection against assaults on these items is contingent upon people and businesses taking precautions.

10.2.1 Consequences of cyber attacks

In 2016, a botnet known as Mirai, which consisted of 400 thousand linked IoT devices, launched a number of devastating cyberattacks. For a few hours, hundreds of websites were down, including Twitter and Netflix, because service provider Dyn was taken down. With an impact of 1.1 Gb per second, the network assaulted a French web provider. Approximately 145,000 cameras were responsible for the assault. Houses and the well-being of both humans and animals have been observed through the use of Internet-connected webcams. However, these vulnerabilities aren't limited to saturated saturation assaults; they were also employed in the 2016 Mirai attack. These cameras have been taken over by unauthorized operators who are now monitoring the owners of the devices.

10.2.1.1 The Stuxnet worm attack

A uranium enrichment plant was among 14 Iranian industrial locations targeted by the Stuxnet computer worm in 2010, which had a size of 500 kb. In its first stages, the worm replicated on systems running Microsoft Windows. Its intended audience was CPSs in manufacturing (PLC). In a Supervisory Control and Data Acquisition (SCADA) system, a PLC regulates the input and output of digital and analog signals, and it may communicate with other equipment, such as the IoT, in this capacity (Maleh, 2021). The worm was able to spy on linked industrial systems and spin centrifuges at such a pace that they tore apart after it eventually got control of the PLC. According to Thomas M. Chen and Saeed Abu-Nimeh, this assault was successful: "The fact that malware can infect devices other than computers was brought to light by Stuxnet. Malware has the potential to impact software-controlled vital physical infrastructures. That dangers might manifest in the physical world are genuine".

10.2.2 IoT challenges

There are many different types of frameworks. It can, therefore, be difficult to evaluate devices. Communication protocols vary widely, for example:

- Wi-Fi
- BLE
- Cellular/Long-Term Evaluation (LTE)
- Zigbee
- Z-wave
- 6LoWPAN
- LORA
- CoAP
- Sigfox
- MOTT
- AMQP
- LoRaWAN

A wide range of resources is required to deal with the potential wide diversity of procedures. To record and analyze BLE packets, for instance, you'd need an Ubertooth One; for Zigbee, you'd need an Atmel RzRaven; and so on.

10.2.2.1 Typical IoT vulnerabilities

The following questions should form the basis of the assessment procedure for device safety: Tell me what this gadget can do. Could you tell me what data the gadget can access? These two considerations allow us to make reasonable estimations about the likelihood and severity of any security issues.

10.2.2.1.1 Appliances and devices

An IoT project cannot be complete without devices. Hardware risks will be covered in this subpart, whereas the frame will be discussed in a separate area. When it comes to protecting sensitive data, physical security is paramount. Devices without proper protection pose a serious risk to your infrastructure (Figures 10.2). Attacks on exposed devices are easy. For instance, a hacker might easily get device information online by utilizing publicly available datasheets and the necessary details on most popular and widely used devices. A website like https://openwrt.org allows users to get extensive information on various devices, including gateways and routers:

This is a risky step since attackers can use the hardware information to find the interfaces utilized, which can give them root access. For example, if an attacker finds the PINS (TX, RX, and GND) using a multimeter in continuity mode, they can connect to the device without a power supply (Figures 10.3):

The USB cable pins are shown in the preceding illustration. Finding the baud rate is essential; it's synonymous with the bit rate (bits per second) but describes the rate of signal changes per second. Put another way, it's the rate at which data expressed as symbols change in a certain time period. To find out what a device's baud rate is, you may use a script that Craig Heffner made and found on GitHub:

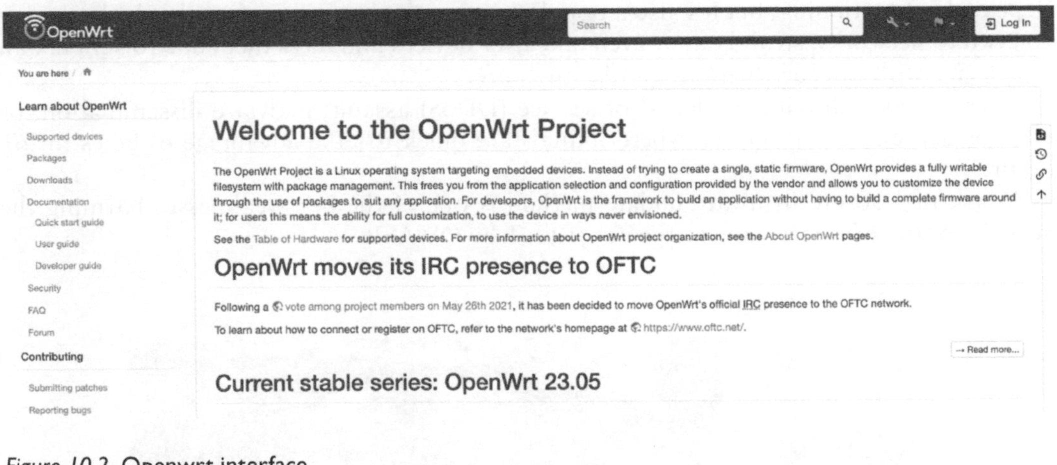

Figure 10.2 Openwrt interface.

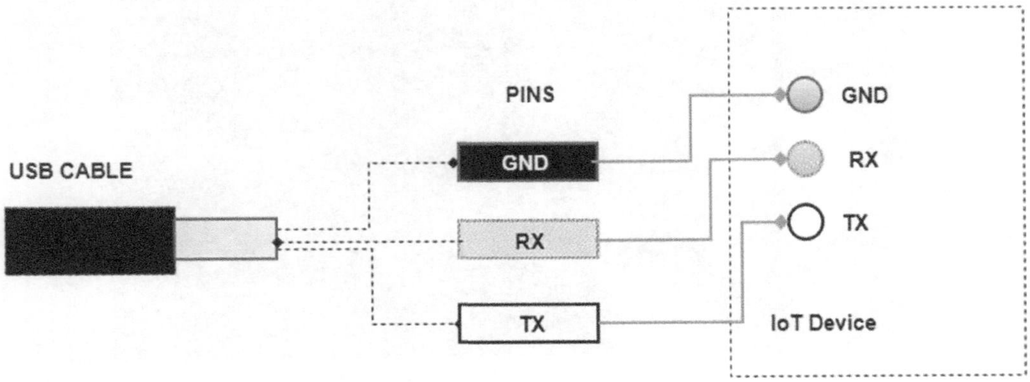

Figure 10.3 UART architecture.

https://github.com/devttys0/baudrate/blob/master/baudrate.py.

Once you've set the appropriate baud rate, you can connect to the device.

10.2.2.1.2 IoT hardware analysis steps

One of the challenges faced by the tester is clearing flash memory. In many cases, flash memory is used to store the device's bootloader and firmware. This is the central component of the device that a hacker will attempt to modify. There are two typical ways to clear the memory of this component. You can connect to the chip itself and use a tool to clear the flash memory, as shown in the screenshot. The other method is to remove the chip and use a chip removal method, which is, of course, intrusive and risky.

The top screenshot shows the test clip connection. The bottom screenshot shows the entire test configuration used for memory dumping using a tool known as Bus Pirate http://dangerousprototypes.com/docs/Bus_Pirate.

The screenshot is taken from a blog article on embedded hardware hacking published by FireEye.

10.2.2.1.3 Firmware attacks

Keeping your fleet of connected IoT devices' firmware up-to-date is critical for their safe and dependable functioning, but it's also a popular unprotected attack surface that bad guys use to get into networks, steal data, or even take over devices and mess them up. To sum up, an IoT device with insecure firmware is also insecure.

Whether it's a distributed denial of service (DDoS) assault, malware dissemination, or data breach and compromise, cybercriminals are quick to take advantage of holes in IoT security.

A lack of a secure firmware update method is one of the key weaknesses harming the security of the IoT, according to the Top 10 IoT by OWASP:

Figure 10.4 An SOIC/SOP test clip connecting the bus interface to a memory chip.

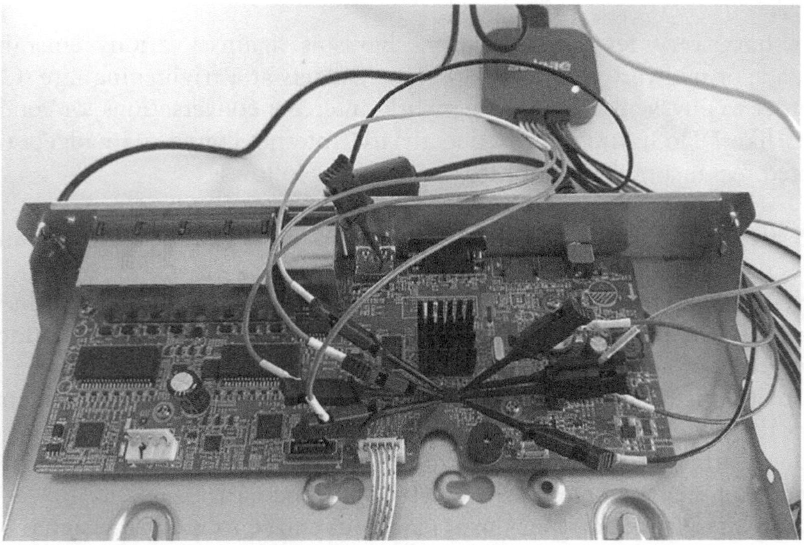

Figure 10.5 Test configuration for memory dumping.

Lack of ability to securely update the device. This includes the absence of firmware validation on the device, the absence of secure delivery (not encrypted in transit), the absence of anti-return mechanisms and the absence of notification of security changes due to updates.

In order to address this risk, enterprises must guarantee that IoT firmware can be safely and reliably updated via over-the-air updates on a regular basis. Updating the firmware of IoT devices might compromise their security, though, and these include:

- **Signature compromise:** Threat actors can install harmful updates on seemingly trustworthy devices by impersonating trusted persons and gaining unauthorized access to code signing keys or firmware signing mechanisms. Prior to enabling code execution on the device, it is crucial to fortify security measures surrounding your code signing keys and validate firmware signatures.
- **Insecure coding:** Buffer overflows are possible due to programming vulnerabilities in devices. In order to compromise security, attackers search for these code errors, which can cause applications to behave erratically or crash. Criminals can employ buffer overflows to launch DDoS attacks and acquire remote access to devices. They use techniques such as malicious code injection or DDoS assaults to exploit these vulnerabilities.
- **Insecure software supply chain:** To build IoT devices, software supply chains and heavy use of open-source components are essential. Due to the lack of supply chain security safeguards, attackers are swift to exploit vulnerabilities in open-source components. The recent incident at SolarWinds highlights the risks associated with software supply chains that are not safeguarded.
- **Forgotten test services in production devices:** No developer should ever move their credentials and debugging services to the final production version of an IoT device while they are still in the testing phase. This is because this gives attackers simple access to the device.

10.2.2.1.4 IoT threats

IoT devices have very few protection mechanisms against various emerging threats. Malicious code or malware may infiltrate these gadgets at a frightening rate (Casola et al., 2019). In order to physically harm the network, intercept conversations, or conduct disruptive assaults like DDoS, attackers frequently use Internet-connected gadgets that are not well protected. An incomplete list of IoT attacks is as follows:

- **DDoS attack:** To render a system or server inaccessible, an attacker can transform devices into a legion of botnets.
- **Attack on HVAC systems:** Attackers take advantage of HVAC system vulnerabilities to steal sensitive data such as user credentials and launch other network attacks.
- **Rolling code attack:** An intruder scrambles and sniffs the signal in order to decipher the code that is transmitted to the receiver of a car. With this code, the intruder may unlock the vehicle and make off with it.
- **BlueBorne attack:** Criminals target adjacent devices by connecting to them and taking advantage of security holes in the Bluetooth protocol.
- **Jamming attack:** An adversary uses malicious traffic to block the signal between the sender and receiver, preventing them from communicating.
- **Backdoor remote access:** An IoT device becomes a backdoor when hackers take advantage of security flaws to infiltrate a company's network.
- **Sybil attack:** An adversary can disrupt communication between nodes and nearby networks by convincing them of a severe case of traffic congestion using a plethora of fake identities.
- **Exploit kits:** Attackers exploit inadequately patched vulnerabilities in IoT devices by using malicious scripts.
- **Man-in-the-middle attack:** In this type of attack, the perpetrator masquerading as the sender gains control of all communications between the sender and the receiver.
- **Repeat attack:** Criminals launch a DoS attack or crash the target device by intercepting legal messages in a lawful transmission and repeatedly sending the intercepted message to the device.
- **Fake malicious device:** Once they get physical access to the network, attackers can replace legitimate loT devices with malicious ones.
- **Side-channel attack:** By monitoring the signals sent by loT devices, known as "side channels," attackers are able to obtain information regarding encryption keys.
- **SQL injection attack:** Criminals launch SQL injection attacks by taking advantage of security holes in the web or mobile apps that manage IoT devices. This gives them access to the devices, which they may then employ for other malicious activities.
- **SDR-based attack**: An intruder can deliver spam messages to linked devices in the IoT network by analyzing communication signals sent by devices using a software-based radio communication protocol.
- **Fault injection attack:** To undermine the security of an IoT device, an attacker can launch a fault injection attack by intentionally introducing improper behavior into the device.
- **Network pivoting:** To steal sensitive information, an attacker employs a rogue smart device to connect to a closed server. Then, they utilize this connection to pivot other devices and network connections to the server.
- **DNS rebinding attack:** One way to compromise a user's router is by DNS rebinding, which involves inserting malicious JavaScript code into a website.

10.3 METHODOLOGY

10.3.1 PT methodology for IoT

This section describes the three-step penetration testing (PT) process presented in the paper PATRIoT: A systematic and agile IoT Pentesting Process (Süren et al., 2023) more thoroughly, followed by how each step is applied in this work.

- **Planning.** To learn more about the product's capabilities, this phase aims to assess the whole infrastructure. Following this, a comprehensive threat assessment, including the existence of vulnerabilities, for various components will be provided. This is accomplished by collecting relevant data, doing threat modeling, and reviewing relevant literature. Although vulnerability analysis is a component of this phase, it will not be carried out at this stage because the chapter already contains a list of potential vulnerabilities that will be used in the threat modeling step.
- **Exploitation.** All of the vulnerabilities and threats that were selected in the previous stage are now being exploited in this step. For the best outcomes, we match vulnerabilities with helpful tools.
- **Reporting.** Finally, all results are provided after being summarized.

10.3.2 Threat modeling

A device's operation and its associated technical parts can be better understood through threat modeling. By breaking down the parts into their component parts, these models can find security restrictions and ways to circumvent them. After that, you may choose where to apply PT by evaluating the device's attack surface in conjunction with threat detection methodologies like STRIDE. As part of the modeling process, we also evaluate the seriousness of the danger. In order to prioritize efforts, evaluation systems like DREAD or NCSC rank the severity of potential dangers. To visualize the threat modeling process, you may utilize tools like draw.io or the Microsoft13 threat modeling tool. Smartphones and IoT devices, such as machine monitoring sensors and cellular service provider networks, are evaluated using the threat modeling approach (Casola et al., 2019).

Depending on the approach chosen, threat modeling may be executed in many ways. A CAD tool created for corporate cyber security management, securiCAD, is one such approach (Ekelund, 2021). The user may model the architecture of a current or future system using it and then use it to simulate assaults and find vulnerabilities using attack graphs. The user is presented with a heat map where various colors indicate the possibility of an attacker using a certain attack type to access that asset. The ability to examine the paths an attacker must take to reach a certain point—represented by red arrows—is another feature offered by securiCAD. To make securiCAD work, you need to use a language that is based on MAL (Johnson et al., 2018). Cyber threat modeling and attack simulations for specific systems in a domain are made possible by MAL, a framework for designing domainspecific attack languages. Among the domainspecific languages (DSLs) developed with MAL, coreLang stands out (Katsikeas et al., 2020). Built for the abstract world of information technology, it allows for the modeling of IT systems and the identification of vulnerabilities linked to typical assaults. The majority of the threat modeling could have been automated and saved time if securiCAD had been integrated with coreLang. Each stage is described thoroughly below.

1. **Identify assets:** In order to find and record important items, reconnaissance is the initial stage. Looking at it from a temporal standpoint, we want to know where to concentrate our most likely assaults. Finding assets with public vulnerabilities can also greatly reduce the amount of effort spent exploiting the system (Johari et al., 2020).
2. **Create an architecture overview:** The second step is to uncover the many technologies that will be used in the product and to conduct a thorough documentation examination of its architecture, subsystems, functionality, etc. (Yadav et al., 2020). This is useful for visualizing potential attacks on the product and how to get unauthorized access. The goal is to find problems with the design and the execution. This stage involves carrying out three separate tasks.
   - **Features and functions documentation:** The purpose of documenting use cases is to aid in comprehending the capabilities and functionalities of the device, its operation, and potential abuse of the technology.
   - **Sketch up the product's framework:** An architectural diagram that details the product's structure, composition, subsystems, and physical characteristics.
   - **Name the technologies that are being utilized:** With the goal of shifting attention to more technology-specific risks later in the process, we identify distinct technologies that are utilized in the implementation. The best and most accurate mitigation strategies can then be identified with this information.
3. **Break down the item:** At this point, we dissect the product's architecture in order to build a security profile that addresses the most prevalent weak spots. 4. Points of entry, limits of trust, and data flow are all discovered during this stage. In order to find vulnerabilities, it is necessary to understand the product's inner workings well.
4. **Detect dangers:** The next step is to conduct a threat assessment using the identified access points. This will help identify any potential risks to the system and its assets.
5. Various methodologies and/or models can be employed for this purpose; one of them is the famous and long-standing STRIDE model. Also included in PATRIOT were typical IoT security measures: To find security flaws, we will use an agile and methodical IoT (IoT) pentesting process and the OWASP top ten list of IoT vulnerabilities (Süren et al., 2021). Note potential dangers. As a last step in threat modeling, documenting each selected threat according to a set of important qualities is carried out in the middle stage (Guzman & Gupta, 2017). Describe the danger, identify its target, outline the attack methods, and list the countermeasures. These are the features that make up a threat document.
6. **Rate and choose potential dangers:** In the last step of threat modeling, known as a risk assessment, all threats are ranked according to the danger they represent. This way, the developer may prioritize the most dangerous risks and address them in order of severity. Because it may not be feasible, either financially or in terms of time, to handle all identified dangers, it is possible that certain threats will be disregarded entirely, even if they do not constitute a significant threat.

10.3.3 Firmware

To some extent, firmware acts as a go-between for both hardware and software. The logic for a device's operation is often provided by instructions that make up firmware. What we mean is that software actually controls the hardware of a gadget at the most basic level (IoT). Sometimes, a so-called real-time operating system, which is higher-level software, is also used (RTOS). The firmware of a device is often kept in read-only memory (ROM), which means that it is allocated on-chip or on-board, typically in flash memory, and may be updated or overwritten. Fixing faults, adding new features, or bolstering security are all goals of firmware updates.

10.4 FIRMWARE PENETRATION TESTING

10.4.1 Binwalk

To extract data from firmware images and analyze them, we utilized Binwalk. This allowed us to find files, code, and other information that was encoded in the firmware binary image. Binwalk is able to analyze executable binaries more efficiently since it uses libmagic and a custom magic signature file.

Installation link: https://github.com/ReFirmLabs/binwalk

Installation

The Kali Linux distribution comes with it already installed. It is recommended to remove the previous version of Binwalk before installing the current version to avoid any API conflicts, as the old version is not compatible with the latest releases.

A python3 interpreter must be installed before it can be installed on a Linux machine.

```
sudo apt-get update
sudo apt-get install python3
```

The next step is to install Binwalk by downloading the binary file from the aforementioned website, then navigating to the unzipped folder and running the following command:

```
$ sudo python3 setup.py install
```

The command below displays all the options:

```
$ binwalk -h
```

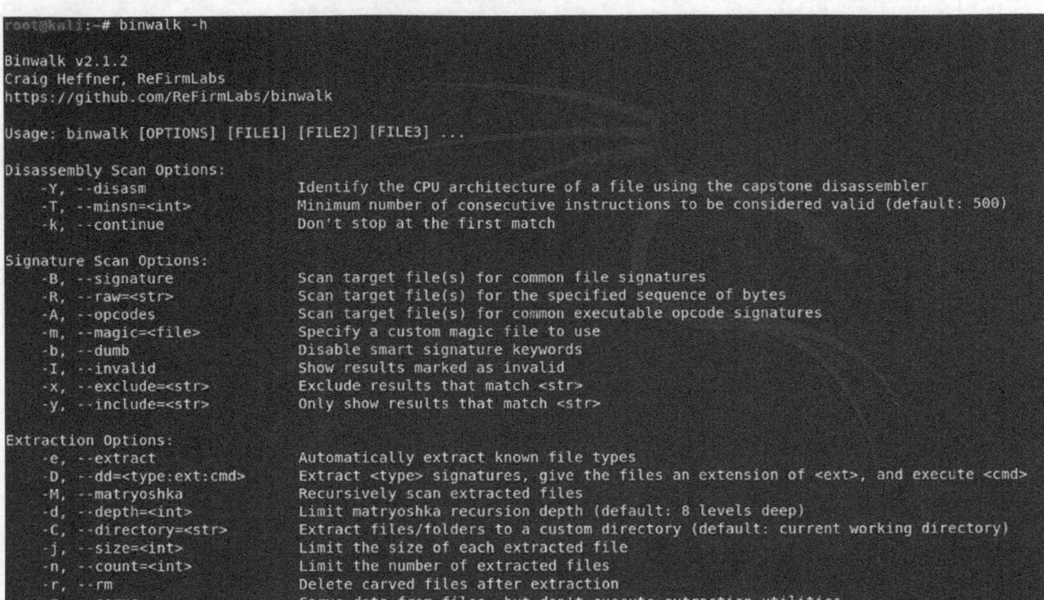

Figure 10.6 Binwalk tool help menu.

Examples of tool use:

Scan for codes, files, and other information

```
$ binwalk <firmware-image>
```

Figure 10.7 Example of using the Binwalk tool.

- **Extract firmware files**

```
$ binwalk -e <firmware-image>
```

- **Recursive file extraction from firmware**

```
$ binwalk -Me <firmware-image>
```

Top 10 firmware security vulnerabilities

- **Generate differences between firmware images**

```
$ binwalk -W <firmware1-image><firmware2-image><firmware3-image>
```

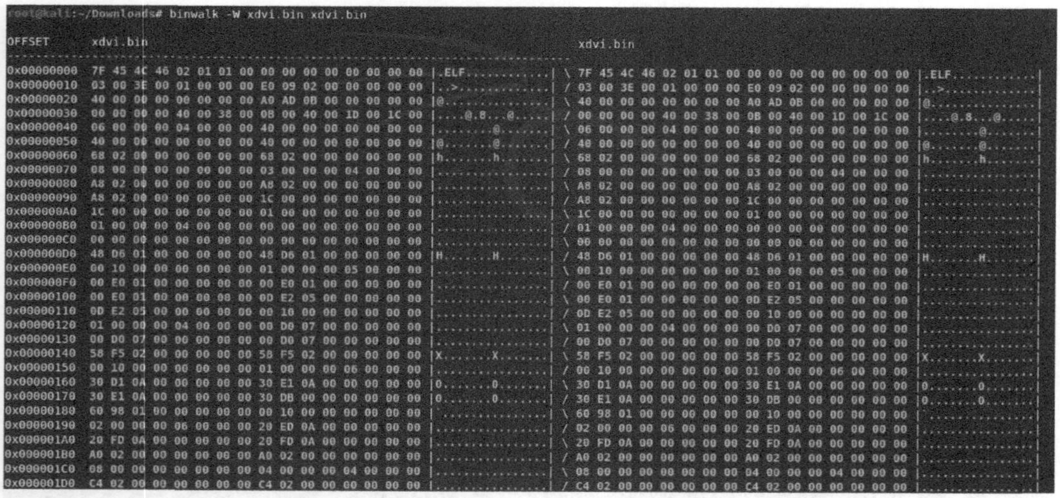

Figure 10.8 Differences between firmware images.

- **Signature analysis**

```
$ binwalk -B <firmware-image>
```

- **Entropy analysis**

```
$ binwalk -E <firmware-image>
```

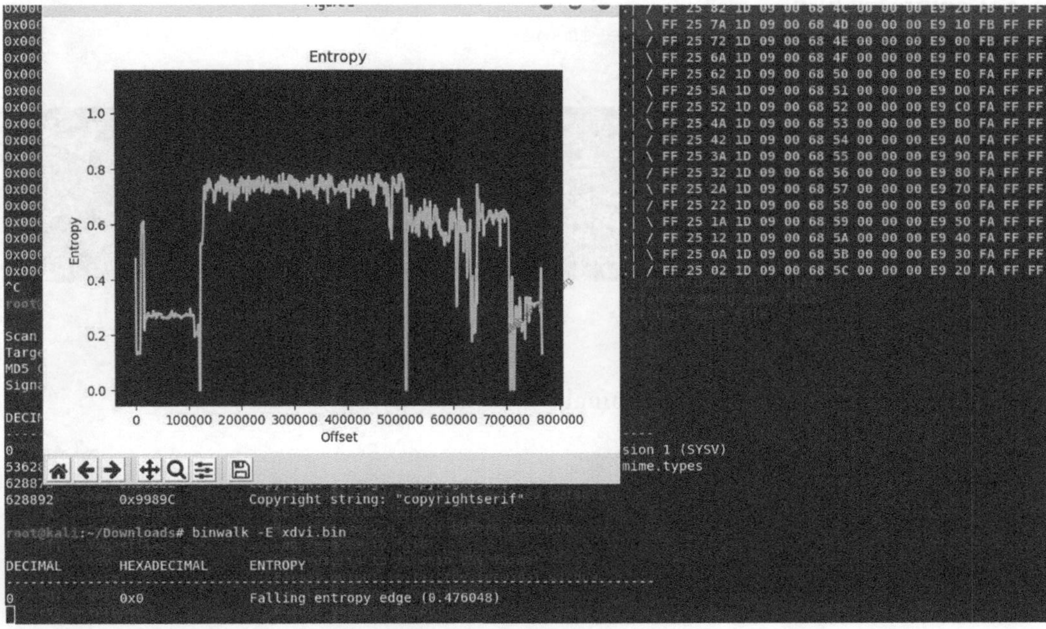

Figure 10.9 Signature analysis.

A high entropy indicates that an encryption technique is available, whereas a low entropy indicates that it may not be applied.

- **Update to the latest version**

```
$ sudo binwalk -u
```

- **Wordy output**

```
$ binwalk --verbose <firmware-image>
```

Figure 10.10 Verbose output.

- **Capture log files**

```
$ binwalk -f file.log <firmware-image>
```

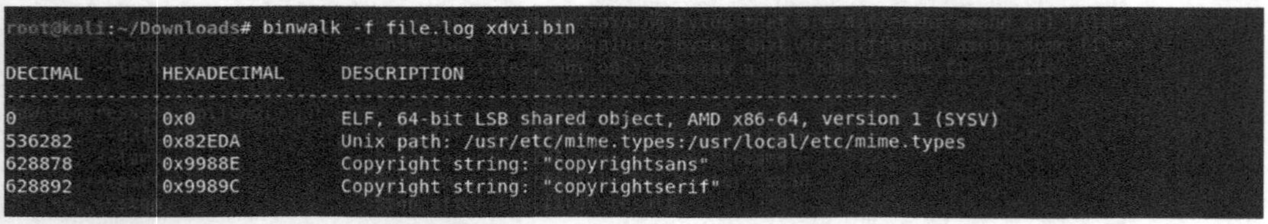

Figure 10.11 Capturing log files.

- **Output format to a current terminal**

```
$ binwalk -t <firmware-image>
```

- **To display the file system of a binary file**

```
$ binwalk -y 'filesystem' <firmware-image>
```

- **To display the CPU architecture of a binary system**

```
$ binwalk --disasm <firmware-image>
```

- **To display the CPU architecture of a binary system**

```
$ binwalk -y "endian"
```

- **To extract the firmware recursively and decompress the file**

```
$ binwalk -reM
```

10.4.2 Operation

The tools you'll need:

- AttifyOS VM or any other Linux-based image
- Firmware analysis toolbox (https://github.com/attify/firmware-analysis-toolkit)
- A firmware you wish to emulate (for example, Netgear WNAP320)

Installation

The first thing to do after acquiring the aforementioned three items is to set up a Firmware Analysis Toolkit.

To automate the process of mimicking new firmware, the Firmware Analysis Toolkit just acts as a wrapper around the actual Firmadyne. Follow these steps to recursively clone the git repository and install FAT:

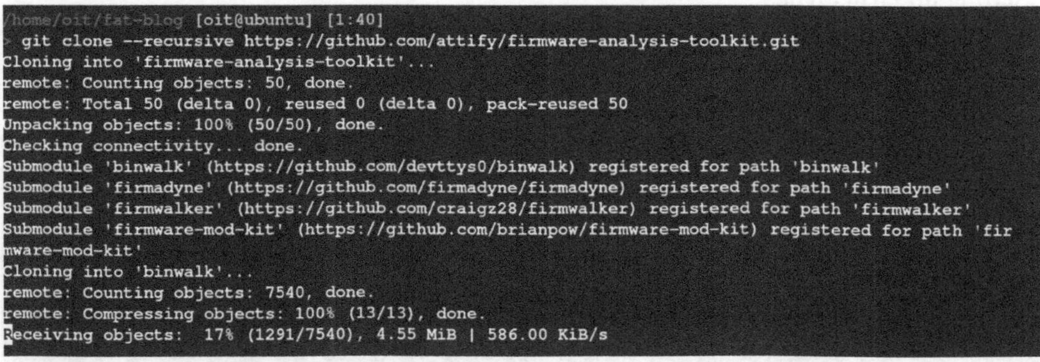

Figure 10.12 Firmware analysis toolkit.

```
git clone --recursive https://github.com/attify/
firmware-analysis-toolkit.git
```

Next, we need to configure the various tools such as BinwalkFirmadyne and Firmware- Mod-Kit.

Setting up Binwalk

Just install the dependencies in the order given below, and then install the following to setup Binwalk:

```
cd firmware-analysis-toolkit/binwalk sudo ./deps.sh
sudo python setup.py install
```

If everything is in order, you may run Binwalk and get the output below.

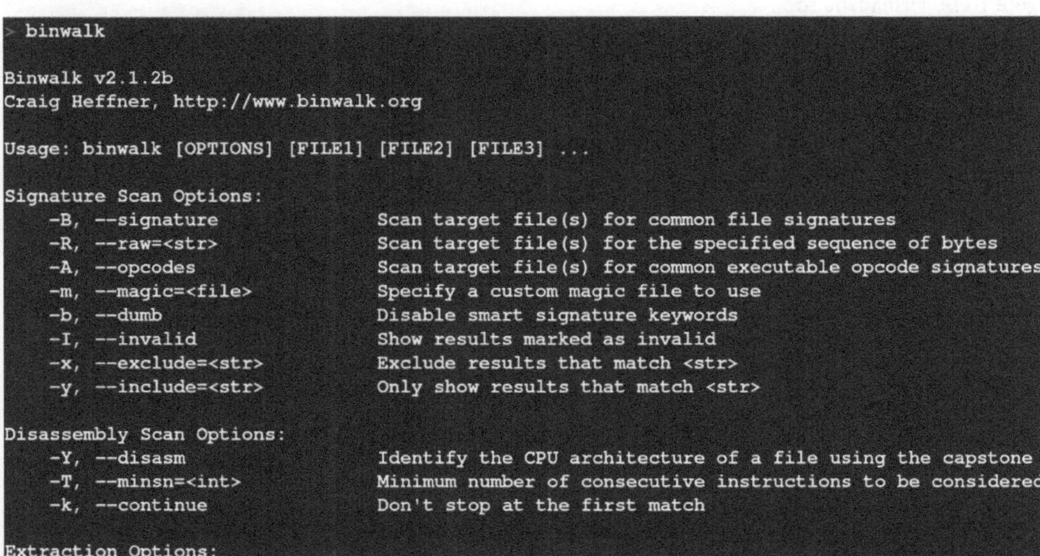

Figure 10.13 Binwalk interface.

Setting up Firmadyne

Get to the Firmadyne folder, then open firmadyne.config to set up Firmadyne. A picture like the one below should do it.

```
#!/bin/sh

# uncomment and specify full path to FIRMADYNE repository
#FIRMWARE_DIR=/home/vagrant/firmadyne/

# specify full paths to other directories
BINARY_DIR=${FIRMWARE_DIR}/binaries/
TARBALL_DIR=${FIRMWARE_DIR}/images/
SCRATCH_DIR=${FIRMWARE_DIR}/scratch/
SCRIPT_DIR=${FIRMWARE_DIR}/scripts/

# functions to safely compute other paths

check_arch () {
    ARCHS=("armel" "mipseb" "mipsel")

    if [ -z "${1}" ]; then
        return 0
    fi

    match=0
    for i in "${ARCHS[@]}"; do
```

Figure 10.14 Firmadyne tool.

Make sure the address matches the existing Firmadyne path and uncomment the line that says FIRMWARE DIR=/home/vagrant/firmadyne/. Here is how the revised line seems to me.

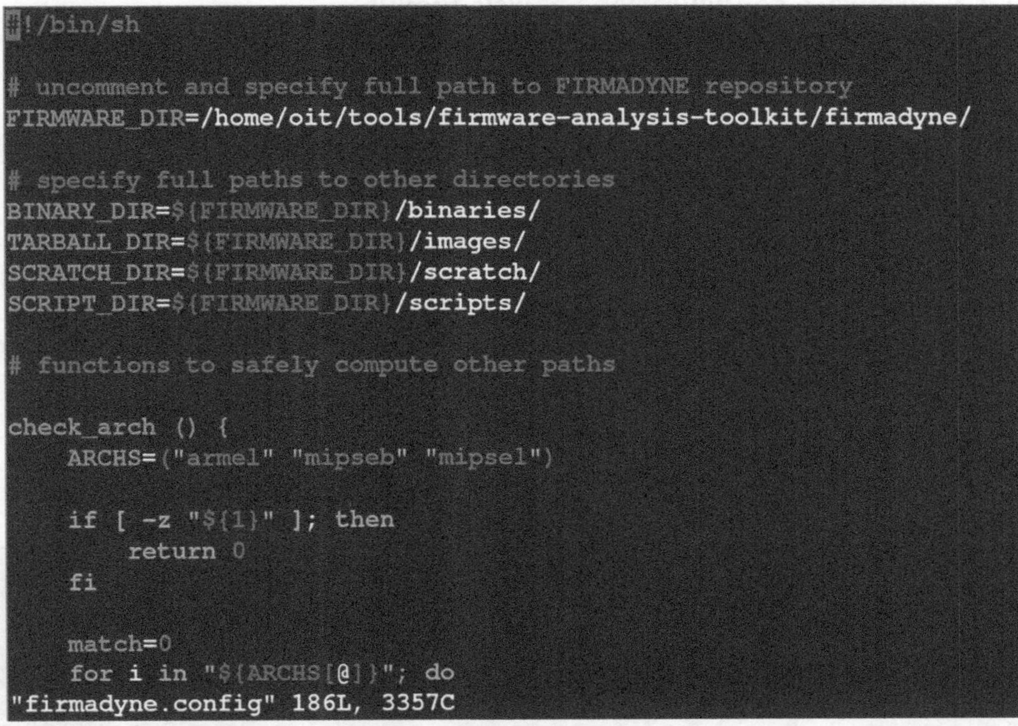

```
#!/bin/sh

# uncomment and specify full path to FIRMADYNE repository
FIRMWARE_DIR=/home/oit/tools/firmware-analysis-toolkit/firmadyne/

# specify full paths to other directories
BINARY_DIR=${FIRMWARE_DIR}/binaries/
TARBALL_DIR=${FIRMWARE_DIR}/images/
SCRATCH_DIR=${FIRMWARE_DIR}/scratch/
SCRIPT_DIR=${FIRMWARE_DIR}/scripts/

# functions to safely compute other paths

check_arch () {
    ARCHS=("armel" "mipseb" "mipsel")

    if [ -z "${1}" ]; then
        return 0
    fi

    match=0
    for i in "${ARCHS[@]}"; do
"firmadyne.config" 186L, 3357C
```

Figure 10.15 Firmadyne configuration.

Downloading the extra binaries needed to run Firmadyne is the next step after updating the path.

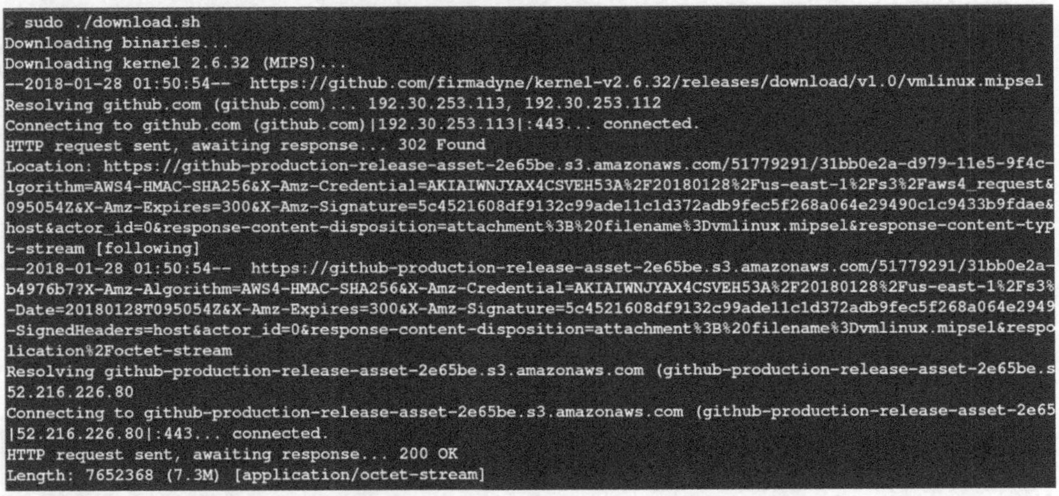

```
> sudo ./download.sh
Downloading binaries...
Downloading kernel 2.6.32 (MIPS)...
--2018-01-28 01:50:54--  https://github.com/firmadyne/kernel-v2.6.32/releases/download/v1.0/vmlinux.mipsel
Resolving github.com (github.com)... 192.30.253.113, 192.30.253.112
Connecting to github.com (github.com)|192.30.253.113|:443... connected.
HTTP request sent, awaiting response... 302 Found
Location: https://github-production-release-asset-2e65be.s3.amazonaws.com/51779291/31bb0e2a-d979-11e5-9f4c-
lgorithm=AWS4-HMAC-SHA256&X-Amz-Credential=AKIAIWNJYAX4CSVEH53A%2F20180128%2Fus-east-1%2Fs3%2Faws4_request&
095054Z&X-Amz-Expires=300&X-Amz-Signature=5c4521608df9132c99ade11c1d372adb9fec5f268a064e29490c1c9433b9fdae&
host&actor_id=0&response-content-disposition=attachment%3B%20filename%3Dvmlinux.mipsel&response-content-typ
t-stream [following]
--2018-01-28 01:50:54--  https://github-production-release-asset-2e65be.s3.amazonaws.com/51779291/31bb0e2a-
b4976b7?X-Amz-Algorithm=AWS4-HMAC-SHA256&X-Amz-Credential=AKIAIWNJYAX4CSVEH53A%2F20180128%2Fus-east-1%2Fs3%
-Date=20180128T095054Z&X-Amz-Expires=300&X-Amz-Signature=5c4521608df9132c99ade11c1d372adb9fec5f268a064e2949
-SignedHeaders=host&actor_id=0&response-content-disposition=attachment%3B%20filename%3Dvmlinux.mipsel&respo
lication%2Foctet-stream
Resolving github-production-release-asset-2e65be.s3.amazonaws.com (github-production-release-asset-2e65be.s
52.216.226.80
Connecting to github-production-release-asset-2e65be.s3.amazonaws.com (github-production-release-asset-2e65
|52.216.226.80|:443... connected.
HTTP request sent, awaiting response... 200 OK
Length: 7652368 (7.3M) [application/octet-stream]
```

Figure 10.16 Firmadyne upgrade.

10.4.3 Firmware configuration analysis toolkit

You should begin by transferring fat.py and reset.py to the Firmadyne directory.

After that, enter fat.py and replace the root password with your own. This will prevent the script from requesting a password when executed. Then, follow the steps below to provide the path to Firmadyne.

That is all for the installation. Make sure your postgresql database is up and running.

-Emulating a firmware image

To imitate firmware, just run./fat.py and pass in the name of the firmware. For this example, we will declare that we are using the firmware WNAP320.zip. Since the brand is only utilized in the database, you are free to choose any brand. Here is the expected outcome:

An IP address will be assigned to you after the firmware's initial configuration procedure is finished. It should be possible to access the web interface, communicate with the firmware using SSH, and execute other network-based actions if the firmware makes use of a web server.

```
#!/usr/bin/env python2.7

import os
import pexpect
import sys

# Put this script in the firmadyne path downloadable from
# https://github.com/firmadyne/firmadyne

#Configurations - change this according to your system
firmadyne_path = "/home/oit/tools/firmware-analysis-toolkit/firmadyne"
binwalk_path = "/usr/local/bin/binwalk"
root_pass = "attify123"
firmadyne_pass = "firmadyne"


def show_banner():
    print """

"fat.py" 122L, 4194C
```

Figure 10.17 Moving Firmadyne files.

```
/home/oit/tools/firmware-analysis-toolkit/firmadyne [git::95a4030 *] [oit@ubuntu] [11:0
5]
> sudo service postgresql start
[sudo] password for oit:
 * Starting PostgreSQL 9.3 database server                                    [ OK ]
```

Figure 10.18 Starting the postgresql database.

```
> sudo ./fat.py

  __        _
 / _|      | |
| |_  __ _ | |_
|  _|/ _` || __|
| | | (_| || |_
|_|  \__,_| \__|

                Welcome to the Firmware Analysis Toolkit - v0.2
    Offensive IoT Exploitation Training  - http://offensiveiotexploitation.com
                By Attify - https://attify.com  | @attifyme

[?] Enter the name or absolute path of the firmware you want to analyse : wnap320.zip
[?] Enter the brand of the firmware : netgear
[+] Now going to extract the firmware. Hold on..
[+] Firmware : wnap320.zip
[+] Brand : netgear
[+] Database image ID : 1
```

Figure 10.19 Fat tool interface.

```
[+] Identifying architecture
[+] Architecture : mipseb
[+] Storing filesystem in database
[+] Building QEMU disk image
[+] Setting up the network connection, please standby
[+] Network interfaces : [('brtrunk', '192.168.0.100')]
[+] Running the firmware finally
[+] command line : sudo /home/oit/tools/firmware-analysis-toolkit/firmadyne/scratch/1/r
un.sh
[*] Press ENTER to run the firmware...
```

Figure 10.20 Information generated by fat.

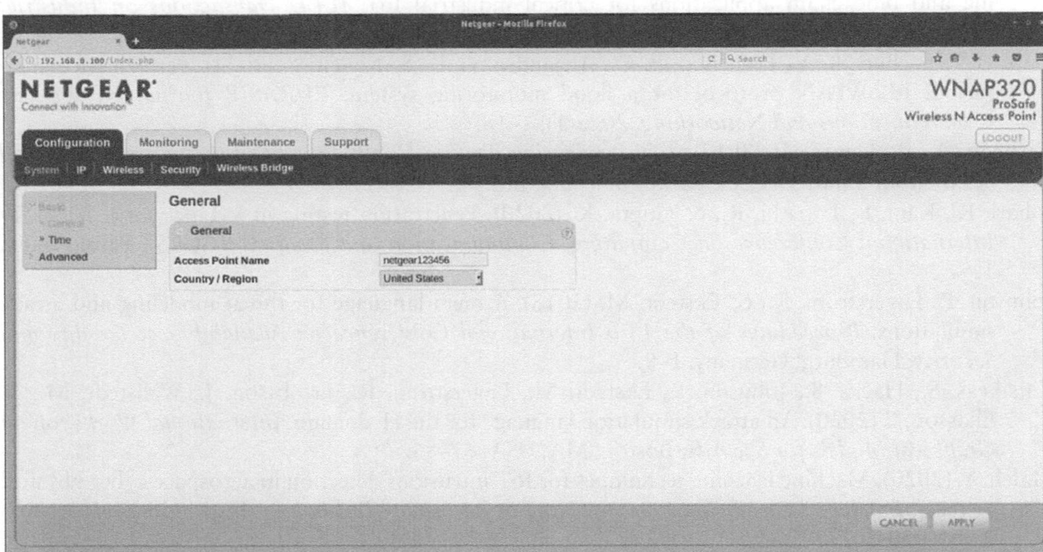

Figure 10.21 Netgear interface.

Let's open Firefox and see if we can access the web interface.

We succeeded in creating a firmware emulator that could contact the web server, which was initially designed for the MIPS Big Endian architecture.

10.5 CONCLUSION

This chapter serves as a comprehensive guide tailored to aid developers, manufacturers, and penetration testers in the development and evaluation of IoT projects. It is initiated by elucidating the intricacies of the IoT ecosystem and delineating the diverse components typically found within an IoT project. In addition, it delineates the array of threats that pose challenges to IoT projects, alongside elucidating the requisite steps for fortifying an IoT environment against potential vulnerabilities.

Moving forward, future iterations of this chapter will delve deeper into emerging trends and advanced techniques for IoT security. This will encompass exploring cutting-edge methodologies for threat detection and mitigation, as well as addressing the evolving landscape of IoT vulnerabilities. Furthermore, forthcoming sections will delve into the integration of artificial intelligence and machine learning algorithms to enhance the resilience of IoT systems against sophisticated cyber threats. Through continuous updates and refinement, this chapter aims to remain an indispensable resource for navigating the complexities of IoT development and security.

REFERENCES

Alaoui, E. A. A., Tekouabou, S. C. K., Maleh, Y., & Nayyar, A. (2022). Towards to intelligent routing for DTN protocols using machine learning techniques. *Simulation Modelling Practice and Theory*, *117*, 102475. https://doi.org/10.1016/j.simpat.2021.102475

Casola, V., De Benedictis, A., Rak, M., & Villano, U. (2019). Toward the automation of threat modeling and risk assessment in IoT systems. *Internet of Things*, *7*, 100056.

Ekelund, J. (2021). *Security Evaluation of Damper System's Communication and Update Process: Threat modeling using vehicleLang and securiCAD. Master Thesis, KTH ROYAL INSTITUTE OF TECHNOLOGY, STOCKHOLM, SWEDEN*. URN: urn:nbn:se:kth:diva-305221.

El-Latif, A. A. A., Maleh, Y., Petrocchi, M., & Casola, V. (2022). Guest editorial: Advanced computing and blockchain applications for critical industrial IoT. *IEEE Transactions on Industrial Informatics*, 1–4. https://doi.org/10.1109/TII.2022.3183443

Gabriel, P.-E., Butt, S. A., Francisco, E.-O., Alejandro, C.-P., & Maleh, Y. (2022). Performance analysis of 6LoWPAN protocol for a flood monitoring system. *EURASIP Journal on Wireless Communications and Networking*, *2022*(1), 1–18.

Guzman, A., & Gupta, A. (2017). *IoT Penetration Testing Cookbook: Identify Vulnerabilities and Secure Your Smart Devices*. Packt Publishing Ltd.

Johari, R., Kaur, I., Tripathi, R., & Gupta, K. (2020). Penetration testing in IoT network. *2020 5th International Conference on Computing, Communication and Security (ICCCS)*, Patna, India, 1–7.

Johnson, P., Lagerström, R., & Ekstedt, M. (2018). A meta language for threat modeling and attack simulations. *Proceedings of the 13th International Conference on Availability, Reliability and Security*, Hamburg, Germany, 1–8.

Katsikeas, S., Hacks, S., Johnson, P., Ekstedt, M., Lagerström, R., Jacobsson, J., Wällstedt, M., & Eliasson, P. (2020). An attack simulation language for the IT domain. *International Workshop on Graphical Models for Security*, Boston, MA, USA, 67–86.

Maleh, Y. (2020). Machine learning techniques for IoT intrusions detection in aerospace cyber-physical systems. *Machine Learning and Data Mining in Aerospace Technology*. https://doi.org/10.1007/978-3-030-20212-5_11

Maleh, Y. (2021). IT/OT convergence and cyber security. *Computer Fraud & Security*, *2021*(12), 13–16. https://doi.org/10.1016/S1361-3723(21)00129-9

Maleh, Y., & Ezzati, A. (2016). Towards an efficient datagram transport layer security for constrained applications in internet of things. *International Review on Computers and Software*, *11*(7), 611–621. https://doi.org/10.15866/irecos.v11i7.9438

Maleh, Y., Ezzati, A., & Belaissaoui, M. (2018). *Security and Privacy in Smart Sensor Networks*. IGI Global. https://doi.org/10.4018/978-1-5225-5736-4

Maleh, Y., Lakkineni, S., Tawalbeh, L., & AbdEl-Latif, A. A. (2022). Blockchain for cyber-physical systems: Challenges and applications. In Y. Maleh, L. Tawalbeh, S. Motahhir, & A. S. Hafid (Eds.), *Advances in Blockchain Technology for Cyber Physical Systems* (pp. 11–59). Springer International Publishing. https://doi.org/10.1007/978-3-030-93646-4_2

Maleh, Y., Sahid, A., & Belaissaoui, M. (2021). Optimized machine learning techniques for IoT 6LoWPAN cyber attacks detection. In A. Abraham, Y. Ohsawa, N. Gandhi, M. A. Jabbar, A. Haqiq, S. McLoone, & B. Issac (Eds.), *Proceedings of the 12th International Conference on Soft Computing and Pattern Recognition (SoCPaR 2020)* (pp. 669–677). Springer International Publishing.

Maleh, Y., Sahid, A., Ezzati, A., & Belaissaoui, M. (2018). Key management protocols for smart sensor networks. *Security and Privacy in Smart Sensor Networks* (pp. 1–23). IGI Global.

Süren, E., Heiding, F., & Lagerström, R. (2021). PATRIoT: A systematic and agile vulnerability research process for IoT. *Unpublished Paper*.

Süren, E., Heiding, F., Olegård, J., & Lagerström, R. (2023). PatrIoT: Practical and agile threat research for IoT. *International Journal of Information Security*, *22*(1), 213–233.

Yadav, G., Paul, K., Allakany, A., & Okamura, K. (2020). Iot-pen: A penetration testing framework for iot. *2020 International Conference on Information Networking (ICOIN)*, Barcelona, Spain, 196–201.

Chapter 11

A fuzzy logic-based trust system for detecting selfish nodes and encouraging cooperation in Optimized Link State Routing protocol

Fatima Lakrami, Ouidad Labouidya, Najib El Kamoun, Hind Sounni, Hicham Toumi, Youssef Baddi, and Zakariaa Jamal

11.1 INTRODUCTION

In ad hoc networks, attacks against routing are a major concern due to these networks' decentralized and dynamic nature. Attackers can exploit weaknesses in the routing process to disrupt communication between nodes. Among the most common attacks are denial-of-service (DoS) attacks, black hole, gray hole and wormhole attacks, and so on. In addition to malicious attacks, the bad behavior of nodes can also compromise routing operations. Although misbehaving nodes are commonly regarded as a security concern, selfishness is often considered a distinct concept in literature and has garnered significant research focus.

The Optimized Link State Routing (OLSR) protocol [1] was initially designed for ad hoc networks, functioning as a proactive routing protocol. Its core mechanism involves identifying specific nodes, such as Multipoint Relays (MPRs), which optimize flooding during neighbor discovery and route calculation. The importance of MPRs selection in the OLSR protocol lies in its critical role in optimizing flooding efficiency during neighbor discovery and route calculation processes. MPRs are designated nodes tasked with forwarding broadcast traffic exclusively from their selectors, significantly reducing the overhead associated with flooding in the network. By carefully selecting MPRs, OLSR can establish proactive routes opposite to the routing path, enhancing the overall performance and stability of the network. Therefore, the process of MPR selection is pivotal in ensuring efficient and reliable communication within the OLSR network.

However, the selection of selfish nodes as MPRs can significantly disrupt the overall performance of the routing protocol. Selfish MPRs prioritize their interests over network efficiency, leading to suboptimal routing decisions, increased congestion, and reduced efficiency. Despite recognizing the importance of addressing this issue, previous efforts to enhance MPR selection within OLSR have predominantly focused on technical improvements rather than directly addressing selfish node behavior. This research gap underscores the need for innovative strategies that explicitly tackle the challenge of selfish behavior in MPR selection within OLSR. Such approaches can greatly enhance the protocol's resilience and effectiveness in practical scenarios. Our recent research publication illuminates the profound impact of selfish behavior within an ad hoc network employing the OLSR protocol. This newly published paper thoroughly investigates the consequences of selfishness on network dynamics, specifically within the framework of the OLSR protocol [2].

This paper concentrates on trust models built upon reputation as a fundamental concept. The primary goal of reputation-based approaches is to foster a community of entities with established trustworthiness. Our approach introduces a fuzzy logic-based trust system

 DOI: 10.1201/9781032714806-13

centered on reputation and direct trust measurement, tailored explicitly for the OLSR protocol. In our proposed system, each node constructs its trust community with direct neighbors by quantifying their trustworthiness. We augment the MPR selection mechanism and routing algorithm in OLSR to identify selfish nodes and incentivize cooperation by continuously monitoring each node's trust evolution to determine suitable rewards or penalties. This endeavor culminates in an extension of OLSR termed STFOLSR (Security-based Trust Model and Fuzzy Logic in OLSR Protocol). STFOLSR integrates willingness, cooperation rate, and reputation metrics to compute global trust. Willingness is determined based on node stability and lifetime, and when combined with the cooperation rate, it evaluates direct trust using a fuzzy system. Reputation is calculated by aggregating recommendations from all legitimate nodes as an indirect trust metric.

A secondary fuzzy system establishes the overall trust level, considering inputs such as direct trust, reputation, and historical global trust values of all direct neighbors. To ensure compatibility and avoid introducing additional control messages or data to routing packets, STFOLSR maintains the operational functions of the protocol, including packet formats. The subsequent sections of this chapter are organized as follows: Section 11.2 reviews the related works. Section 11.3 outlines the research methodology. Section 11.4 presents simulation results, analyzes the performance of STFOLSR, and finally, Section 11.5 offers concluding remarks.

11.2 RELATED WORKS

Reducing and eliminating selfish nodes is crucial for ensuring packet accessibility and stabilizing links for end-to-end (E2E) communication.

The investigation into routing protocols and selfish behavior has attracted considerable interest, with numerous research articles delving into this subject. In this section, we give a survey of notable studies published during the last 5 years.

Authors in [3] introduce a novel Energy-Efficient Protected Optimal Path-Routing Protocol system based on Fuzzy C-Means Clustering. The proposed system utilizes the TACIT technique, based on AES-based encryption to detect and contend with the problem of malicious nodes. In [4], the authors propose to integrate the Interval Type-2 Fuzzy Logic Controller and Modified Dingo optimization (MDO) (IT2FLC-MDO) algorithm in OLSR to optimize the routing process through a node ranking mechanism based on their energy. In [5], the authors propose a fuzzy logic-based method for assessing the trustworthiness of vehicles. This novel approach incorporates three trust factors represented by fuzzy sets, aimed at characterizing potential malicious behaviors exhibited by a vehicle.

The work of [6] minimizes selfish nodes in MANET using the Reputation-Based Technique. This technique is used to control node-to-node communication by knowing and observing its neighbors node. The determination of the study proposed by [7] is designing an algorithm for detecting misbehavior node for tracking the attacker using an IDS method based on communication that is E2E between the source and the recipient.

In [8], the authors propose a new extension of OLSR called A-OLSR based on an adaptive neuro-fuzzy inference system (ANFIS) that uses node energy as a metric for selecting MPRs. In [9], the authors propose OLSR Fuzzy Cost (OLSR–FC) as an extension of the OLSR protocol based on Fuzzy logic for detecting selfish nodes and enhancing the throughput in MANETS. The authors in [10] propose to combine the Genetic Algorithm with Fuzzy logic to improve the performance of OLSR. This solution is used to Detect and Mitigate the gray hole attack over an OLSR network. In [9], a combination of a fuzzy-based system with an

OLSR routing protocol called PRO-OLSR is proposed. PRO-OLSR uses the fuzzy system's metrics as input membership functions: average delay, packet delivery ratio, direct trust, and attack suspected ratio. The simulation is done in MATLAB, and the obtained results prove the proficiency of the PRO-OLSR technique. However, this technique has the cost of inducing a high overhead to the control messages. In [11], the authors propose extending the multipath OLSR routing protocol that implements a trust-based fuzzy logic rules prediction method. The authors aim to isolate the nodes qualified as "untrustworthy" to build reliable and secure paths. The simulation result confirms the new extension's effectiveness; however, the proposed algorithm's complexity increases, which may deplete node energy rapidly.

11.3 RESEARCH METHOD

This section introduces a three-dimensional model, which incorporates three functions to establish a trust relationship with nodes designated as MPRs. In formulating the proposed model, we operate under the following assumptions:

- no central coordination or database,
- no global view of the system,
- nodes are autonomous,
- nodes can be selfish or malicious (misbehaving), and
- a single ID identifies each entity.

The original OLSR updates through Hello messages and TC messages the following sets:

- {**Link_Set:LS**}: Association of all neighbors and their link's type,
- {**Neighbor_set:NS**}: Set of all direct 1-hop-neighbors,
- {**2_Hop_Neighbor_Set:2NS**}: Set of all 2-hop-neighbors,
- {**MPR_Set:MPS**}: Set of nodes selected as MPRs,
- {**MPR_Selector_Set:MSS**}: Set of MPR selectors, and
- {**Topology_Information_Set:TIS**}: Set of topology information.

In this study, the fundamental sets of basic OLRS remain unchanged. However, to enhance the efficiency of resource allocation in trust calculation, each node evaluates the trust only for its 1-hop-symmetric neighbors. We add the **Sym_NS{}** to regroup nodes in **NS{}** with link_ type=Symmetric. The **MPS{}** contains all symmetric nodes from **Sym_NS{}** selected as MPRs. Figure 11.4 resumes evaluating the trust relationship between two nodes i and j. We present first an overview of different annotations used in this document:

- i: The node performing the MPR selection,
- j: A neighbor of i with a symmetric link,
- k: A current/previous MPR selector of j.

The following fields are recorded in **Sym_NS{}**:

- **NODE_ID**: Each node has a unique ID,
- **HELLO_SSJ**: The signal strength of the received Hello message from neighbor j,
- **Δ_HELLO_SSJ**: Difference of the signal strength of two successive Hello packets, used to accumulate the signal strength variation. This avoids unexpected links breaking:

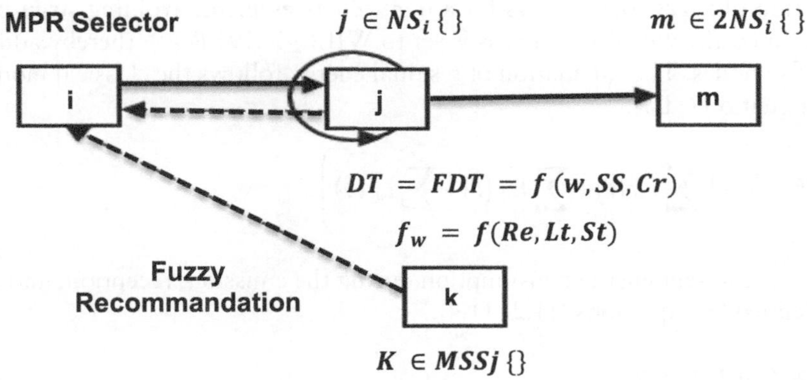

Figure 11.1 A summary of the STFOLSR trust relationship establishment process.

- DTi_j: Direct trust value between i and the neighbor j,
- W_j: The willingness received from a neighbor j,
- HW_j: The past unexpired value of node j,
- Cr_j: Cooperation rate of node j,
- GT_j : The global trust of node j,
- Tcj: Expresses if the node j belongs to the trust community of i, and the decision about its punishment or reward,
- Re_j: The reputation of node j,
- HGT_j: the past unexpired value of global trust of node j.

Each entry includes a timestamp labeled "Column_Name_Current_Time" to indicate whether the global trust record has expired. To assess the reliability of a recommendation, each node "i" maintains a list of the MPR selectors of node "j" denoted as MSSj{}, along with the following details for each node "k" in this set: its recommendation for node "j," its reputation, and the timestamp for the last update of the recording (as illustrated in Figure 11.1). In addition, we have adjusted the NS{} and 2NS{} structures by incorporating an additional field for recording nodes' reputation, a modification which will be elaborated on later.

11.4 THE DEGREE OF COOPERATION OF A NODE

In the original OLSR protocol, the concept of "willingness" denotes a node's capacity to act as an MPR. By default, the willingness is set to a predefined value labeled WILL_DEFAULT (with a value of 3). A value of WILL_NEVER (0) signifies a node's reluctance to forward traffic, often due to selfish or malicious reasons. Conversely, a value of WILL_ALWAYS (7) indicates a node's eagerness to assume the role of an MPR. This study evaluates willingness through a fuzzy system, considering factors such as node lifetime and stability. Both stability and lifetime are pivotal in maintaining link availability among nodes and their MPR/MPR_Selectors in highly mobile networks. In addition, they contribute to establishing trust, particularly when a node's availability is uncertain or incomplete. We categorize mobile nodes into three stability levels (very stable, stable, unstable) and six energy levels (Very Low (VL), Low (L), Relatively Medium (RM), Medium (M), Relatively High (RH), High (H)).

The utilization of fuzzy logic allows for a nuanced assessment, avoiding strict binary classifications, especially when willingness is set to WILL_NEVER (0), thereby addressing the issue of selfish nodes. The calculation of residual energy follows the classical model outlined in [12] by Equation (11.1).

$$RE_i = E_i(0) - \left(\sum_t E_i t(t) + \sum_t E_i r(t) + \sum_t E_i i(t) \right) \tag{11.1}$$

Et, Er, and Ei represent energy consumption during the emission, reception, and idle stages. They are defined by Equations (11.2–11.4).

$$\text{Et} = \text{Pcs} \times \text{Packet\_size} \tag{11.2}$$

$$\text{Er} = \text{Pcr} \times \text{Packet\_size} \tag{11.3}$$

$$\text{Ei} = \text{Pci} \times \text{Packet\_size} \tag{11.4}$$

Pcs, Pcr, and Pci represent power consumption levels during emission, reception, and idle stages. We did not consider the fixed channel access cost to simplify the calculation.

The node lifetime is expressed by Equation (11.5):

$$\text{Node\_Lifetime}_i = \frac{RE_i}{\text{Transmission\_Rate}_i} \tag{11.5}$$

Each summation over time t is conducted within a time interval of duration *T*, where *T* denotes the monitoring and updating interval for various records. The value of *T* is subject to adjustment, varying in accordance with the rate of change in a trust metric. For instance, if the signal power of HELLO packets increases while the willingness, particularly the stability of a node, decreases, the observation interval is shortened to prevent the node from relocating and exiting the coverage area of its selector node. Node stability is determined using an indirect mobility metric. We conducted numerous experiments to assess the relevance of various indirect mobility metrics reported in the literature in recent years. Our focus was on metrics that calculate the rate of change in the neighborhood, as the operation of the OLSR protocol predominantly relies on nodes' relationships with their 1-hop neighborhood. Equation (11.6), as developed in [13], yields precise outcomes in detecting and quantifying node stability, demonstrating its effectiveness within an OLSR architecture without GPS. Nodes implementing this function can acquire real-time information regarding their position and velocity. The stability of a node is determined by equations (11.6) and (11.7) provided below.

$$\text{Stab}_i(t) = \frac{1}{\text{Instab}_i(t)} \tag{11.6}$$

$$\text{Instab}_i(t) = k_1 . \frac{\sum_{m}^{n} \text{Neighbors}_{im}^{in}}{\text{Total\_Neighbors}_i^{in}} + k_2 . \text{Instab}_i(t-1) \tag{11.7}$$

$\text{Neighbors}_{im}^{in}$: Associations of all the links of a node *i* with direct neighbors *m*.

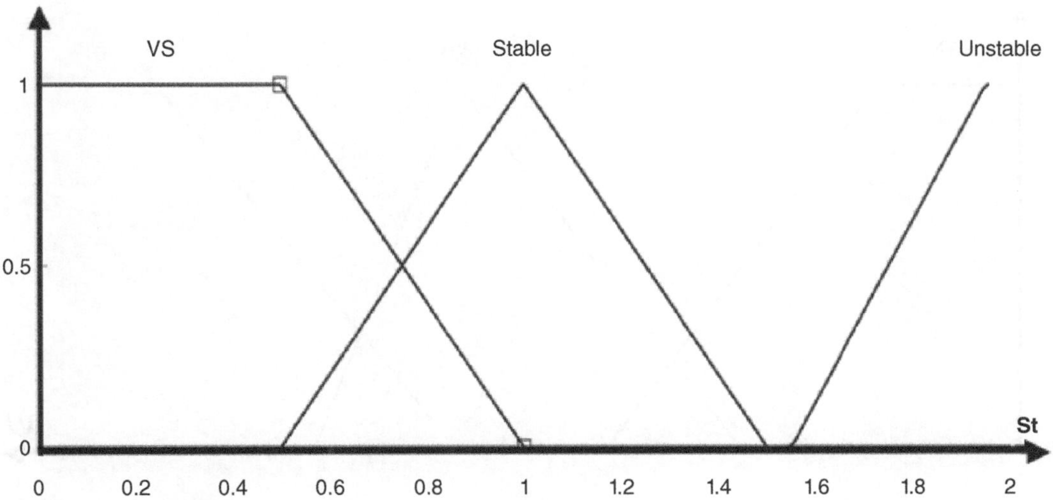

Figure 11.2 Membership functions of fuzzy set Lf (node's lifetime).

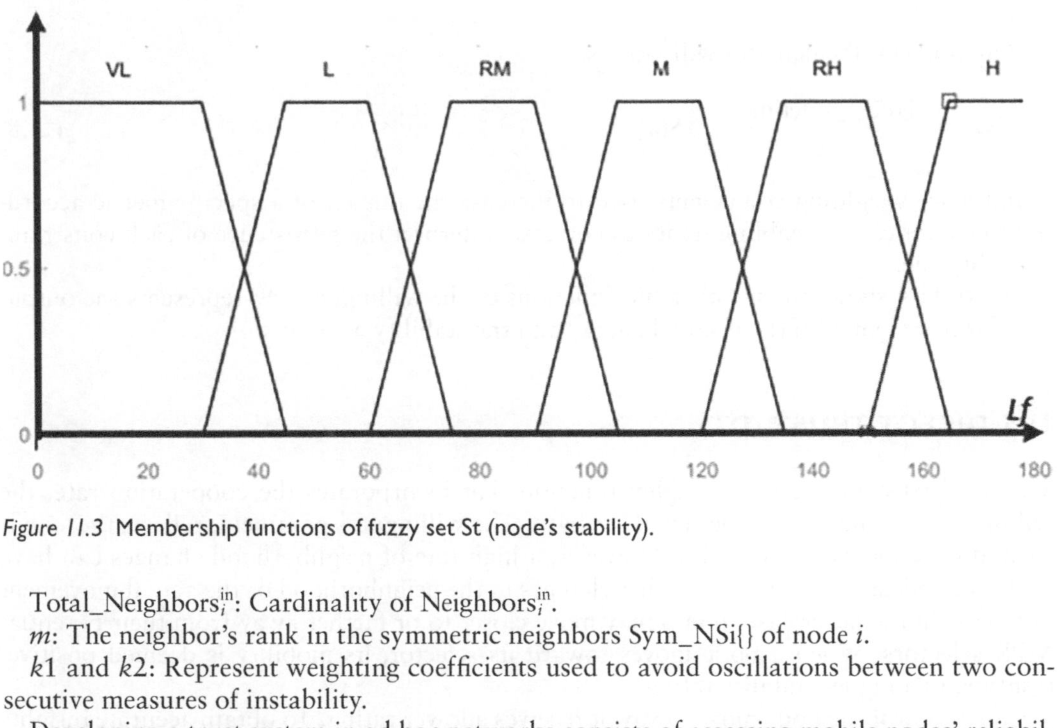

Figure 11.3 Membership functions of fuzzy set St (node's stability).

$\text{Total\_Neighbors}_i^{in}$: Cardinality of Neighbors_i^{in}.

m: The neighbor's rank in the symmetric neighbors Sym_NSi{} of node *i*.

*k*1 and *k*2: Represent weighting coefficients used to avoid oscillations between two consecutive measures of instability.

Developing a trust system in ad hoc networks consists of assessing mobile nodes' reliability by collecting and analyzing information on their current and historical behavior. Fuzzy logic allows each node to have a relative level of membership. It also allows the development of a conceptual model that is tolerant and flexible regarding nodes' behavior concerning data imprecision. In STFOLSR, second-hand information is avoided since it maximizes the overhead to the network. The membership functions of a node's lifetime and stability are presented, respectively, by Figures 11.2 and 11.3:

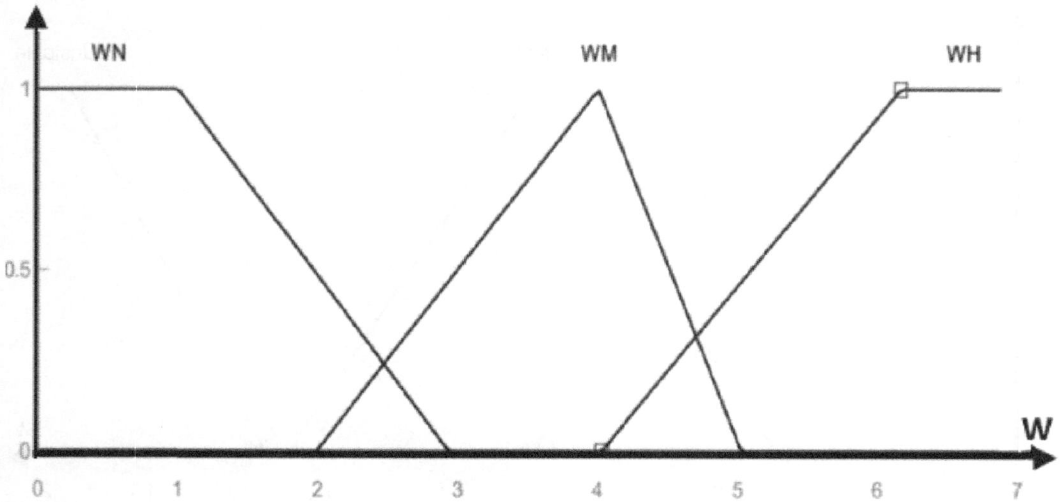

Figure 11.4 Membership functions of fuzzy set W (the willingness).

Equation (11.8) calculates willingness:

$$W_i = \alpha.\frac{\text{Node\_Lifetime}_i}{60} + \beta.\text{Stab}_i \tag{11.8}$$

α and β are weighting coefficients used to increase the impact of a specific metric according to its criticality, enabling nodes to correctly interpret the persistence of each constraint (mobility/energy).

Figure 11.4 shows the membership functions of the willingness. W represents the output of a fuzzy system with the residual energy and the stability as inputs:

11.5 DIRECT TRUST (DT)

The local trust metric is a complex function that incorporates the cooperation rate, the willingness of a node, and the signal strength. The willingness parameter reflects the stability and residual energy of nodes. However, a high rate of neighborhood changes can have various implications, indicating either changes in the neighborhood or the actual movement of nodes. In the latter case, nodes may move closer to or farther away from their potential MPR selectors. When a node moves toward its selector, its mobility is deemed positive, resulting in a higher stability rating.

Conversely, if the node moves away, it receives a lower rating. To obtain accurate insights into mobility patterns, tracking the evolution of signal strength by measuring the signal power of HELLO packets received from symmetric neighbors is essential. This measured signal strength is then compared to predefined thresholds (Threshold_High and Threshold_Low), categorizing it into one of three levels: Low, Medium, or High.

To calculate the cooperation rate, nodes monitor the packet transmission activities of their symmetric neighbor set (Sym_NS{}), determining the percentage of co-transmitted Topology Control (TC) packets relative to the total number of TC messages generated and received via their 1-hop neighbors. HELLO packets cannot fulfill this role due to their limited coverage, which extends only to the 1-hop neighborhood, whereas TC messages are broadcast throughout the entire network. The cooperation rate (%Forwarded TC) is computed according to Equations (11.9–11.11).

$$Cr_j(\%) = \frac{1}{2} \times \left(FTCr_j(\%) + FTCs_j(\%)\right) \tag{11.9}$$

$$FTCr_j(\%) = \frac{\sum_t TCr_i(t)}{\sum_t TCd_i(t)} \times 100 \tag{11.10}$$

$$FTCr_j(\%) = \frac{\sum_t TCs_i(t)}{\sum_t TCg_i(t)} \times 100 \tag{11.11}$$

- *i*: The node performing the evaluation
- j: Symmetric neighbor of *i*
- $Cr_j(\%)$: Percentage of cooperation rate of a node *j*
- $FTCr_j(\%)$: Percentage of forwarded TC packet to be received by *i* through *j*
- $FTCr_j(\%)$: Percentage of forwarded TC packet sent by *i* to be forwarded by *j*
- TCr_i: TC messages received by *i* through *j*
- TCd_i: TC messages destined node *i*
- TCs_i: TC messages sent by node *i*
- TCg_i: TC messages generated by node *i*

The direct trust is evaluated only for 1-hop-symmetric-neighbors through a fuzzy system with the following inputs: The willingness disseminated by nodes in HELLO messages, the signal strength of received HELLO, and the cooperation rate expressed by the percentage of forwarded TC messages. Figures 11.5–11.7 represent membership functions of SS (signal strength), Cr (cooperation rate), and DT (direct trust).

Table 11.1 represents the fuzzy inference engine for determining the direct trust level:

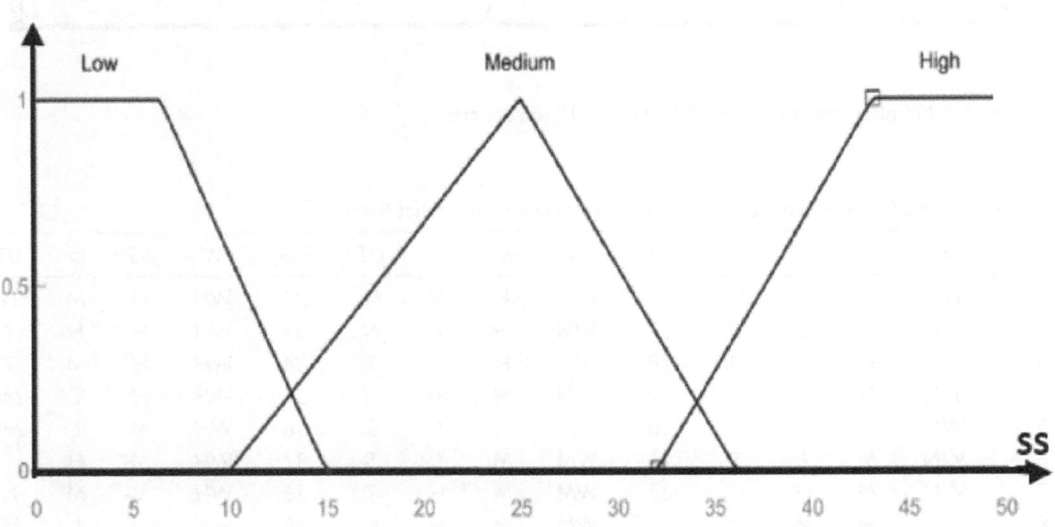

Figure 11.5 Membership functions of fuzzy set SS (single strength).

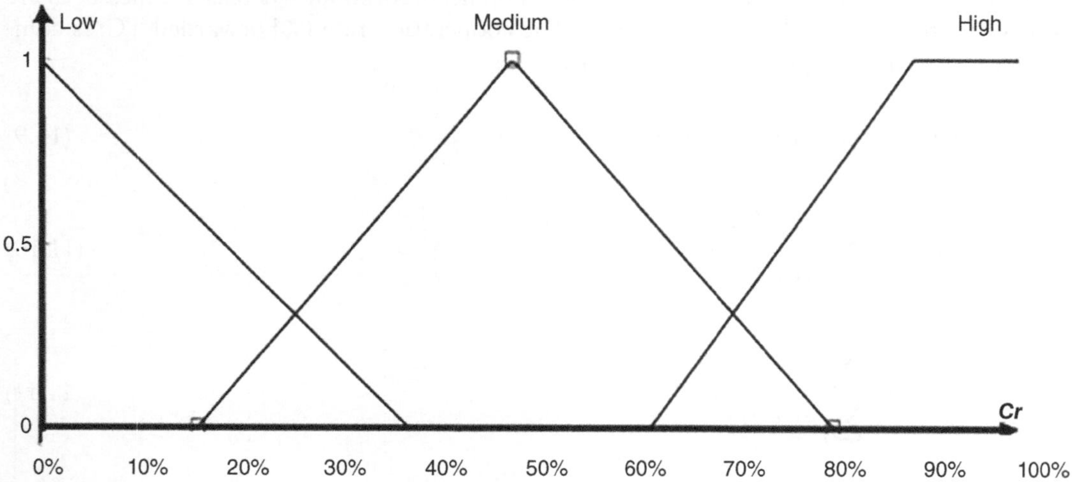

Figure 11.6 Membership functions of fuzzy set Cr (cooperation rate).

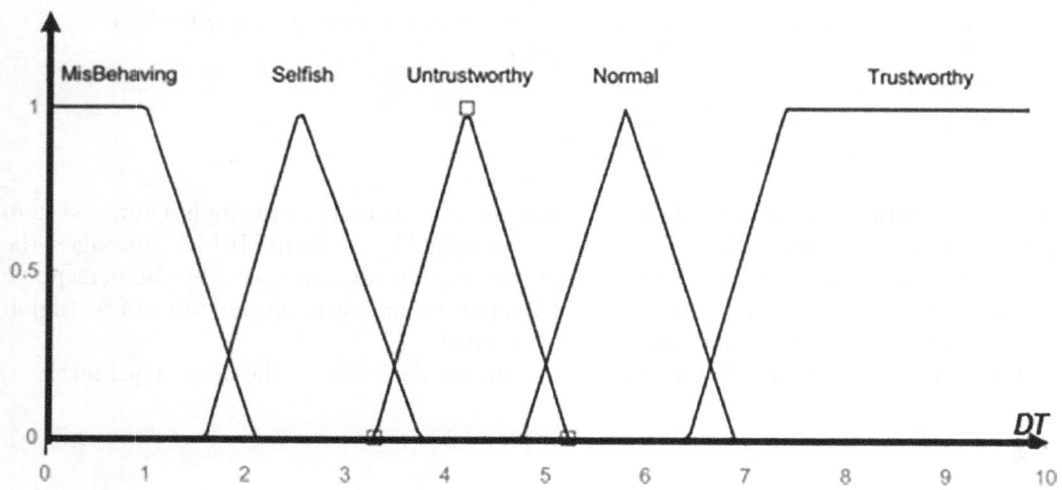

Figure 11.7 Membership functions of fuzzy set DT (direct trust).

Table 11.1 The fuzzy inference engine for determining direct trust level

Rule	*W*	*SS*	*Cr*	*DT*	*Rule*	*W*	*SS*	*Cr*	*DT*	*Rule*	*W*	*SS*	*Cr*	*DT*
1	WN	*H*	*M*	*MI*	*16*	*WM*	*H*	*M*	*MI*	*31*	*WH*	*H*	*M*	*N*
2	WN	*H*	*L*	*S*	*17*	*WM*	*H*	*L*	*N*	*33*	*WH*	*H*	*H*	*T*
3	WN	*H*	*H*	*MI*	*18*	*WM*	*H*	*H*	*T*	*34*	*WH*	*H*	*M*	*T*
4	WN	*H*	*M*	*S*	*19*	*WM*	*H*	*M*	*N*	*35*	*WH*	*H*	*L*	*MI*
5	WN	*H*	*L*	*S*	*20*	*WM*	*H*	*L*	*S*	*36*	*WH*	*M*	*L*	*MI*
6	WN	*M*	*L*	*S*	*21*	*WM*	*M*	*L*	*S*	*37*	*WH*	*M*	*H*	*T*
7	WN	*M*	*H*	*U*	*22*	*WM*	*M*	*H*	*T*	*38*	*WH*	*M*	*M*	*N*
8	WN	*M*	*M*	*U*	*23*	*WM*	*M*	*M*	*N*	*39*	*WH*	*M*	*L*	*MI*

(*Continued*)

Table 11.1 (Continued) The fuzzy inference engine for determining direct trust level

Rule	*W*	*SS*	*Cr*	*DT*	*Rule*	*W*	*SS*	*Cr*	*DT*	*Rule*	*W*	*SS*	*Cr*	*DT*
9	WN	*M*	*L*	*U*	*24*	*WM*	*M*	*L*	*S*	*40*	*WH*	*L*	*H*	*T*
10	WN	*L*	*H*	*N*	*25*	*WM*	*L*	*H*	*U*	*41*	*WH*	*L*	*M*	*N*
11	WN	*L*	*M*	*N*	*26*	*WM*	*L*	*M*	*U*	*42*	*WH*	*L*	*L*	*U*
12	WN	*L*	*L*	*M*	*27*	*WM*	*L*	*L*	*MI*	*43*	*WH*	*H*	*M*	*N*
13	WN	*H*	*M*	*S*	*28*	*WM*	*H*	*M*	*N*	*44*	*WH*	*H*	*L*	*S*
14	WN	*H*	*L*	*S*	*29*	*WM*	*H*	*L*	*S*	*45*	*WH*	*M*	*L*	*S*
15	WN	*M*	*L*	*MI*	*30*	*WM*	*M*	*L*	*MI*					

11.6 THE REPUTATION OF THE NODE: INDIRECT TRUST

The reputation of a node, denoted as node j, encapsulates a collective assessment derived from recommendations furnished by other nodes that have interacted with node j either presently or in the past. The process of evaluating reputation primarily hinges on two factors: the credibility of the received recommendations and the timeliness of the information, ensuring it remains current and not expired. Within the framework of STFOLSR, the initial phase entails gathering recommendations pertaining to node j from nodes that have directly interacted with it and whose experiences are still relevant. Subsequently, these recommendations are consolidated by leveraging the Ordered Weighted Averaging (OWA) operator, as delineated in Equations (11.12) and (11.13), to synthesize the inputs received through Transmission Control (TC) messages from various nodes.

$$\mathrm{Re}_j(t) = \frac{1}{n}.\sum_{k=1}^{n} w_k.\mathrm{Rc}_{kj}(t) \tag{11.12}$$

$$\mathrm{Re}_j = \frac{1}{2}.\left(\mathrm{Re}_j(t) + \mathrm{Re}_j(t-1)\right) \tag{11.13}$$

If recommendations originate from nodes already designated as MPRs for node i, their weighting will be determined by their established global trust scores for each respective MPR. Should the recommending node not hold the status of an MPR but possesses a reputation value within the Neighboring Set (NS{}) or the 2-Hop Neighboring Set (2NS{}), its recommendation will be weighted accordingly by this reputation value. Conversely, if the recommending node lacks validation, its weighting coefficient will assume a lower value ($w_k \leq 0.5$) to mitigate its influence. The concept of reputation here is an indirect measure of trust, considering both past and current reputations as outlined in Equation (11.13). The fixed weighting coefficients aim to balance past and present reputations. In our scenario, unlike certain other approaches, we assume that a node's propensity to cooperate may fluctuate with a probability of 50%. For a newly encountered node, the default trust value will be contingent upon the specific application context in which the network operates. Over time, trust levels will evolve based on feedback received from other nodes, reflecting the dynamic nature of trust within the network.

11.7 GLOBAL TRUST

The global trust of a node j is assessed through a fuzzy system utilizing three primary inputs: the direct trust (DTj) associated with node j, its reputation (Rej), and the historical value of the node j's global trust (HGTj). This evaluation entails delineating eight distinct fuzzy levels to characterize global trust, ranging from very low (VL) to very high (VH). The resultant trust value is constrained within the objective range of 0–1. A summary of the STFOLSR global trust evaluation algorithm is presented in Figure 11.8.

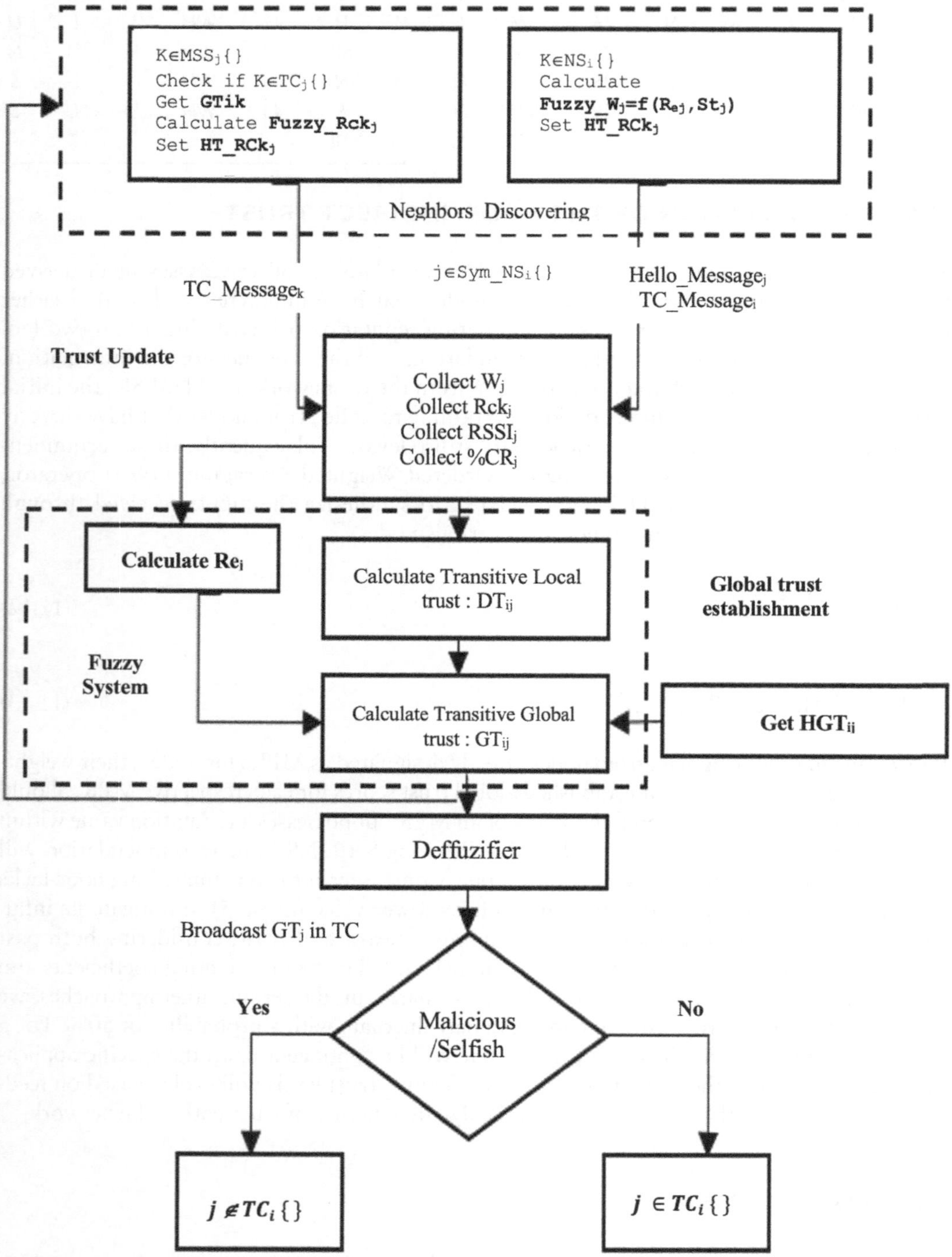

Figure 11.8 STFOLSR global trust evaluation.

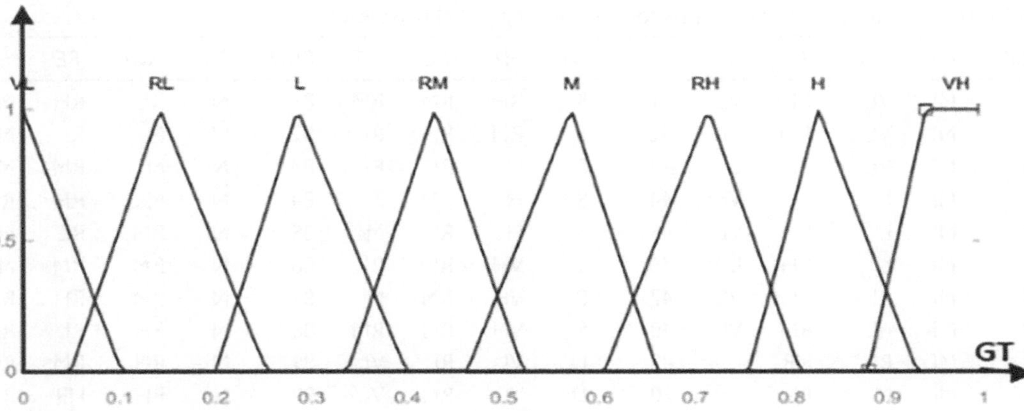

Figure 11.9 Membership functions of the global trust (GT) fuzzy set.

Figure 11.9 presents the fuzzy memberships for the global trust output:

Based on binary logic, trust models tend to categorize nodes into two states: trustworthy or untrustworthy. However, such rigid classifications can harm the network, as many nodes may be unfairly labeled as untrustworthy. It is imperative to consider past and present trust relationships, as entities may exhibit varying degrees of trust toward each other over time. Continuously monitoring trust progression is essential to preempt any unexpected misbehavior that could compromise network integrity.

To incentivize selfish nodes to cooperate, nodes that refuse to engage in cooperative behavior are penalized by being denied the relay of packets originating from or destined for these nodes. Any node classified as malicious or selfish based on its global trust value will face punitive measures. For instance, if GTi is less than or equal to 0.5, node i will not be selected as an MPR by any other node, and consequently, no node will agree to serve as its MPR. The Trust Community (Tc) indicator is periodically updated, where only nodes with Tc greater than 0.5 are deemed trustworthy. An increase in Tc's value results in rewards for the node, such as being entrusted to forward its traffic and potentially becoming an MPR. Conversely, if Tc decreases below 0.5, the node faces punitive actions.

Table 11.2 outlines the fuzzy inference engine to determine the global trust level. In scenarios where multiple nodes cover the same 2-Hop Neighboring Set (2NS{}) with identical global trust values, the selection of the MPR adheres to the following rules:

$$\mathrm{GT}_{ij} = f(\mathrm{Re}_j, \mathrm{DT}_{ij}, \mathrm{HGT}_j)$$

$$D_+ = \left\{\max(\mathrm{GT}_J \mid J = 1,2......m) / j \in MPS_i\{\ \} \,\&\, j \in Tc_+\right\}$$

$$D_- = \left\{\min(\mathrm{GT}_J \mid J = 1,2......m) / j \in MPS_i\{\ \} \,\&\, j \in Tc_-\right\}$$

In this scenario, the decision-making process relies on the evolution of their trust indicators (Tc), reflecting the extent of their affiliation with a trusted community. The node demonstrating a consistent increase in trust over time will be designated as the MPR.

Table 11.2 The fuzzy inference engine for determining global trust level

Rule	*DT*	*HGT*	*RE*	*GT*	*RULE*	*DT*	*HT*	*RE*	*GT*	*RULE*	*DT*	*HGT*	*RE*	*GT*
1	MI	VL	RL	VL	41	S	RH	RM	RM	81	N	RL	RH	M
2	MI	VL	RM	VL	42	S	RH	RH	RM	82	N	M	RL	M
3	MI	VL	RH	L	43	S	H	RL	RL	83	N	M	RM	M
4	MI	L	RL	VL	44	S	H	RM	RL	84	N	M	RH	RM
5	MI	L	RM	VL	45	S	H	RH	M	85	N	RM	RL	M
6	MI	L	RH	L	46	S	VH	RL	RL	86	N	RM	RM	M
7	MI	RL	RL	VL	47	S	VH	RM	M	87	N	RM	RH	RH
8	MI	RL	RM	VL	48	S	VH	RH	RM	88	N	RH	RL	RM
9	MI	RL	HR	L	49	U	VL	RL	VL	89	N	RH	RM	RM
10	MI	M	RL	L	50	U	VL	RM	VL	90	N	RH	HR	RH
11	MI	M	RM	RL	51	U	VL	RH	L	91	N	H	RL	RM
12	MI	M	HR	RL	52	U	L	RL	VL	92	N	H	RM	RM
13	MI	RM	RL	L	53	U	L	RM	VL	93	N	H	RH	RH
14	MI	RM	RM	L	54	U	L	RH	L	94	N	VH	RL	RM
15	MI	RM	RH	M	55	U	RL	RL	VL	95	N	VH	RM	RH
16	MI	RH	RL	L	56	U	RL	RM	VL	96	N	VH	RH	H
17	MI	RH	RM	RL	57	U	RL	RH	VL	97	T	VL	RL	VL
18	MI	RH	RH	RM	58	U	M	RL	VL	98	T	VL	RM	RM
19	MI	H	RL	L	59	U	M	RM	VL	99	T	VL	RH	RH
20	MI	H	RM	L	60	U	M	RH	RL	100	T	L	RL	L
21	MI	H	RH	RM	61	U	RM	RL	L	101	T	L	RM	M
22	MI	VH	RL	RL	62	U	RM	RM	L	102	T	L	RH	RM
23	MI	VH	RM	M	63	U	RM	RH	RL	103	T	RL	RL	RL
24	MI	VH	RH	M	64	U	RH	RL	VL	104	T	RL	RM	M
25	S	VL	RL	L	65	U	RH	RM	L	105	T	RL	RH	RM
26	S	VL	RM	RL	66	U	RH	RH	RL	106	T	M	RL	RM
27	S	VL	RH	RL	67	U	H	RL	VL	107	T	M	RM	M
28	S	L	RL	VL	68	U	H	RM	L	108	T	M	RH	RH
29	S	L	RM	VL	69	U	H	RH	RL	109	T	RM	RL	M
30	S	L	RH	L	70	U	VH	RL	M	110	T	RM	RM	RM
31	S	RL	RL	L	71	U	VH	RM	M	111	T	RM	RH	H
32	S	RL	RM	L	72	U	VH	RH	RM	112	T	RH	RL	RM
33	S	RL	RH	RL	73	N	VL	RL	L	113	T	RH	RM	RH
34	S	M	RL	L	74	N	VL	RM	RL	114	T	RH	RH	H
35	S	M	RM	RL	75	N	VL	RH	RH	115	T	H	RL	RH
36	S	M	RH	M	76	N	L	RL	VL	116	T	H	RM	RH
37	S	RM	RL	VL	77	N	L	RM	L	117	T	H	RH	VH
38	S	RM	RM	VL	78	N	L	RH	M	118	T	VH	RL	RH
39	S	RM	RH	M	79	N	RL	RL	RL	119	T	VH	RM	VH
40	S	RH	RL	RL	80	N	RL	RM	RL	120	T	VH	RH	VH

11.8 SIMULATION AND RESULTS

Our proposal is implemented and evaluated using Network Simulator 3 (NS 3.17) [14], which provides a recent implementation of the OLSRv2 module. This work implements three algorithms: the 2ACK mechanism, STOLSR [15], and our proposal STFOLSR.

Table 11.3 Simulation parameters

Parameter	*Value*
Field size	$1{,}500 \times 1{,}500\,m^2$
Number of nodes	200
Traffic type	CBR
Packet size	1024 bytes
Number of selfish nodes	5%, 15%, 30%, 40%, 50%
Mobility model	Random waypoint
Pause time	$0 \rightarrow 10$ seconds
Node placement	Random
Maximum node speed	15 m/s
Default trust value	0.5
Trust calculation interval	15 seconds
Transmission range	50–150 m
Transport protocol	UDP
Mac protocol	802.11n
Propagation radio model	Two ray
Routing protocol	OLSRv2
Simulation time	900 seconds

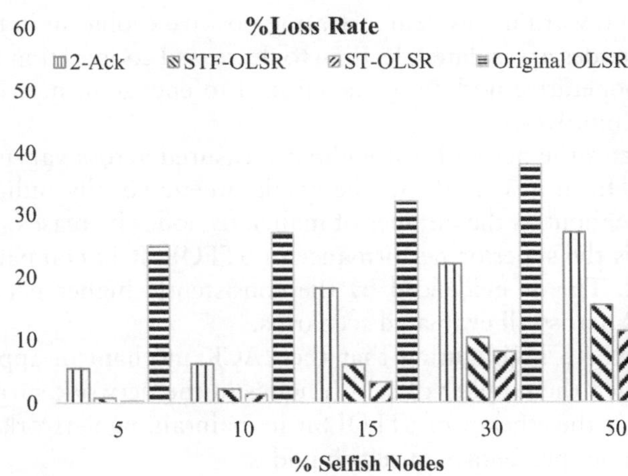

Figure 11.10 Percentage of loss rate for a variable percentage of selfish nodes and a fixed velocity.

The network contains a variable number of selfish and mobile nodes during all simulations. Table 11.3 summarizes different simulation parameters:

The 2ACK scheme finds good behavior using a new type of acknowledgment bundle, termed 2ACK. This technique is based on receiving an acknowledgment to verify that a packet has been forwarded. We implement in NS3 the 2ACK extension presented by [16]. In this work, the acknowledgment is explicitly sent to two upstream hops by nodes in the opposite direction of routing to verify their cooperation. A 2ACK is used to ensure the successful receiving of data and identify nodes' misbehavior. STOLSR is an OLSR extension that implements a basic trust model based on cooperation degree on nodes using crisp logic.

Figure 11.10 illustrates the loss rate as it fluctuates across varying percentages of selfish nodes within the network, ranging from 5% to 50%. The velocity of nodes remains fixed at 5 m/s, with a pause time of 50 seconds. It is evident from the graph that as the proportion of malicious nodes escalates, the loss rate proportionally increases across all evaluated algorithms.

Notably, a discernible trend emerges wherein STFOLSR consistently exhibits a lower loss rate than 2ACK and STOLSR. For instance, when the network comprises 50% selfish nodes, the loss rate peaks at 27% for 2ACK, 15% for STOLSR, and notably decreases to 10% for STFOLSR.

This observed reduction in loss rate with STFOLSR can be attributed to its enhanced trust-based mechanisms, which effectively identify and mitigate the impact of selfish nodes on network performance. By dynamically adjusting trust levels and imposing penalties on non-cooperative nodes, STFOLSR fosters a more resilient and reliable network environment, resulting in a reduced loss rate even in the presence of a higher percentage of selfish nodes.

Figure 11.11 depicts the observed percentage of loss rate across various nodes velocities for the three investigated solutions. In this scenario, 25% of nodes are identified as selfish, with half being mobile. Node velocities range from 1 to 15 m/s. It is evident that as nodes' mobility, indicated by their speed, increases, the percentage of loss rate also escalates.

STFOLSR exhibits a higher delivery packet ratio than the other mechanisms, indicating its superior performance in mitigating packet loss. STOLSR and STFOLSR incorporate a trust model that integrates stability as a key trust metric. This enables these extensions to manage changes in the neighborhood, ensuring network stability effectively.

Furthermore, within the 2ACK system, the percentage of collusion rate tends to rise, potentially leading to packet loss due to signal fading. Colluding nodes may fail to receive an acknowledgment, compromising normal nodes' reliability. Conversely, STFOLSR implements a mechanism to reward nodes demonstrating positive evolution in their global trust. Conversely, punished nodes are isolated, leading to decreased cooperation from neighboring nodes. As a result, cooperative nodes are encouraged to engage in network operations as their trustworthiness improves.

Figure 11.12 illustrates the network throughput measured across varying percentages of selfish nodes, ranging from 5% to 50%. The results unequivocally indicate a downward trend in network throughput as the number of malicious nodes increases.

Of particular note is the superior performance of STFOLSR in comparison to STOLSR and the 2ACK model. This is evidenced by the consistently higher network throughput achieved by STFOLSR across all evaluated scenarios.

Furthermore, it is worth highlighting that the 2ACK mechanism appears to reach its performance limits when the number of selfish nodes in the network surpasses 15%. This observation underscores the efficacy of STFOLSR in maintaining network throughput even in the presence of a higher percentage of selfish nodes.

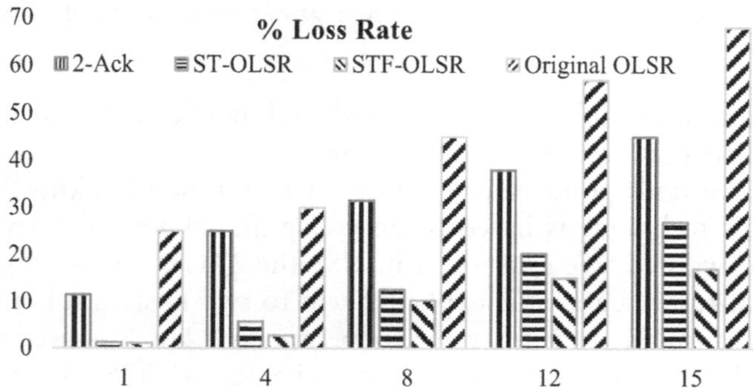

Figure 11.11 Percentage of loss rate for variable velocities and a fixed percentage of selfish nodes.

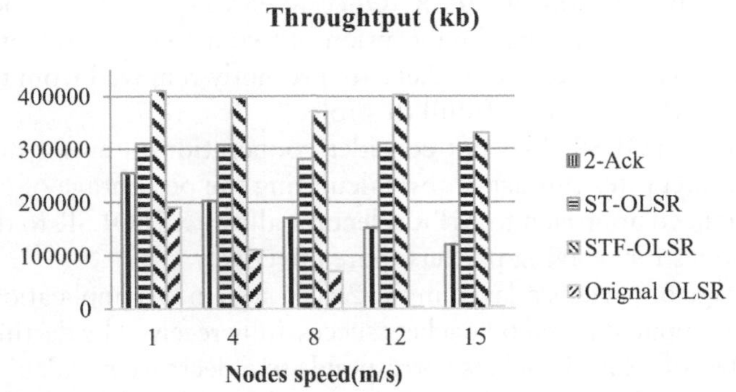

Figure 11.12 Throughput for a variable percentage of selfish nodes and a fixed velocity.

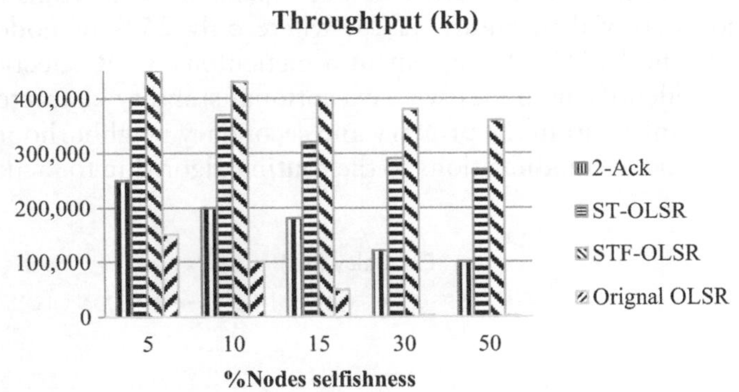

Figure 11.13 Throughput for variable velocities and a fixed percentage of selfish nodes.

This superior performance of STFOLSR can be attributed to its robust trust-based mechanisms, which effectively identify and mitigate the impact of selfish behavior on network performance. By dynamically adjusting trust levels and incentivizing cooperative behavior, STFOLSR fosters a more resilient and efficient network environment, ultimately leading to higher network throughput.

Figure 11.13 illustrates a scenario where throughput is plotted against increasing nodes' mobility. As node speed escalates, there is a noticeable decline in network throughput. This trend is evident across various node velocities ranging from 1 to 15 m/s, with the selfish nodes' percentage constant at 25%.

A key observation is the considerable disparity in performance between STFOLSR and 2ACK. While STOLSR and STFOLSR yield comparable results, the difference becomes apparent when assessing the throughput. This disparity can be attributed to the fact that STOLSR and STFOLSR incorporate stability as a metric for MPR selection, ensuring efficient management of network topology changes.

STFOLSR stands out as the superior performer, demonstrating higher throughput compared to the other implemented models. This superior performance is attributed to its robust trust-based mechanisms, which effectively mitigate the impact of selfish behavior on network performance. Conversely, 2ACK exhibits diminishing efficiency as nodes' speed increases significantly, resulting in a notable decline in throughput.

In Figure 11.14, a notable observation is the significantly low percentage of data packets relayed in the 2ACK system. Under the 2ACK solution, MPR selection follows the same algorithm as the native version of OLSR. Once MPRs are selected, nodes monitor their neighbors' cooperation by tracking the reception of two acknowledgments. Any nodes failing to cooperate by relaying received packets are promptly removed from the MPR list, with new nodes subsequently selected to fulfill this role.

In contrast, while STOLSR does not consider cooperation rate as a metric for selecting MPRs, STFOLSR integrates this aspect by calculating the percentage of co-transmitted TC messages. This nuanced approach to MPR selection allows STFOLSR to discern and prioritize cooperative nodes for relaying packets more effectively.

The low percentage of packets relayed in the 2ACK system has implications for delay calculations, as delay is computed based on packets successfully received by destinations. Therefore, reducing the number of relayed packets corresponds to a decrease in calculated delay.

Despite the observed low percentage of data packets relayed in the 2ACK system, the OLSR selfish delay remains relatively moderate compared with alternative approaches. This is primarily attributable to the inherently low percentage of retransmitted packets within selfish OLSR, contributing to a more efficient and expedient packet transmission process.

In the scenario portrayed in Figure 11.15, where only 25% of nodes exhibit selfish behavior, STOLSR and STFOLSR implement a meticulous MPR selection process. This selection hinges on identifying nodes with exceptional stability characteristics, ensuring robust and resilient links within the primary and secondary neighborhoods. Furthermore, these algorithms introduce modifications to the routing algorithm to facilitate the creation

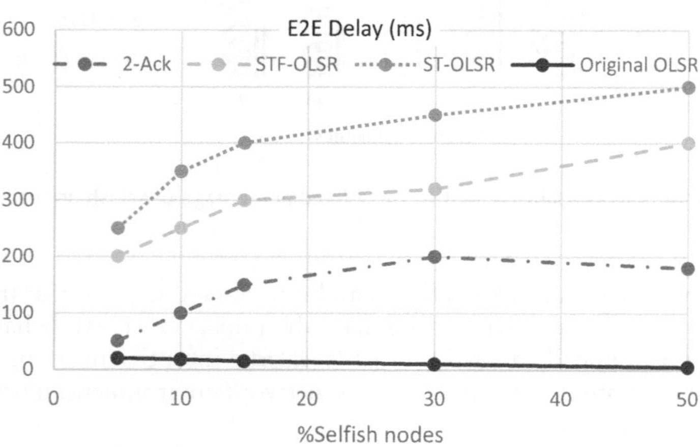

Figure 11.14 E2E delay for a variable percentage of selfish nodes and a fixed velocity.

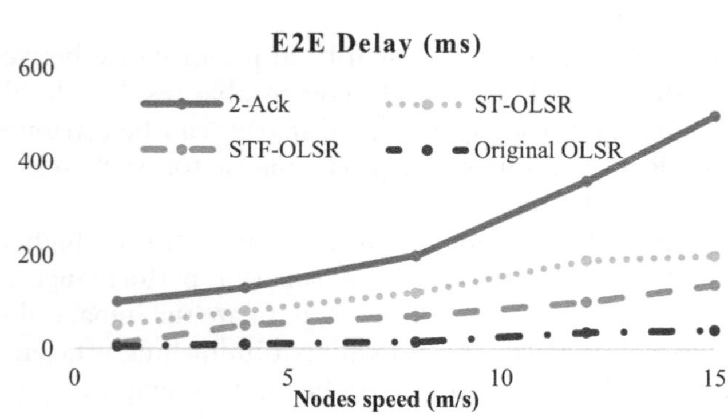

Figure 11.15 E2E delay for variable velocities and a fixed percentage of selfish nodes.

of paths in the direction opposite to MPR selection. This adjustment considers the intricate transitive trust relationships between nodes and their assigned MPR selectors. By leveraging this approach, the selected routes boast an elevated degree of trust and stability, thus enhancing communication's overall performance and reliability within the network. This strategic routing ensures that data transmission occurs along pathways with the highest levels of trust, thereby minimizing the likelihood of disruptions or packet loss (Figures 11.16 and 11.17).

The concept of overhead in network operations encompasses the additional computational and communication burden incurred during trust evaluation and dissemination processes. Within this context, the 2ACK mechanism introduces a notable overhead due to incorporating new messages, specifically acknowledgments, to verify the trustworthiness of neighboring nodes. Consequently, the utilization of 2ACK results in a heightened routing (data) overhead compared to alternative protocols.

In environments characterized by high node density, exchanging acknowledgments within the 2ACK mechanism can lead to network saturation and an increase in collision rates. This phenomenon occurs as the network becomes inundated with acknowledgment messages, potentially causing congestion and impeding the smooth data transmission flow.

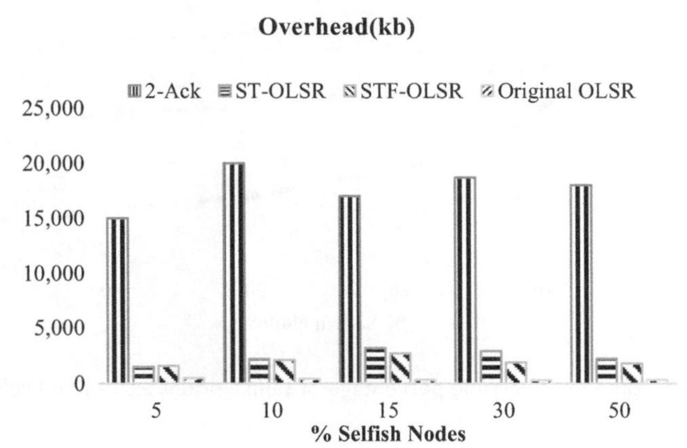

Figure 11.16 Network overhead for a variable percentage of selfish nodes and a fixed velocity.

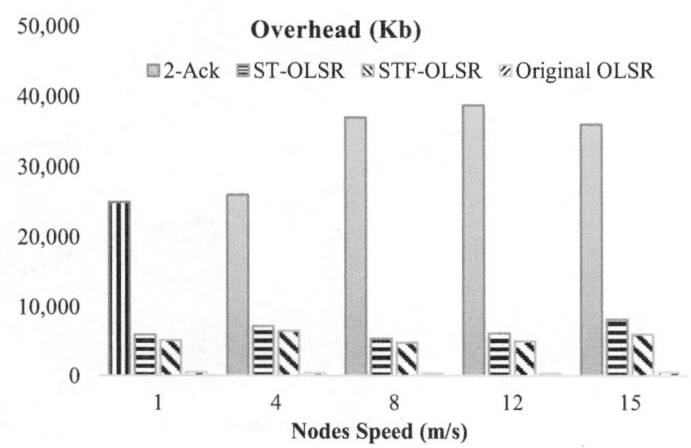

Figure 11.17 Network overhead for variable velocities and a fixed percentage of selfish nodes.

Despite the substantial amount of information STFOLSR broadcasts regarding node trustworthiness and stability, its overhead closely aligns with that of STOLSR. This suggests that STFOLSR efficiently manages the dissemination of trust-related data without significantly increasing the network overhead.

Conversely, the 2ACK mechanism introduces a disproportionately large overhead due to its reliance on additional acknowledgment messages. This excess overhead can strain network resources and adversely impact overall network performance, particularly in high-density scenarios.

Based on the insights gleaned from Figures 11.18 and 11.19, it becomes evident that STFOLSR outperforms other extensions by offering a significantly higher network lifetime. The underlying rationale behind this enhanced performance lies in the algorithm's approach to MPR selection. Specifically, STFOLSR prioritizes nodes as MPRs if they exhibit higher

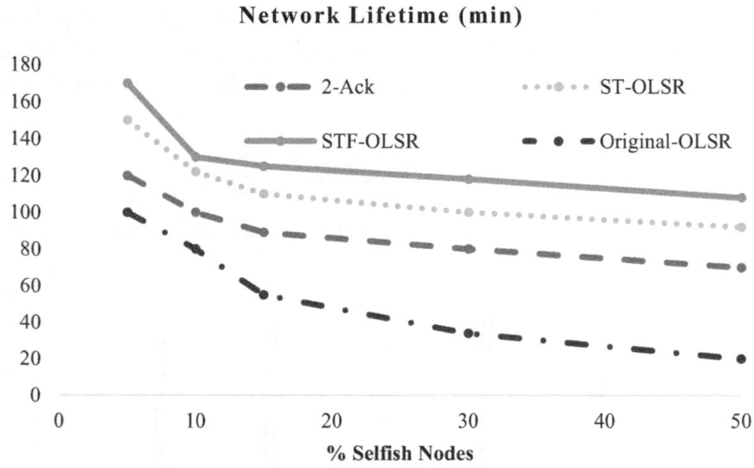

Figure 11.18 Network lifetime for a variable percentage of selfish nodes and a fixed velocity.

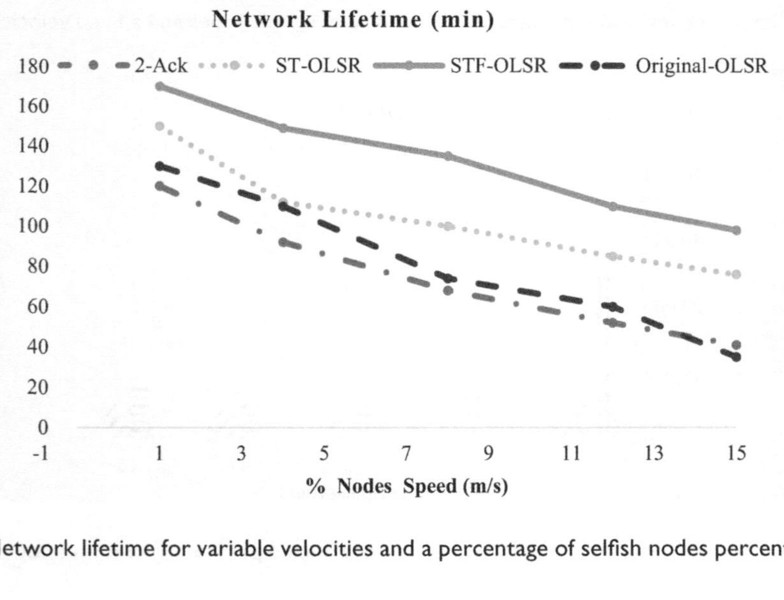

Figure 11.19 Network lifetime for variable velocities and a percentage of selfish nodes percentage.

stability or remain within the coverage area of their MPR selectors. By opting for local MPRs, STFOLSR mitigates the need for nodes to expend excessive power in message transmission and reception, thereby preserving their energy and prolonging their operational lifetime.

In contrast, selecting remote MPRs would necessitate nodes to invest greater power in communication activities, resulting in quicker depletion of energy resources and reduced network lifetime. STFOLSR's emphasis on local MPR selection thus contributes significantly to the prolonged sustainability of the network.

Moreover, to foster a trust relationship conducive to high Quality of Service (QoS), the trust mechanism embedded within STFOLSR incorporates considerations beyond merely nodes' behavioral attributes. It also considers the stability and reliability of links, which are crucial for ensuring the durability and longevity of network services. By accounting for these additional parameters, STFOLSR enhances network resilience and robustness, ultimately leading to an extended network lifetime and improved overall performance.

11.9 CONCLUSION

This chapter introduces STFOLSR, a novel fuzzy logic-based trust model designed to enhance the performance of the OLSR routing protocol. The primary objective of this model is to mitigate the adverse effects of nodes' selfish behavior on network performance by refining the process of MPR selection and routing mechanisms within OLSR. The proposed solution diverges from traditional OLSR routing by implementing a multilevel process with various design constraints. It integrates a trust function that evaluates several factors, including the willingness of nodes based on their residual energy and stability, the direct trust derived from nodes' cooperation, and the reputation synthesized from recommendations received from peers. The model uses fuzzy logic to imbue decision-making with greater precision and adaptability in addressing nodes' selfish behavior. It achieves this by offering nuanced degrees of punishment and reward based on node behavior, thereby incentivizing cooperative behavior.

In addition, fuzzy logic enables the categorization of nodes while yielding more refined outputs, facilitating less radical decision-making. Simulation results demonstrate the efficacy of STFOLSR compared to the original OLSR protocol and two alternative extensions, namely 2ACK and STOLSR. STFOLSR consistently outperforms these solutions, showcasing improved network performance and reliability. Furthermore, the routing algorithm is enhanced to incorporate trust evaluation in the direction of MPR selectors by considering the average trust of MPRs forming the pathway.

Future endeavors will focus on implementing the proposed solution in a real-world testbed to assess its feasibility and practical deployment considerations. This real-world deployment will provide valuable insights into the solution's effectiveness and potential impact on real network environments.

REFERENCES

1. Clausen, T. H. The optimized link state routing protocol v2, 2003. inria-00471712.
2. Lakrami, F., et al. A secure based trust model for Optimized Link State Routing protocol (OLSR). *2023 10th International Conference on Wireless Networks and Mobile Communications (WINCOM),* Istanbul, Turkiye. IEEE, 2023.

3. Purushothaman, V., Sivasamy, J., Karthik, S., et al. “Fuzzy C-means clustering based energy-efficient protected optimal path-routing protocol for MANET.” *International Journal of Intelligent Systems and Applications in Engineering*, 2024, vol. 12, no 2s, pp. 453–466.
4. Mamatha, C. R., Ramakrishna, M. An energy efficient model for node ranking and routing path computation using optimal type-2 fuzzy logic controller, 2023. DOI: 10.21203/rs.3.rs-1652397/v1
5. Hasan, M. M., Jahan, M., Kabir, S. A trust model for edge-driven vehicular ad hoc networks using fuzzy logic. *IEEE Transactions on Intelligent Transportation Systems*, 2023. DOI: 10.1109/TITS.2023.3305342
6. Teklu, A., Biratu, G. History-aware selfish node detection on mobile ad-hoc network, 2023. DOI: 10.21203/rs.3.rs-2528219/v1
7. Sridhar, B., Raji, K., Nandhini, S., et al. A hierarchical OLSR wireless sensor network attack detection using machine learning approach. *2023 2nd International Conference on Applied Artificial Intelligence and Computing (ICAAIC)*. IEEE, Salem, India, 2023, pp. 1206–1211.
8. Sharma, V., Alam, B., Doja, M. N. A-OLSR: ANFIS based OLSR to select multipoint relay. *International Journal of Electrical & Computer Engineering*, 2019, vol. 9, no 1. DOI: 10.11591ijece.v9i1.pp646-651
9. Gurjinder, K., Navpreet, K. Detection and mitigation of Gray hole attack over OLSR protocol in MANET using GA and Fuzzy. *International Journal of Advanced Trends in Computer Applications (IJATCA)*, 2017, vol. 4, no 7, pp. 7–12.
10. J. Diógenes, Neto, R. B., Patto, V. S., et al. OLSR Fuzzy Cost (OLSR-FC): An extension to OLSR protocol based on fuzzy logic and applied to avoid selfish nodes. *Revista de Informática Teórica e Aplicada*, 2019, vol. 26, no. 1, pp. 60–77.
11. Singh, A., Harpreet, K. Improved security protocol using fuzzy logic with multi-parametres in ad-hoc networks. *International Journal for Research in Applied Science & Engineering Technology (IJRASET)*, 2017, vol. 5, no VIII. DOI: 10.22214/ijraset.2017.8136
12. Xiangyun, Z., Wang, C. Trust-based multipath OLSR routing protocol using analytic hierarchy process and fuzzy logic. *Computer Applications and Software*, 2015, vol. 3, p. 67.
13. Allard, G., Minet, P., Nguyen, D.-Q., et al. Evaluation of the energy consumption in MANET. *International Conference on Ad-Hoc Networks and Wireless*. Springer, Berlin, Heidelberg, 2006, pp. 170–183.
14. Henderson, T., Lacage, M., Riley, G. F. Network simulations with the ns-3 simulator. *Modeling and Tools for Network Simulation*. Springer, Berlin, Heidelberg, 2010, pp. 15–34.
15. F. Lakrami, Kamoun, N. E. L., Labouidya, O., et al. Analysis and evaluation of cooperative trust models in ad hoc networks: Application to OLSR routing protocol. *International Conference on Artificial Intelligence and Symbolic Computation*. Springer, Cham, 2019, pp. 38–48.
16. Wankhade, S. V. 2ACK-scheme: Routing misbehavior detection in manets using OLSR. *International Journal of Advanced Research in Computer Engineering & Technology (IJARCET)*, 2012, vol. 1, no 5, pp. 1–7.

Chapter 12

Collaborative Cloud–SDN architecture for IoT privacy-preserving based on federated learning

Anas Harchi, Hicham Toumi, and Mohamed Talea

12.1 INTRODUCTION

Internet of Things (IoT), as defined by the Internet Engineering Task Force (IETF), is the network of physical objects or "things" embedded with electronics, software, sensors, actuators, and connectivity to enable objects to exchange data with the manufacturer, operator, and/or other connected devices [1]. The IoT refers to a system with devices that are often constrained in communication and computation capabilities, now becoming more commonly connected to the Internet or at least to an IP network and to various services built on top of the capabilities these devices jointly provide. This development is expected to usher in more machine-to-machine communication using the Internet with no human user actively involved [1]. While IoT offers numerous benefits and opportunities, it also presents several security challenges that need to be addressed to ensure the safe and reliable operation of connected devices and the protection of user data.

While software-defined networks (SDNs) have proven effective in combating cyber threats and attacks, they are not a panacea for cybersecurity in the IoT environment. Moreover, using machine and deep learning algorithms in cybersecurity introduces unique privacy concerns requiring careful management and consideration.

This chapter proposes a scalable solution integrating federated learning (FL)—a collaborative machine learning technique across decentralized devices—with SDN to create a robust intrusion detection system (IDS). This system can adapt to evolving security threats and enhance network user protection. The SDN component facilitates granular control over IoT device communications, ensuring secure and efficient data exchange. SDN controllers enforce access policies and monitor traffic dynamically, thwarting unauthorized access and preserving data integrity.

In the proposed architecture, FL plays a crucial role by allowing model training to occur locally on IoT devices, mitigating the risk of centralizing IoT data, and minimizing the potential for data breaches. Only model updates are shared with a central server, preserving data privacy.

Initially, the primary Fundamental Knowledge will be introduced. Following that, a review of related works will be conducted in the subsequent section. In the ensuing section, emphasis will be placed on the approach, and a subsequent discussion will touch upon future work.

12.2 THEORETICAL BACKGROUND AND RELATED WORKS

12.2.1 IoT and Cloud security challenges

In the past few years, we have witnessed a surge in the frequency and intensity of assaults aimed at IoT devices and Cloud infrastructure. Such attacks represent substantial threats to individuals, businesses, and essential infrastructure.

DOI: 10.1201/9781032714806-14

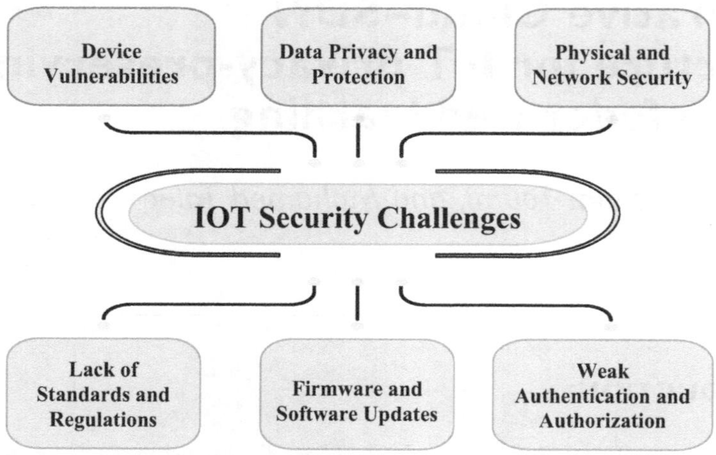

Figure 12.1 IoT security challenges.

IoT devices such as smart home appliances, wearables, and industrial sensors often lack robust security measures. This makes them attractive targets for cybercriminals. Moreover, attackers can exploit vulnerabilities in these devices to gain unauthorized access, launch distributed denial of service (DDoS) attacks, or even use them as entry points to compromise other devices or networks.

Cloud computing environments have also become prime targets for cyberattacks due to their widespread adoption and the valuable data they store. Attackers may attempt to breach Cloud infrastructure to gain unauthorized access to sensitive information, disrupt services, or launch attacks against other targets hosted in the Cloud [2].

Cyberattacks are complicated procedures as they use advanced techniques to penetrate systems and computer networks. These attacks can bypass firewalls and antivirus programs to steal sensitive information.

Within the domain of IoT, a myriad of challenges exist, particularly in the realm of security. To offer a more comprehensive understanding, the compilation of the most prominent challenges is presented in Figure 12.1 for a visual representation:

IoT raised the cybercrime cost. As more people and businesses connect online, it provides more opportunities for cybercriminals to target unsuspecting individuals and exploit vulnerabilities. The global cost of cybercrime is expected to surge in the next 5 years, as shown in the estimated cost of cybercrime worldwide realized by Statista's Cybersecurity Outlook in Figure 12.2 [3,4].

12.2.2 SDN for IoT security and privacy improvement

SDN is an innovative approach to design, implement, and manage networks that separate the network control (control plane) and the forwarding process (data plane) for a better user experience. This network segmentation offers numerous benefits in terms of network flexibility and controllability. On the one hand, it allows combining the advantages of system virtualization and Cloud computing and, on the other hand, creates an implementation of a centralized intelligence that enables clear visibility over the network for the sake of easy network management and maintenance as well as enhanced network control and reactivity [5].

SDN has come to light in recent years. However, the concept of this approach has been evolving since the mid-1990s. Ethane (management architecture) and OpenFlow have given

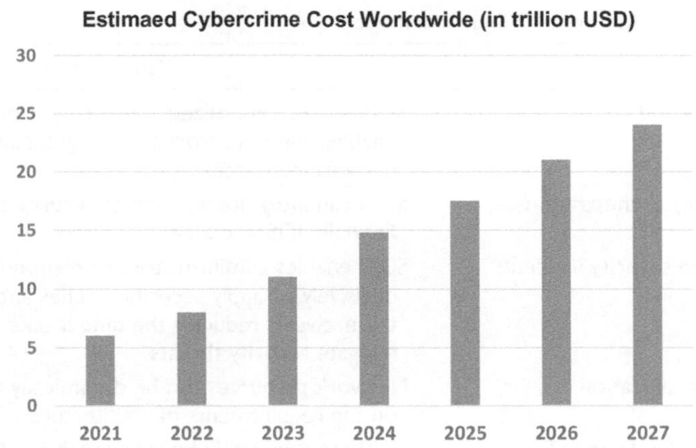

Figure 12.2 Cybercrime cost worldwide.

Table 12.1 SDN versus Classical networking

Characteristics	*SDN architecture*	*Classical architecture*
Programmability	✓	
Centralized control	✓	
Error-prone configuration		✓
Complex network control		✓
Network flexibility	✓	
Improved performance	✓	
Easy implementation	✓	
Efficient configuration	✓	
Enhanced management	✓	

birth to a real implementation of SDN [5]. OpenFlow protocol provides a standardized way of managing traffic and describes how a controller communicates with network devices. The devices supporting OpenFlow consist of two logical components: a flow table that defines how to process and forward packets within the network and an exposed OpenFlow application programming interface (API) that handles the exchanges between switch/router and controller. In Table 12.1 presented below, a comparison is depicted, highlighting distinctions between SDN architecture and Classical architecture:

The demand for Cloud services in its various forms is increasing drastically. Although these services are centralized in Data Centers, they pose important challenges for service providers. With the rapid growth of the client's demands, the operator is required to respond accordingly by considering additional servers, network components, high quality of service, and secure architecture abiding by the standards. This generally comes at the cost of non-negligible effort in facing new challenges within the core network where SDN leads and governs.

SDN stands as a pivotal player in fortifying the security of IoT deployments. SDN serves as a critical enabler for enhancing the security of IoT deployments, providing the flexibility, scalability, and control necessary to protect against evolving cyber threats effectively. By harnessing the capabilities of SDN, IoT ecosystems can significantly enhance their levels of privacy and security, as illustrated in the insights presented in Table 12.2.

Table 12.2 SDN benefits for an IoT environment

Capability	*Description*
Network visibility and control	SDN allows centralized control and management and enables administrators to have granular control over network resources.
Security and policy orchestration	SDN can integrate with other security technologies like firewalls, IDS, and SIEM.
Rapid response to security incidents	SDN enables administrators to respond quickly and decisively or apply security patches and updates across the network, reducing the time it takes to detect and mitigate security threats.
Efficient resource allocation	Network resources can be dynamically allocated based on the requirements of architecture.
Integration with Cloud and IoT	SDN can seamlessly integrate with the Cloud infrastructure. It can provide connectivity between IoT devices and Cloud resources.
Scalability and flexibility	SDN provides scalability and flexibility, making deploying and managing large-scale IoT environments easier.

Overall, SDN's abilities provide organizations with better protection for their IoT infrastructure from evolving cyber threats and mitigate potential risks associated with IoT device connectivity.

12.2.3 Intrusion detection with federated learning

IDS detects attacks by capturing and analyzing network packets. Listening on a network segment or switch, one network-based IDS can monitor the network traffic affecting multiple hosts connected to the network segment. From the deployment-based IDS perspective, IDS is further subclassified as host-based IDS (HIDS) or NIDS. HIDS is deployed on a single information host. Its task is to monitor all the activities on this single host and scan for its security policy violations and suspicious activities. The main drawback is its deployment on all the hosts that require intrusion protection, which results in extra processing overhead for each node and ultimately degrades the performance of the IDS. In contrast, NIDS is deployed on the network to protect all devices and the entire network from intrusions. The NIDS will constantly monitor the network traffic and scan for potential security breaches and violations [6].

In recent years, we have seen a lot of use of artificial intelligence to improve the detection rate of an intrusion system. John McCarthy defines AI as the science and engineering of making intelligent machines. It is related to the similar task of using computers to understand human intelligence, but AI does not have to confine itself to biologically observable methods. The goal of AI is to develop machines that behave as though they are intelligent [7].

In 1959, Arthur Samuel defined machine learning as a field of study that allows computers to learn without being explicitly programmed. ML algorithms are mainly divided into four categories: supervised learning, unsupervised learning, semisupervised learning, and reinforcement learning [8–10].

Centralized machine learning refers to the traditional approach where data is collected from multiple sources and sent to a central server or a Cloud-based system to process and train ML models. In this setup, all the data is combined in one location, and the models are trained on the centralized dataset. The trained models are then deployed to make predictions or perform tasks [11]. Decentralized ML, conversely, involves training ML models directly on local devices or edge devices rather than sending data to a central server. Each device performs local model training using its data, and the models may collaborate or exchange information to improve their performance [11].

FL is an effective solution for decentralized systems that require on-device training without compromising data privacy [11]. Recently, FL has become one of the most extensively employed solutions for maintaining the privacy and integrity of data with low latency [12,13]. FL overcomes the disadvantages of centralized paradigms and outperforms traditional ML techniques to maintain data privacy while sharing information with other systems. FL models use an exceptional strategy that allows them to share a trained ML model with multiple devices, and the trained ML model will help these devices learn from the environment using the available computational resources with its superior features and operational concepts.

IDS with FL is a combination that leverages the power of collaborative machine learning while preserving data privacy. In the context of IDS, FL can be used to improve the accuracy and effectiveness of intrusion detection by combining data from multiple sources while maintaining privacy.

12.2.4 Related works

Within the scope of this work, the primary focus has been on exploring and analyzing current literature and research on implementing FL techniques for improving intrusion detection systems. Subsequent sections will investigate anomaly-based IDSs, focusing on those tailored for use in IoT settings.

Placing the IDS is an important design decision in IoT systems compared to computer networks. These included centralized IDSs, central IDSs, and mixed IDSs [14–17]. Recently, with the advancement of FL [18], FL-based IDSs [19–21] have become increasingly popular. FL allows distributed training to achieve better generalization performance by exploiting diverse training data sets from many IoT devices. The Cloud server then aggregates these local models into a global model. Since then, several articles (for example, [20,21]) have investigated the use of the FL framework to enable distributed edge devices to learn an anomaly detection model using only the local data at each edge device. In anomaly-based IDS, machine learning mechanisms have been widely explored in the literature.

Other studies have been introduced to address the current problems of the IoT. The authors of [22] proposed a detection system based on SDN using blockchain-based FL on a three-layer architecture. In contrast, the authors of [23] proposed an AI-enabled architecture to secure the physical layer of the IoT using modules dedicated to one use case. The authors of [24] present DeepFed, a deep FL detection system tailored to cyber–physical systems (CPSs). It maintains a centralized architecture. In contrast, FLchain [25] exploits the distributed aspects of the blockchain to remove the single point of failure.

12.3 DESIGN OF THE PROPOSED SCHEME

12.3.1 Objectives of a federated learning module for Cloud computing security

FL can be applied in many fields. Google has presented a TensorFlow FL version in their input method where many mobile phones keep their user data locally but jointly learn a machine learning model to predict the next typing word, FL offers several advantages, as discussed in Table 12.3:

These distinctive advantages of FL make it one of the most widely used techniques in various IoT applications. Although FL has been extensively researched in previous literature, a lack of dedicated research signifies its application for IoT security. Therefore, this work highlights the application of IDS with FL to improve IoT security.

Table 12.3 FL advantages for Cloud computing security

Advantages	*Description*
Privacy enhancement	FL does not require raw data to train the model. Therefore, the likelihood of leaking confidential information to a third party is very low, this privacy enhancement mechanism makes FL a suitable candidate for developing a robust security approach.
Low-latency communication	By eliminating the need to transmit the IoT data to the server, the use of FL helps to reduce the communication latency caused by data offloading. FL also reduces load and minimizes the use of network resources.
Quality of learning improved	FL can improve the convergence rate and learning quality to achieve the desired accuracy [26], which is not possible with traditional ML techniques. In addition, the distributed learning capability of FL improves the scalability of IoT networks.

12.3.2 Model architecture

This chapter introduces a collaborative Cloud–SDN framework to preserve IoT privacy, leveraging FL for intrusion detection. Recent advancements have facilitated the sharing of peers' insights while safeguarding privacy.

In the proposed model, decentralization of machine learning tasks occurs through the relocation of the training of IDS models to the devices, eliminating the need for engagement with a centralized entity to which the devices transmit their data. As shown in Figure 12.3, only model parameters are communicated to a centralized entity responsible for disseminating global improvements to the model, making the latter better for all participants, as the learning process takes place on the device and data is never shared externally.

The adaptability of the learning process is evident in its hybrid nature, offering the flexibility to employ either the IoT device or the IoT gateway as a local training model. This decision is contingent upon various factors, with processing capabilities pivotal in determining the optimal choice. Power efficiency, network bandwidth, and real-time requirements further shape the selection process, highlighting the intricate balance between resource utilization and performance optimization in IoT-based learning environments.

12.3.3 Exploring the main contributions of the proposed model

The proposed scheme aims to achieve detection accuracy approaching or exceeding that of central IDSs while maintaining the privacy and security of the collected data and reducing the communication effort. This chapters's contributions are summarized below:

- **Federated learning for intrusion detection:** The chapter utilizes the FL approach to perform intrusion detection in IoT networks. FL enables the collaborative training of a machine learning model across multiple IoT devices without sharing raw data. This approach helps preserve data privacy while improving the accuracy of intrusion detection.
- **Privacy-preserving IoT architecture:** The chapter introduces a collaborative Cloud–SDN architecture designed for IoT environments. This architecture considers and addresses the privacy concerns associated with IoT data by implementing privacy-preserving techniques.
- **Enhanced security:** Our collaborative architecture enhances the security of IoT networks by integrating SDN capabilities. SDN enables centralized network control, facilitates dynamic traffic monitoring, and responds rapidly to potential intrusions. This helps mitigate security threats in real time and enhances overall network security.

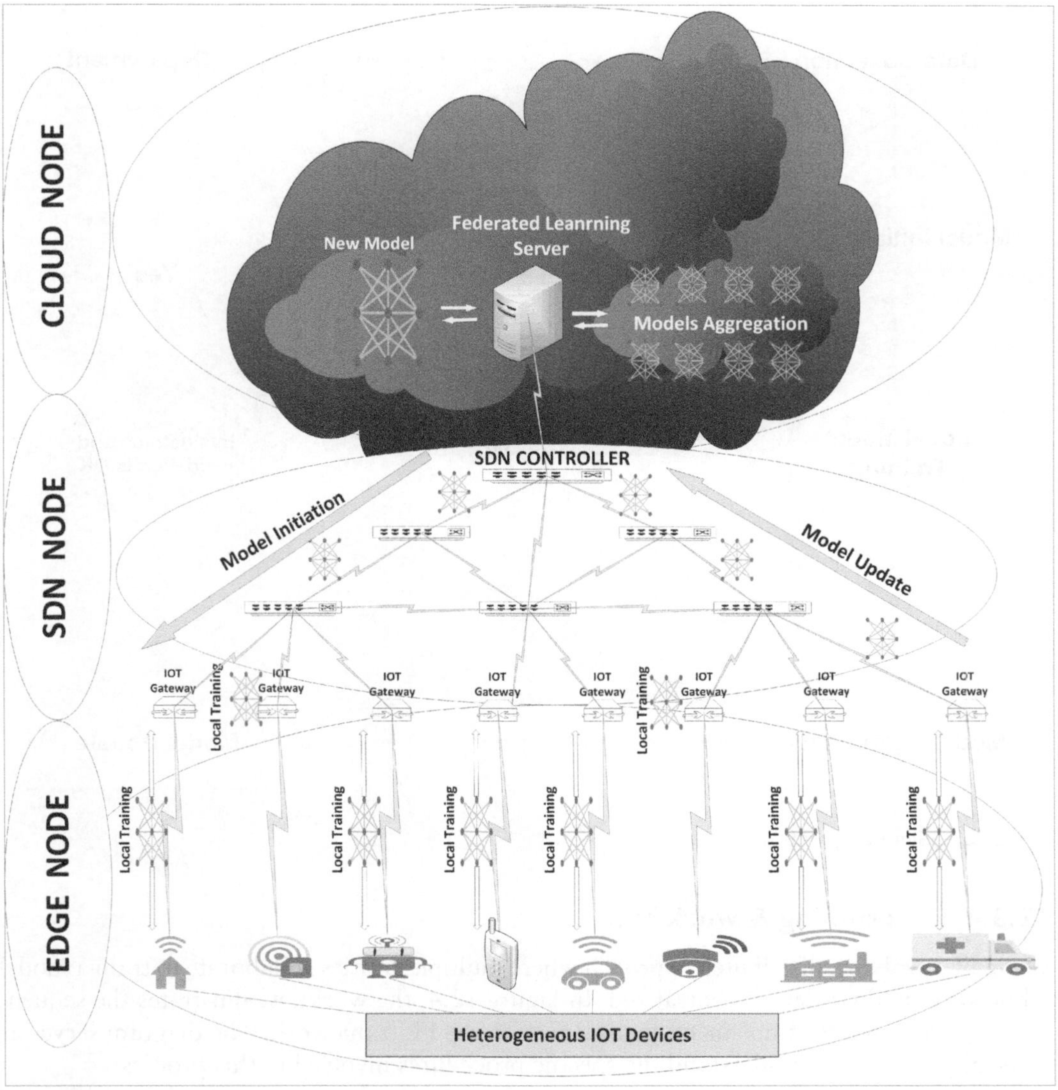

Figure 12.3 Proposed model.

- **Collaborative intrusion detection:** The proposed architecture fosters collaboration among IoT devices by allowing them to participate in the training process collectively. Instead of relying solely on a centralized entity, each IoT device contributes its local knowledge to the global intrusion detection model.

Overall, the contributions of this work lie in developing a collaborative Cloud–SDN architecture that leverages FL for privacy-preserving intrusion detection in IoT environments. By addressing privacy concerns and fostering collaboration among IoT devices, this architecture offers an effective solution for enhancing the security of IoT networks while preserving data privacy.

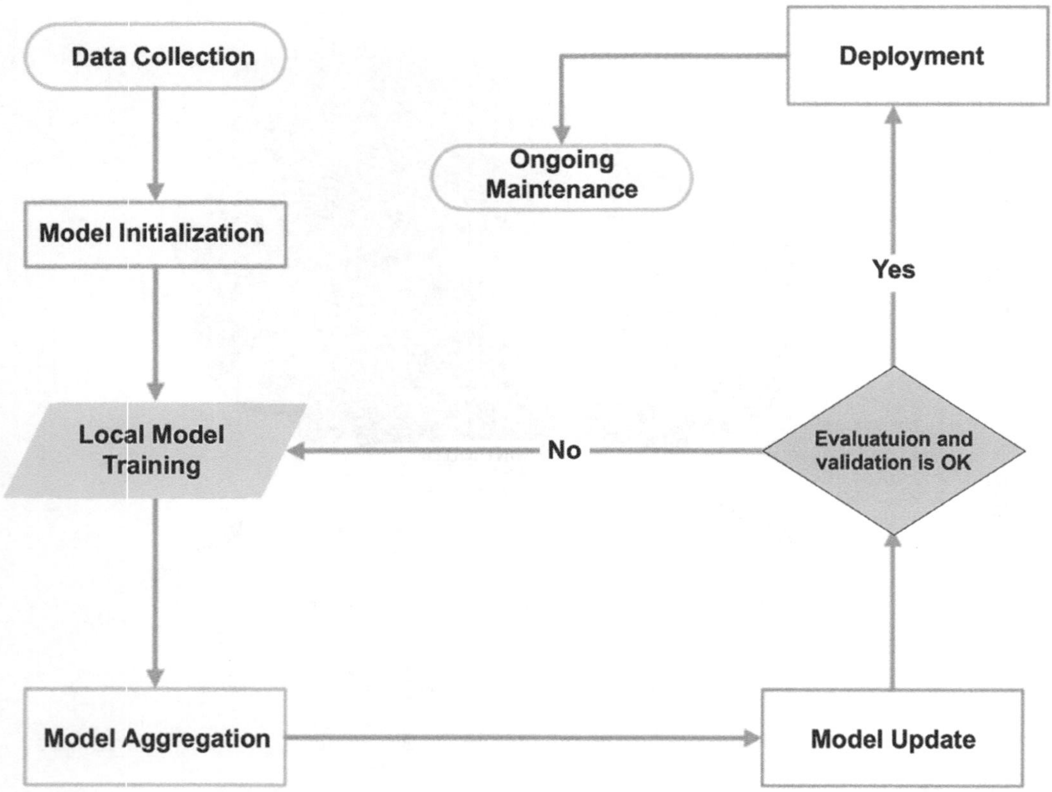

Figure 12.4 General workflow.

12.3.4 Functioning & workflow

Using FL involves a distributed approach where multiple entities collaborate to train a model while keeping their data decentralized. In Figure 12.4, the workflow illustrates the sequential steps required for implementing an IDS with the FL framework. The diagram serves as a visual guide to help understand the specific procedures involved in this process:

Step 0- Define the model's goal: This model's primary objective is to enhance the privacy and security of IoT devices and their data by implementing a collaborative approach that combines Cloud computing, SDN, and FL for intrusion detection.

Step 1- Data collection: This data could include network traffic logs, system logs, or any relevant information for intrusion detection.

Step 2- Model initialization: An initial machine learning model is created and shared with each participating entity. This model serves as the starting point for training.

Step 3- Local model training: Each entity independently trains the shared model using their local data. They operate their computational resources and retain control over their data.

Step 4- Model aggregation: The locally trained models from each entity are sent back to the central coordinator. To create a global model, the coordinator aggregates the models' using techniques like model averaging or weighted averaging.

Step 5- Model update: The global model is shared with each participating entity. They update their local models with the newly aggregated global model.

Step 6- Iterative training: Steps 4–6 are repeated for multiple iterations, allowing the model to learn from the combined knowledge of all participating entities while preserving data privacy.

Step 7- Evaluation and validation: Periodically, the performance of the global model is evaluated using a validation dataset or metrics defined for intrusion detection. This step helps identify the effectiveness of the IDS and potential areas for improvement.

Step 8- Deployment: Once the desired performance is achieved, the trained global model can be deployed for real-time intrusion detection. Each entity applies the deployed model to its local environment, making predictions based on its own data.

Step 9- Ongoing maintenance: The IDS system continues to operate, and the FL process can be periodically repeated to incorporate new data, update the model, and adapt to evolving intrusion patterns.

In Figure 12.5, an effort has been made to construct a workflow that depicts the sequential steps in the proposed model's initial iteration. After collecting data and initializing the FL model, each IoT device independently trains the initial machine learning model using its locally stored and encrypted data. After local training, the IoT devices send their updated models to the Cloud server. The Cloud server aggregates these models while preserving privacy. The updated global model is then sent back to the IoT devices for further local training. This process is repeated iteratively to refine the global model. Once the global model has achieved satisfactory performance, each device uses the deployed model to analyze network

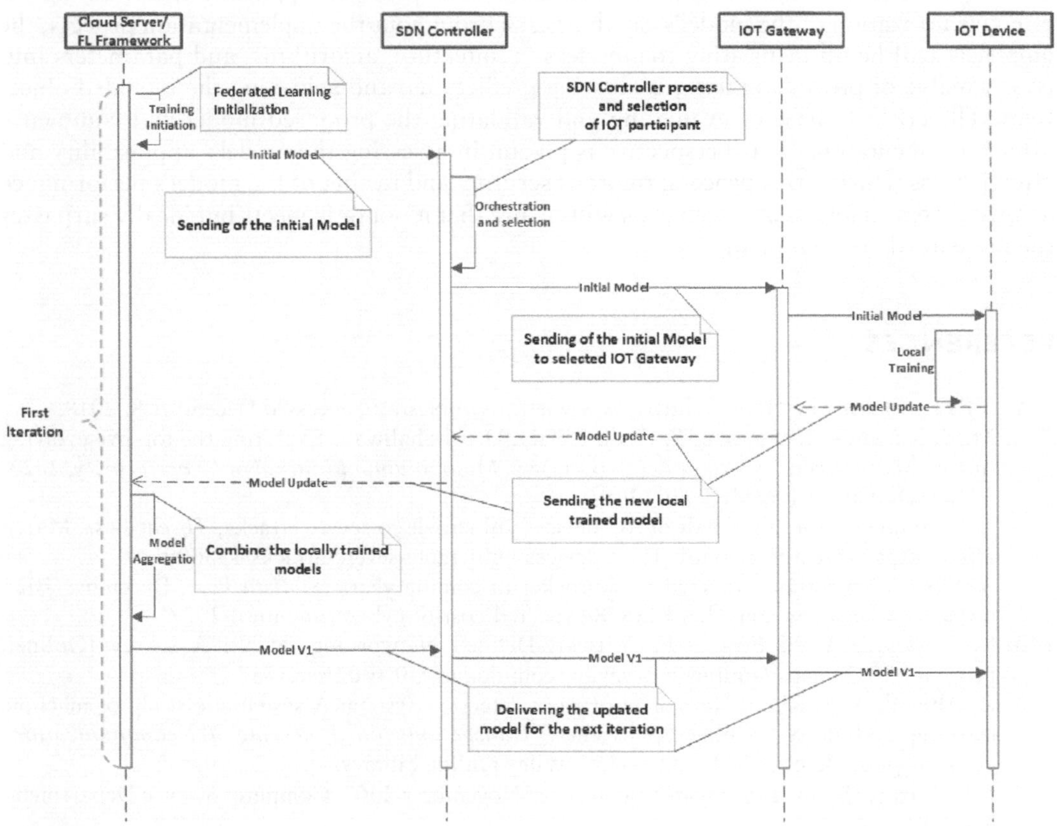

Figure 12.5 First iteration workflow.

traffic and detect potential intrusions or anomalies. Based on the detected intrusions or anomalies, appropriate response actions are taken, such as notifying network administrators, blocking malicious traffic, or triggering automated remediation processes.

12.4 CONCLUSION AND FUTURE WORK

Enhancing security with the explosion of IoT is a challenging mission for the whole computer information industry. This paper explores a novel approach that contributes to a common objective. Emphasis is placed on providing an overview of each technology incorporated in the proposed model and key considerations for integrating these technologies. The proposed architecture presents a promising approach for privacy-preserving intrusion detection in IoT environments. By combining the strengths of Cloud computing, SDN, and FL, this architecture addresses the challenges of intrusion detection while safeguarding the privacy of IoT devices and their users. However, it is important to note that challenges and considerations may be associated with implementing this architecture. These may include addressing communication and synchronization issues between IoT devices and the Cloud, ensuring the security of model updates during aggregation, and addressing potential performance overhead due to the distributed nature of FL.

In the upcoming phase of work, the primary focus will be on the practical execution of the proposed model. This phase involves transforming the conceptual model, which has been meticulously designed and fine-tuned, into a fully functional software system or framework. The objective is to bridge the gap between theory and practical application, facilitating the effective utilization of the model's capabilities. Throughout the implementation process, the emphasis will be on translating the model's architecture, algorithms, and parameters into code capable of processing data, generating predictions, and achieving the intended objectives. The critical phase of evaluating and validating the proposed model will commence after implementation. This perspective is pivotal in assessing the model's applicability and effectiveness. During this process, rigorous scrutiny and testing of the model's performance against diverse datasets and scenarios will ensure that it not only meets but ideally surpasses the predefined criteria for success.

REFERENCES

1. "IETF - Internet of Things"; https://www.ietf.org/topics/iot/, accessed December 8, 2018.
2. M.M. Mijwil, O.J. Unogwu, Y. Filali, I. Bala, H. Al-Shahwan. Exploring the top five evolving threats in cybersecurity: An in-depth overview. *Mesopotamian Journal of Cybersecurity*, 2023, 57–63. DOI:10.58496/MJCS/2023/010
3. L. O'Donnell. More than half of IoT devices vulnerable to severe attacks, *Threat Post*, March 2020. https://threatpost.com/half-iot-devices-vulnerable-severe-attacks/153609/
4. A. Fleck. Cybercrime expected to skyrocket in coming years, *Statista Post*, December 2022 https://www.statista.com/chart/28878/expected-cost-of-cybercrime-until-2027/
5. K. Benzekki, A. El Fergougui. Software-Defined Networking (SDN): A survey [Online], February 2017. https://onlinelibrary.wiley.com/doi/abs/10.1002/sec.1737
6. Z. Ahmad, A. S. Khan. Network intrusion detection system: A systematic study of machine learning and deep learning approaches. *Transactions on Emerging Telecommunications Technologies*, October 2020, 32, e4150. Wiley Online Library.
7. J. McCarthy. *What Is Artificial Intelligence?* Novembrer 2007. Computer Science Department, Stanford University.

8. I.H. Sarker. Machine learning: Algorithms, real-world applications and research directions? *SN Computer Science,* March 2021, 2, 160.
9. J. Han, J. Pei, M. Kamber. *Data Mining: Concepts and Techniques.* Amsterdam: Elsevier, 2011.
10. M. Mohammed, M.B. Khan, B.E. Bashier Mohammed. *Machine Learning: Algorithms and Applications.* CRC Press, 2016.
11. M. Aledhari, R. Razzak, R.M. Parizi, F. Saeed. Federated learning: A survey on enabling technologies, protocols, and applications. *IEEE Access*, 2020, 8, 140699–140725.
12. Z. Chen, N. Lv, P. Liu, Y. Fang, K. Chen, W. Pan. Intrusion detection for wireless edge networks based on federated learning. *IEEE Access*, 2020, 8, 217463–217472. https://doi.org/10.1109/ACCESS.2020.3041793
13. W.Y.B. Lim, N.C. Luong, D.T. Hoang, Y. Jiao, Y.-C. Liang, Q. Yang, D. Niyato, C. Miao. Federated learning in mobile edge networks: A comprehensive survey. *IEEE Communications Surveys and Tutorials*, 2020, 22(3), 2031–2063. https://doi.org/10.1109/COMST.2020.2986024
14. D. Oh, D. Kim, W.W. Ro. A malicious pattern detection engine for embedded security systems in the internet of things. *Sensors*, 2014, 14(12), 24188–24211.
15. A. Ferdowsi, W. Saad. Generative adversarial networks for distributed intrusion detection in the internet of things. In *2019 IEEE Global Communications Conference (GLOBECOM),* Waikoloa, HI, USA, pp. 1–6, IEEE, 2019.
16. S. Raza, L. Wallgren, T. Voigt. Svelte: Real-time intrusion detection in the internet of things. *Ad Hoc Networks*, 2013, 11(8), 2661–2674.
17. J.P. Amaral, L.M. Oliveira, J.J. Rodrigues, G. Han, L. Shu. Policy and network-based intrusion detection system for ipv6-enabled wireless sensor networks. In *2014 IEEE International Conference on Communications (ICC),* pp. 1796–1801, IEEE, Sydney, NSW, Australia, 2014.
18. B. McMahan, E. Moore, D. Ramage, S. Hampson, B. A. Y Arcas. Communication-efficient learning of deep networks from decentralized data. In *Artificial Intelligence and Statistics (AISTATS 17)*, Florida, USA, pp. 1273–1282, 2017.
19. T. D. Nguyen, S. Marchal, M. Miettinen, H. Fereidooni, N. Asokan and A.-R. Sadeghi. DIOT: A federated self-learning anomaly detection system for iot. In *2019 IEEE 39th International Conference on Distributed Computing Systems (ICDCS),* Dallas, Texas, USA, pp. 756–767, IEEE, 2019.
20. Y. Liu, S. Garg, J. Nie, Y. Zhang, Z. Xiong, J. Kang, M.S. Hossain. Deep anomaly detection for time-series data in industrial iot: A communication-efficient on-device federated learning approach. *IEEE Internet of Things Journal*, 2020, 8(8), pp. 6348–6358.
21. V. Mothukuri, P. Khare, R.M. Parizi, S. Pouriyeh, A. Dehghantanha, G. Srivastava. Federated learning-based anomaly detection for iot security attacks. *IEEE Internet of Things Journal*, 2021. https://doi.org/10.1109/JIOT.2021.3077803
22. S. Rathore, B. Wook Kwon, J.H. Park. BlockSecIoTNet: Blockchain-based decentralized security architecture for IoT network. *Journal of Network and Computer Applications*, 2019, 143, 167–177.
23. H. HaddadPajouh, R. Khayami, A. Dehghantanha, K.-K.R. Choo, R.M. Parizi. AI4SAFE-IoT: An AI-powered secure architecture for edge layer of Internet of things. *Neural Computing and Applications*, 2020, 32, 16119–16133.
24. B. Li, Y. Wu, J. Song, R. Lu, T. Li, L. Zhao. DeepFed: Federated deep learning for intrusion detection in industrial cyber-physical systems. *IEEE Transactions on Industrial Informatics*, 2020, 17, 5615–5624.
25. U. Majeed, C.S. Hong. FLchain: Federated learning via MEC-enabled blockchain network. *2019 20th Asia-Pacific Network Operations and Management Symposium (APNOMS),* IEEE, 2019.
26. D.C. Nguyen, M. Ding, Q.-V. Pham, P.N. Pathirana, L.B. Le, A. Seneviratne, J. Li, D. Niyato, H.V. Poor, Federated learning meets blockchain in edge computing: Opportunities and challenges. *IEEE Internet Things Journal*, 2021, 8(16), 12806–12825, https://doi.org/10.1109/JIOT.2021.3072611.

Chapter 13

An adaptive cybersecurity strategy based on game theory to manage emerging threats in the SDN infrastructure

Jihad Kilani, Youssef Baddi, Faycal Bensalah, and Yousra Fadili

13.1 INTRODUCTION

In an ever-evolving digital landscape where information technology is ubiquitous, the security of computer networks has emerged as a major concern. Cyberattacks are becoming increasingly sophisticated and targeted, highlighting the limitations of traditional protection methods. Faced with this reality, this chapter presents an innovative adaptive cybersecurity approach, using game theory [1] as a foundation to proactively detect and respond to emerging threats.

The key to this approach resides in modeling the interaction between attacker and defender in a strategy game. Each party weighs up the potential benefits and risks of its actions, anticipating the actions of the other. This dynamic perspective makes it possible to consider confrontation scenarios as adaptive processes in which players continually adjust their tactics in response to opposing actions [2,3].

The modeling of strategic decisions is based on fundamental game-theoretic concepts such as action, strategy, and outcome. Attackers and defenders develop their tactics by considering their opponents' potential actions. Each side's decisions create an overall outcome that guides the development of the confrontation, creating a framework within which strategies evolve in response to observed actions and responses.

One of the main strengths of this approach is its ability to encourage adaptive strategies. Aware of the volatility of attacks, defenders constantly adjust their tactics to cope with evolving threats. This adaptability enables them to react rapidly to attacks and design effective countermeasures. In addition, this model favors collaboration between security players. This partnership enhances a common comprehension of offensive tactics and best defense practices, thus improving threat preparation and response.

The model also integrates the specific context of each organization. By considering factors such as acceptable risk levels, available resources, and company specifics, security policies can be tailored to the unique needs of each environment.

This adaptive, game-theoretic cybersecurity model offers a new perspective on dealing with emerging threats in an ever-changing digital world. Focusing on strategic decisions, adaptability, and collaboration paves the way for more robust protection methods better adapted to the complex challenges of modern IT security. Its ability to predict competitors' movements and constantly adapt can help strengthen network resilience and create a more secure digital environment for companies facing challenges.

Software-defined network (SDN) architecture [4] is a fundamental component of our adaptive approach to network threat management. With their static structure and decentralized control, traditional networks often struggle to react rapidly to changing threats. In contrast, SDN offers flexibility and centralized control, enabling the network to be

 DOI: 10.1201/9781032714806-15

flexibly reconfigured to meet changing needs. Using the SDN controller, we can orchestrate real-time security actions, dynamically modifying flow rules, security policies, and traffic paths. The ability to adapt network topology and direct traffic according to security needs greatly enhances our adaptive approach to threat management. SDN forms the foundation for strategic modeling, translating decisions into action, and the coordinated application of security policies, creating a proactive and reactive security ecosystem [5].

13.2 RELATED WORKS

Research in the field of cybersecurity and adaptive threat management has produced a variety of contributions. This section reviews several related works that have addressed similar issues and highlights distinct aspects of our approach to innovation.

- **"Game theory for adaptive defensive cyber deception"** [6]:
  This chapter explores the application of game theory to adaptive defensive network fraud. Our approach differs in its integration of game theory but focuses specifically on proactive threat management, enabling attackers to predict actions and make preemptive decisions to counter threats.
- **"Adaptive MTD security using Markov game modeling"** [7]:
  The research team proposed a Markov game model for adaptive mobile target defense (MTD) security. Similarly, our proposal proposes a more holistic approach using game theory to guide the entire threat management process, including detection, decision, and action, and considering more diverse situations.
- **"Adaptive security architecture defense platform based on software defined security"** [8]:
  This work presents a defense based on software-defined security. Our approach adapts to this context but introduces game theory to optimize threat response and defense, resulting in more proactive adaptive management.
- **"ADVICE: Towards adaptive scheduling for data collection and DDoS detection in SDN"** [9]:
  This chapter addresses DDoS detection and data collection scheduling limitations in SDN. Our proposal covers these aspects while using game theory to guide decisions, thus offering an optimized adaptive response to threats.
- **"Adaptive and intelligent data collection and analytics for securing critical financial infrastructure"** [10]:
  This chapter focuses on intelligent data collection to secure critical financial infrastructures. Our approach extends beyond the integration of game theory to proactively manage threats, enabling better prediction of attacks and adaptation of strategies in consequence.

Compared to these related works, our proposal is distinguished by using game theory as a foundation for adaptive cyber-threat management. This proactive approach makes it possible to predict attackers' actions, adjust strategies flexibly, and proactively build resilience in the face of various threats.

13.3 ARCHITECTURE

Adaptive network security implementations based on game theory can use SDN architectures to offer the required flexibility and responsiveness [11]. Here is the recommended architecture describing in detail its main classes and modules.

13.3.1 SDN infrastructure layer

This layer is based on the SDN network infrastructure, comprising switches and controllers. The SDN controller is responsible for overall network management and security policy implementation. They communicate with switches to define data paths and flow rules [12].

13.3.2 Data collection class

This layer collects network monitoring data, including traffic, security events, and unusual behavior. Distributed sensors and probes collect this information in real time for subsequent analysis.

13.3.3 Behavioral analysis

This module uses the data collected to analyze network behavior. It identifies normal traffic patterns and detects potential anomalies. Using statistical and correlation techniques, it determines the actions of attackers and defenders.

13.3.4 Decision modeling module

This module applies the principles of game theory to model the strategic decisions of attackers and defenders. It considers the possible actions of each and calculates the optimal strategies based on the results obtained. These strategies define the actions to be taken to maximize the respective benefits.

13.3.5 Policy management module

This module takes calculated strategic decisions and translates them into specific security policies. These policies are then distributed to the SDN controller for application to the appropriate switches. It can also automatically adjust policies per evolving threats and new data.

13.3.6 Automatic response module

Depending on the policies defined, this module can trigger automatic response actions. This can include isolating compromised devices, blocking suspicious traffic, or reconfiguring data paths to isolate threats.

13.3.7 Contextual smart layer

This layer can integrate contextual information such as user profiles, critical applications, and specific business needs. This information enriches decision-making by adapting strategy according to requirements and priorities.

13.3.8 Class collaboration and information sharing

This layer encourages collaboration between security entities by sharing information on detected attacks, effective tactics, and identified vulnerabilities. This reinforces a common understanding of threats and fosters collaborative responses (Figure 13.1).

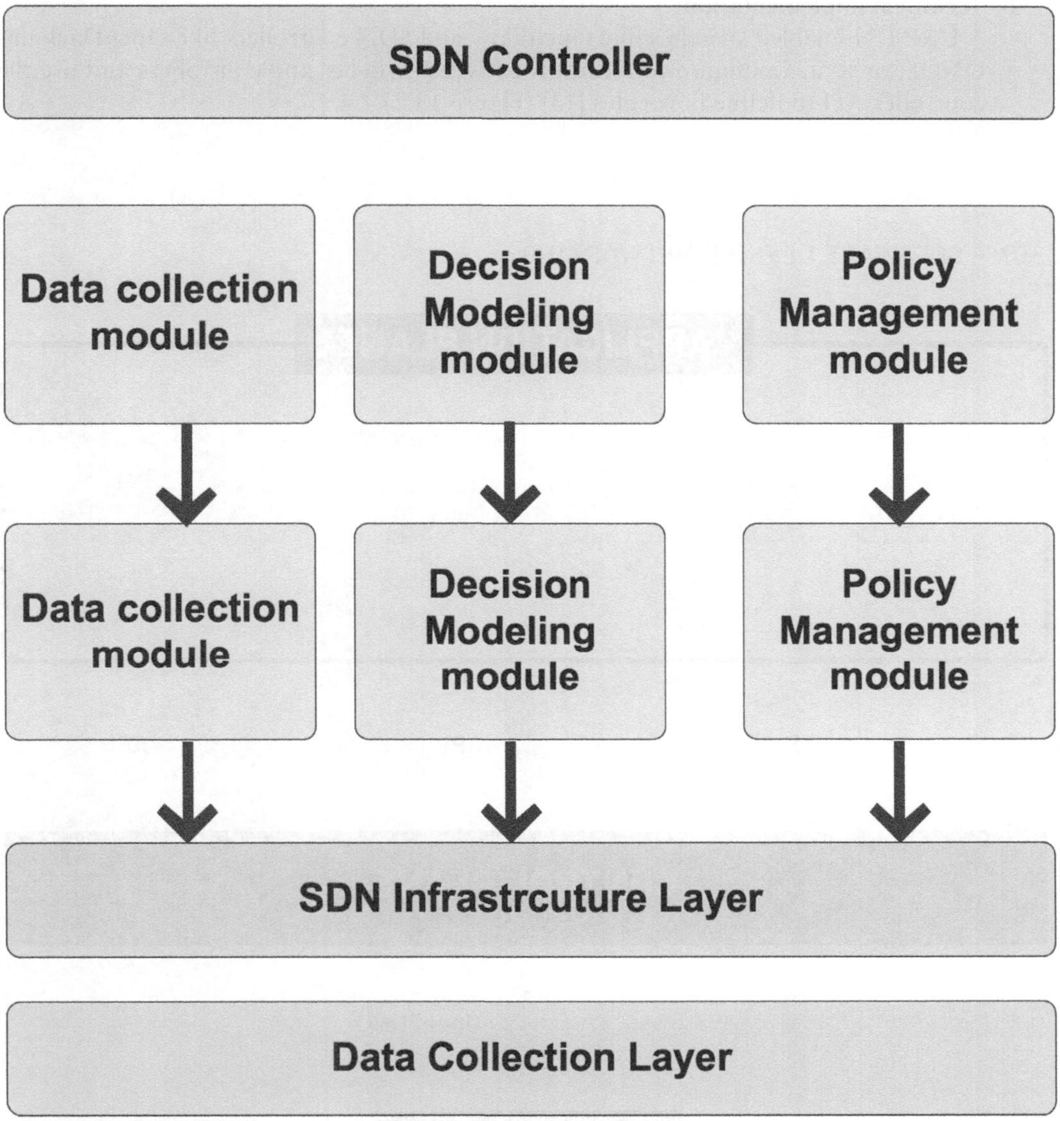

Figure 13.1 Proposed SDN architecture.

13.4 COMPONENTS AND TECHNICAL INSTRUCTIONS

13.4.1 SDN infrastructure layer

This layer covers the entire SDN infrastructure, including SDN switches and controllers. The switch manages network traffic, while the controller provides centralized management and network programmability [12].

a. **Components**
   SDN switch, SDN controller.

b. Technical implementation

Use SDN-enabled switches like OpenFlow and SDN controllers like OpenDaylight, ONOS, or Ryu. Configure connections between switches and controllers and use the controller API to define flow rules [13] (Figure 13.2).

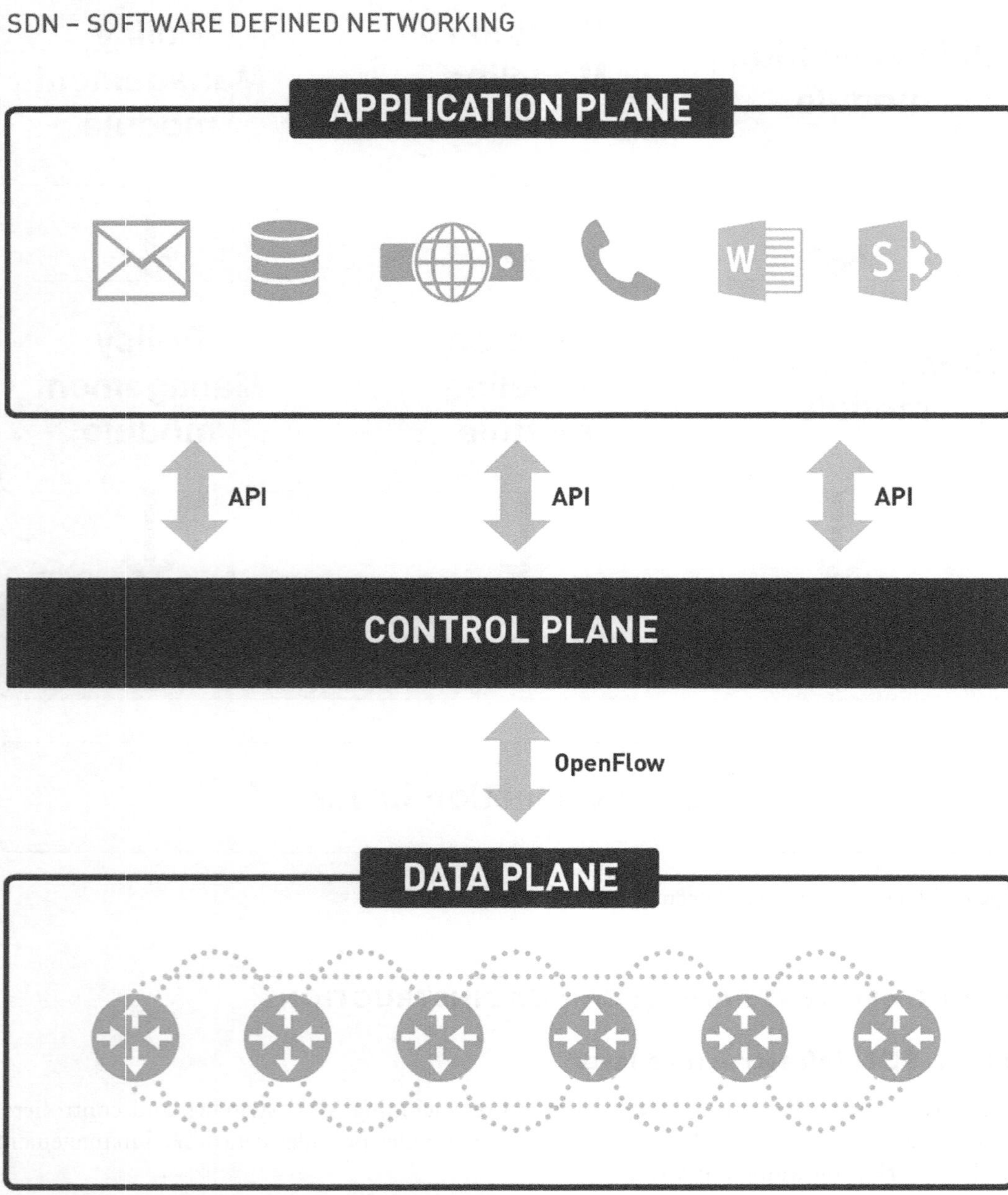

Figure 13.2 SDN infrastructure layer.

13.4.2 Data collection module

The data collection module [14] performs a crucial role in SDN networks by enabling the centralization and aggregation of data from various network elements. This data can then be used for a variety of tasks, such as:

- **Network monitoring:** the data collection module can gather information on network performance, such as latency, throughput, and packet loss. This information can be used to identify performance problems and optimize the network.
- **Anomaly detection:** The data collection module can detect anomalies in network traffic, which can be useful for identifying attacks and intrusions.
- **Traffic analysis:** The data collection module can analyze network traffic and identify trends and patterns. This information can be used to optimize resource allocation and improve network security.
- **Debugging:** The data collection module can collect information useful for debugging network problems.

13.4.2.1 Data collection module operation

The data collection module generally operates in several stages:

- Data collection is collected from various network elements, such as switches, routers, and access points.
- **Data aggregation:** Collected data is then aggregated and stored in a centralized format.
- **Data analysis:** Aggregated data can then be analyzed to identify trends, patterns, and anomalies.
- **Data presentation:** The results of data analysis can be presented in reports, graphs, and other visual formats.

13.4.2.2 Benefits of the data collection module

The data collection module offers several advantages, including:

- **Improved network visibility:** The data collection module provides better network visibility, which can help identify and resolve problems more quickly.
- **Improved network security:** The data collection module can detect attacks and intrusions more quickly.
- **Network optimization:** The data collection module can optimize the network and improve performance.
- **Easier debugging:** The data collection module can be used to collect information useful for debugging network problems.
- The data collection module is an essential component of SDN networks. It collects, aggregates, and analyzes network data, which can be used for various tasks, such as network monitoring, anomaly detection, traffic analysis, and debugging.

13.4.2.3 Components

1. **Probes:**

   Similar to sensors, they can be more specialized and collect more complex or remote data. Examples: network probe, process activity probe, safety probe.

2. **Control points:**
   Represent specific points in a system or process where data is collected, which can be physical (e.g., a test point in a pipeline) or virtual (e.g., an inspection point in software).
3. **Sensors:**
   Sensors refer to devices or software that collect data on the network and transmit it to the central SDN controller. This data can include:
   - Network traffic passing through various points on the network
   - Status of network switches and routers
   - Consumption of network resources (bandwidth, memory, etc.)

   This information is crucial for the SDN controller to:
   - Visualize network status in real time
   - Detect anomalies and performance problems
   - Make dynamic decisions to optimize network traffic and resources

13.4.2.4 Technical implementation

Deploy specialized software or hardware agents as sensors and transducers to collect traffic and event information. Control points can be configured to switch to replicate traffic to specific locations for testing purposes. OpenFlow sensors are installed on network switches and routers.

Software agents deployed on servers or containers to collect data on running applications and services.

13.4.3 Behavioral analysis

This module analyzes collected data to detect anomalies and suspicious activity [15]. The behavioral analysis module for SDN networks uses artificial intelligence and machine learning to analyze network traffic and identify abnormal behavior. It is a powerful tool for detecting threats and attacks on the network and optimizing network performance and security.

13.4.3.1 Operation

The behavioral analysis module works in several stages:

1. **Data collection:**
   The module collects network traffic data from various sources, such as OpenFlow probes, event logs, and monitoring tools.
2. **Data preprocessing:**
   The collected data is then preprocessed to clean and normalize formats and remove redundant or irrelevant information.
3. **Behavioral analysis:**
   Machine learning and artificial intelligence techniques analyze network traffic and identify behavior patterns. This analysis enables detecting abnormal behavior that could indicate a threat or anomaly.
4. **Threat detection:**
   The module uses identified behavior patterns to detect threats and attacks on the network. These may be denial of service (DoS) attacks, intrusions, malware, or other types of malicious activity.

5. **Alert generation:**
   When the module detects a threat, it generates an alert with detailed information on the threat, such as source, destination, type of attack, and severity level.
6. **Incident response:**
   Alerts generated by the module can trigger automatic incident response actions, such as blocking suspicious traffic, quarantining infected devices, or notifying network administrators.

13.4.3.2 Components

Anomaly detection algorithms:

These algorithms analyze data for significant departure points from an established normal behavior pattern. There are many different types, each with its strengths and weaknesses.

Signal processing techniques:

These techniques extract valuable features and information from signals, which anomaly detection algorithms can exploit.

Combining anomaly detection algorithms and signal processing techniques results in a more robust and accurate anomaly identification system. Signal processing prepares the data for analysis, while anomaly detection algorithms exploit the extracted features to identify elements that deviate from normality.

13.4.3.3 Technical implementation

Use machine learning algorithms such as neural networks, decision trees, or methods based on hidden Markov models to detect unusual behavior. Preprocess data using signal processing techniques to normalize and reduce noise.

The anomaly detection algorithms and signal processing techniques selected generate anomaly scores or flags.

These are sent to the SDN controller, which can trigger actions, block suspicious traffic, quarantine infected devices, or reroute traffic via other routes.

It can generate alerts and notify administrators of further investigation and possible manual intervention.

13.4.4 Decision modeling module

This module applies game theory to model the interaction between attackers and defenders is a network traffic management and optimization tool to make strategic decisions [16].

13.4.4.1 Operation

The module gathers information on network status, including network topology (switch and link layout), link load status (traffic level), and current routing policies, as well as the preferences and objectives of different network players.

The module builds a mathematical model of the network as a game (game modeling). In this model, players can be switches, SDN controllers, or other network entities. Each player's strategies represent the actions they can take, such as routing traffic along specific paths. Payouts represent each player's gains or losses based on all players' actions. The module uses game-theoretic algorithms to find equilibria in the game. An equilibrium is a set of strategies where no player is interested in unilaterally changing his strategy. The module translates the calculated equilibrium strategies into concrete actions for the SDN network. These actions can include:

- Modifying routing policies
- Allocation of network resources
- Traffic prioritization

In summary, the game-theoretic decision modeling module is a promising tool for optimizing and managing SDN networks. It enables strategic decisions, considering the interactions between the various network players and drawing on game-theoretic concepts.

13.4.4.2 Components

In game theory, two key concepts are essential for understanding strategic decision-making: **reward matrix** and **game resolution algorithms.**

A reward matrix, also known as a payoff matrix, is a table that represents the results of a game for all the players involved. It shows the payoffs (wins or losses) each player receives for each combination of strategies chosen by all players.

Game-solving algorithms take the information from the reward matrix to find optimal strategies for each player. These algorithms aim to find a Nash equilibrium, i.e., a set of strategies in which no player has any interest in changing strategy, given the strategies chosen by the other players. Each player "does his best" given what the others are doing.

13.4.4.3 Technical implementation

Model the confrontation scenarios as a payoff matrix, where each tile represents the payoff associated with a given offensive and defensive action. Use game-solving algorithms, such as the monomorphic algorithm, to calculate optimal strategies for both sides based on the results obtained.

The technical implementation of the reward matrix and game-solving algorithm may vary according to the application and chosen algorithm.

13.4.5 Policy management module

This module translates strategic decisions into security policies and distributes them to the application's SDN controller [17]. It enables network behavior policies to be defined, deployed, and managed centrally and programmably. These policies dictate how the network should handle traffic according to various criteria, such as source, destination, traffic type, priority, etc.

13.4.5.1 Operation of the policy management module

Network administrators define policies using a high-level declarative language. This language makes it possible to specify the conditions to be met by traffic and the actions to be taken accordingly. Graphical interfaces or command-line tools can be used to facilitate policy definition.

The policy management module translates policies defined by administrators into a format that the SDN controller can understand. This translation may involve compiling policies into specific OpenFlow rules that can be installed on network switches.

The policy management module deploys the translated rules on the appropriate SDN switches. This can be done centrally by the SDN controller or distributed by the switches.

The policy management module enables you to manage the policy lifecycle, including Enable or disable policies as needed, modify existing policies to meet changing network or application requirements, track policy activity, and identify potential problems.

13.4.5.2 Components

Privacy policy: SDN offers many advantages for network management and security but raises privacy issues. SDN's centralized control plane concentrates important information and control over the network, making it crucial to implement a robust privacy policy.

SDN controller API: The SDN controller API is a set of functions and protocols that applications can use to interact with and program the SDN controller. It provides a standardized way of accessing and manipulating the network via the controller.

13.4.5.3 Technical implementation

Use the SDN controller API to define and distribute flow rules based on the security policies generated by the decision modeling module. Policies can include rules for blocking, redirecting, or quarantining traffic. Implementing a privacy policy within an SDN controller API involves taking two key aspects into account, data collection and use and API design and access.

13.4.6 Automatic response module

An automatic response module in SDN is a software component that continuously monitors the network and automatically takes predefined actions in response to specific events or conditions [18]. This automates incident response, improves network security and efficiency, and reduces the need for manual intervention.

This module triggers actions based on defined security policies.

a. **Components**

   Action response, communicate with the SDN controller.

b. **Technical implementation**

   Configure the SDN controller to support response actions. Once a threat situation is identified, the autoresponder module communicates with the respective controller to trigger the actions defined in the security policy.

13.4.7 Contextual smart layer

A context-aware intelligent layer in an SDN is an emerging concept that aims to improve network management by incorporating context and intelligence into decision-making [19]. It builds on traditional SDN architecture by introducing an additional layer between the SDN controller and the physical network infrastructure.

This layer integrates contextual information to personalize decisions and policies.

a. **Components**

   User profile, application information, compliance policy.

b. **Technical implementation**

   Collect and store contextual data such as user characteristics, application requirements, and security preferences. Use this information to adjust policies generated by the decision modeling module and to customize privacy policies.

13.4.8 Class collaboration and information sharing

SDN relies heavily on class collaboration and information sharing between components to achieve efficient, dynamic network management. This layer encourages collaboration between security players [20].

Based on these principles, SDN enables better network management, increased agility and efficient use of resources, while demanding constant innovation and attention to security.

a. **Components**

   Share information and communicate securely.

b. **Technical implementation**

   Establish secure communication channels to share information about detected attacks, effective defenses, and recent attack tactics. Use encrypted communication protocols and secure platforms to facilitate collaboration.

   Each module and layer work together to create an adaptive, game-theoretic cyber-threat management solution that proactively responds to emerging threats proactively and flexibly. Technical implementations may vary depending on the technology, programming language, and protocol selected by the development team.

13.5 MODELING

```
# Initialization
def init():
    configure_controllers_and_switches()
    set_probes()
# Main loop
def main_loop():
    while True:
        # Recover data
        data =retrieve_monitoring_data()
        # Analyze behavior
        abnormal_behavior =detect_abnormal_behavior(data)
        if behavior_in:
            # Decision model
            gain_matrix =generate_gain_matrix()
            attacker_strategy, defender_strategy =solve_game(hand_matrix)
            # Policy management
            pirate_policies =
translate_strategy_into_policies(pirate_strategy)
            protection_policies=
            translate_strategy_into_policies(protection_strategy)
            distribute_policies(pirate_policies, protection_policies)
        # Automatic response
        threat_identified =surveiller_activites_infra()
        if threat_detected:
            declencher_action_reponse()
        # Inclusion of contextual intelligence
        context =retrieve_context_information()
        adjust_strategies_in_context_function(context)
        # Collaboration and information sharing
        share_threat_information()
        # Monitoring and readjustment
        results_evaluation =evaluate_policy_effectiveness()
        readjust_strategies(results_evaluation)
```

```
# Start algorithm execution
if __name__ == "__main__":
    initialization ()
    main_loop()
```

Algorithm 1.

13.6 EXPLANATION OF THE MODELING ALGORITHM

The algorithm presented is designed to analyze the behavior of a group of individuals and detect abnormal behavior. It is a continuous loop that runs continuously and performs the following tasks.

13.6.1 Initialization

Configuring controllers and switches: This step configures the tools and systems used to collect and analyze data.

Defining probes: Probes are data collection points used to observe the behavior of individuals. They can be physical (e.g., surveillance cameras) or digital (e.g., activity logs).

13.6.2 Main loop

Data recovery: Data are collected from the probes defined during initialization.

Behavior analysis: The algorithm analyses the collected data to detect abnormal behavior. This analysis can use various techniques, such as machine learning or statistical analysis.

Decision: If abnormal behavior is detected, the algorithm proceeds to the next step to determine the best response.

a. **Decision model**

   Payoff matrix: The payoff matrix is a tool for evaluating the various possible response options regarding cost and effectiveness.

   Attacker's strategy and defender's strategy: The algorithm calculates the optimal strategy for the attacker and the optimal strategy for the defender based on the payoff matrix.

   Translating strategies into policies: Strategies are then translated into concrete policies that can counter abnormal behavior.

b. **Automatic response**

   Activity monitoring: Suspicious activity is continuously monitored for potential threats. **Triggering response action:** If a threat is detected, the algorithm triggers an appropriate response action.

c. **Inclusion of contextual intelligence**

   Contextual information retrieval: Contextual information, such as the environment and current events, is collected.

   Context-dependent adjustment of strategies: Detection and response strategies are adjusted according to context for greater effectiveness.

d. **Collaboration and information sharing**

   Sharing threat information: Information on detected threats is shared with other security systems for better coordination.

e. **Monitoring and readjustment**

   Evaluation of results: The effectiveness of applied policies is continuously evaluated.

   Readjustment of policies: Policies are readjusted according to the results of the evaluation for better performance.

The algorithm shown is a simplified example and can be adapted to suit specific contexts and requirements. Including contextual intelligence and collaboration with other security systems can significantly improve the algorithm's effectiveness.

13.7 RESULTS AND DISCUSSION

We have carried out in-depth tests to evaluate the effectiveness of our game-theoretic adaptive network security solution. The tests were conducted in a simulated enterprise network environment, using SDN controllers and various threat scenarios. The results presented below highlight the performance of our solution compared with the conventional security approach.

We measured performance using the following metrics:

- **Threat detection rate (TDR):** The percentage of detected threats concerning the total number of simulated threats.
- **False-positive rate (FPR):** The percentage of alerts generated in error concerning the total number of alerts.
- **Automated response rate (ARR):** The percentage of threats to which an automated response has been deployed.
- **Defense success rate (DSR):** The percentage of threats for which the automated response succeeded in minimizing the impact.

13.7.1 TDR

The first graph shows each threat scenario's threat detection rate (TDM; Figure 13.3). Our game-theoretic adaptive cybersecurity solution showed significantly higher threat detection than conventional methods. This improvement in detection was particularly noticeable in

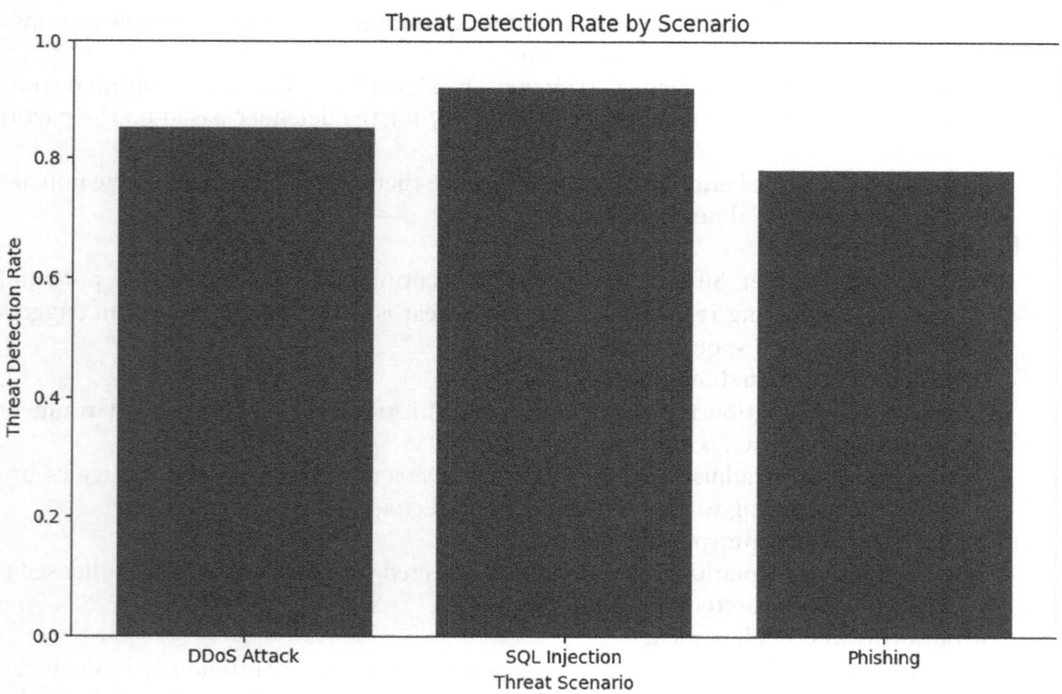

Figure 13.3 Threat detection rate by scenario.

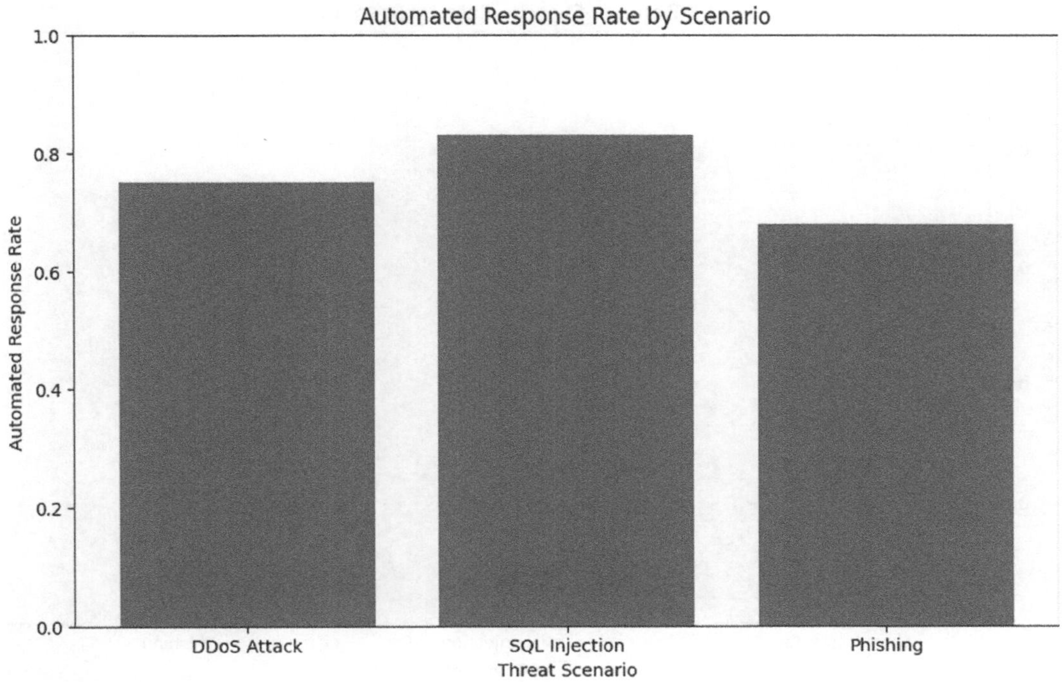

Figure 13.4 Automated response rate by scenario.

the context of SQL injection and phishing attacks, where the TDM was 0.92 and 0.78, respectively. This improvement is due to our solution's ability to predict attackers' actions and dynamically adjust security strategies.

13.7.2 TRA

The second graph shows each threat scenario's ARR (Figure 13.4). It was found that our solution implements the automated response faster than the conventional method. This translates into a higher ARR in all situations, demonstrating that our solution can detect threats quickly and take appropriate defensive measures. The effectiveness of our approach was particularly visible in the context of phishing attacks, where the TRA scored 0.68, indicating a proactive and adaptive response.

13.7.3 TSD

The third graph shows each threat scenario's DSR (Figure 13.5). We found that our solution could successfully mitigate the impact of attacks in all situations, with a high success rate. This highlights the ability of our approach to rapidly adjust defense strategies in response to attackers' actions. The high SDR reflects the growing resilience of our solution in the face of various threats, which is essential for ensuring business continuity in critical network environments.

13.7.4 TFP

The fourth graph shows each threat scenario's FPR (Figure 13.6). The lower FPR indicates that our solution generates fewer false alarms than the conventional method. It is important to reduce these false positives to avoid unnecessary interruptions and ensure efficient use

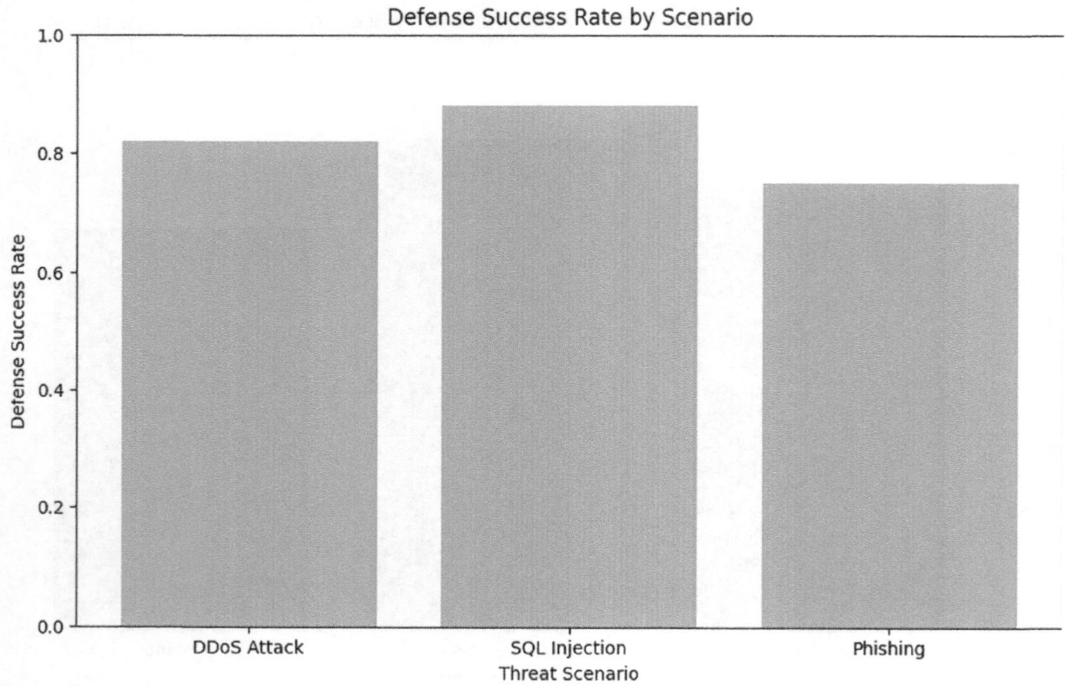

Figure 13.5 Defense success rate by scenario.

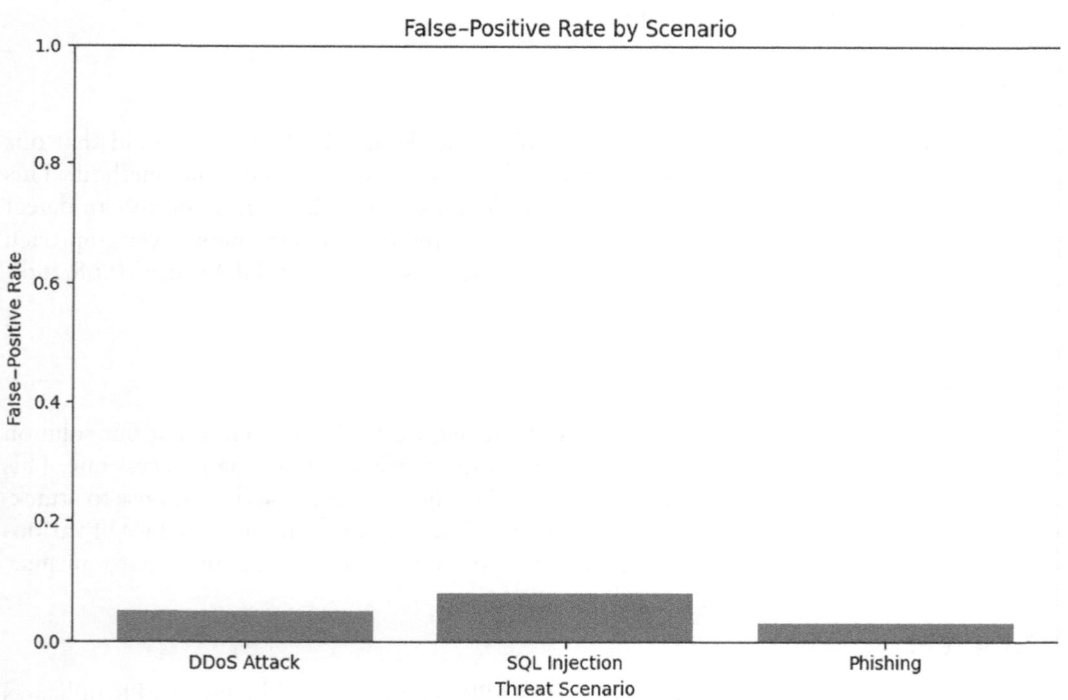

Figure 13.6 False-positive rate by scenario.

of security resources. Our solution has a low FPR in all scenarios, confirming its ability to accurately distinguish real threats from normal events.

Overall, the results of our rigorous testing demonstrate the significant advantages of our game-theoretic adaptive cybersecurity solution over traditional methods. Here is a summary of the main findings of our study:

Improved detection: Our solution significantly improved the threat detection rate (TDM) in all threat scenarios evaluated. Thanks to our proactive decision model based on game theory, our solution can predict attackers' actions and effectively identify potential threats.

Faster-automated response: The results show that our approach enables faster-automated responses to detected threats. The high ARR testifies to our solution's ability to act quickly to minimize the impact of attacks, thus minimizing the impact on the network.

Defensive effectiveness: The ability of our solution to dynamically adapt defense strategies to the actions of attackers is demonstrated by a high DSR. Our solutions effectively combat threats in various scenarios, guaranteeing business continuity.

Reduce false positives: Our solution stands out for its low FPR. This feature demonstrates the accuracy of our method for distinguishing real threats from normal events, minimizing false alarms, and avoiding unnecessary interruptions.

13.8 CONCLUSION

This chapter presents a new game-theoretic approach to adaptive network security. Our solution incorporates a proactive decision model to enhance network resilience in the face of threats. Rigorous experiments demonstrated the superior performance of our solution compared with conventional methods. Using decision modeling based on predictable game scenarios, our solution improves threat detection, accelerates automated responses, and strengthens overall network defenses.

Despite its advantages, our approach has certain limitations. One is the complexity of modeling player interactions, which may require adjustments for flexible threat scenarios. Furthermore, the effectiveness of our solution depends on the quality of the input data and the accuracy of the game model. In the future, we plan to extend our solution by further exploring player interaction in real network environments. In addition, integrating machine learning techniques can help optimize defensive strategies based on real-time data. Adapting our approach to large-scale networks and advanced threat scenarios is a promising prospect. In short, our solution provides a new perspective on adaptive network security by integrating game theory into SDN architecture. The results pave the way for proactive security, based on the predictive actions of attackers, and contribute to building a more secure and resilient cyber ecosystem in the face of ongoing challenges from cyber criminals.

REFERENCES

1. Maschler, M., Zamir, S., & Solan, E. (2020). *Game theory*. Cambridge University Press.
2. Bahnasse, A., Louhab, F. E., Talea, M., Oulahyane, H. A., Harbi, A., & Khiat, A. (2017, November). Towards a new approach for adaptive security management in new generation virtual private networks. In *2017 International Conference on Wireless Networks and Mobile Communications (WINCOM)* (pp. 1–6). Rabat, Morocco: IEEE.
3. Bahnasse, A., Talea, M., Louhab, F. E., Laafar, S., Harbi, A., & Khiat, A. (2017, July). SAS-IMS for smart mobile security in IP multimedia subsystem. In *Proceedings of the 2017 International Conference on Smart Digital Environment* (pp. 35–41). doi: 10.1145/3128128.3128134.

4. Bahnasse, A., Louhab, F. E., Oulahyane, H. A., Talea, M., & Bakali, A. (2018). Novel SDN architecture for smart MPLS traffic engineering-DiffServ aware management. *Future Generation Computer Systems*, *87*, 115–126.
5. Hamza, K. I., Kilani, J., Bensalah, F., & Baddi, Y. (2023). Evaluation and analysis of network safety mechanisms in SDN infrastructure. In *2023 IEEE 6th International Conference on Cloud Computing and Artificial Intelligence: Technologies and Applications (CloudTech)* (pp. 1–6). Marrakech, Morocco.
6. Ferguson-Walter, K., Fugate, S., Mauger, J., & Major, M. (2019, April). Game theory for adaptive defensive cyber deception. In *Proceedings of the 6th Annual Symposium on Hot Topics in the Science of Security* (pp. 1–8). San Diego, CA.
7. Chowdhary, A., Sengupta, S., Alshamrani, A., Huang, D., & Sabur, A. (2019). Adaptive MTD security using Markov game modeling. In *2019 International Conference on Computing, Networking and Communications (ICNC)* (pp. 577–581). Honolulu, HI.
8. Mengyao, L. U., Hongyu, T. A. N. G., Wei, T. A. N. G., & Jian, Z. H. A. N. G. (2020). Adaptive security architecture defense platform based on software defined security. *Telecommunications Science*, *36*(12), 133.
9. Peng, J. C., Cui, Y. H., Qian, Q., Guo, C., Jiang, C. H., & Li, S. F. (2021). ADVICE: Towards adaptive scheduling for data collection and DDoS detection in SDN. *Journal of Information Security and Applications*, *63*, 103017.
10. Abie, H., Boudko, S., Soceanu, O., Greenberg, L., Shribman, A., Gallego-Nicasio, B., ... Aiello, M. (2020). Adaptive and intelligent data collection and analytics for securing critical financial infrastructure. In *Cyber-Physical Threat Intelligence for Critical Infrastructures Security: A Guide to Integrated Cyber-Physical Protection of Modern Critical Infrastructures* (pp. 104–140). Edited by John Soldatos, James Philpot and Gabriele Giunta.
11. Maleh, Y., Qasmaoui, Y., El Gholami, K., Sadqi, Y., & Mounir, S. (2023). A comprehensive survey on SDN security: Threats, mitigations, and future directions. *Journal of Reliable Intelligent Environment*, *9*, 201–239.
12. Anas, I., Bensalah, F., Bahnasse, A., & Talea, M. (2023, March) Evaluating the performance of an SD-WAN network. *Scandinavian Journal of Information Systems*, *35*(1), 628–633.
13. Baddi, Y., Sebbar, A., Zkik, K., Maleh, Y., Bensalah, F., & Boulmalf, M. (2023). MSDN-IoT multicast group communication in IoT based on software defined networking. *Journal of Reliable Intelligent Environments*, *10*(1), 1–12.
14. Borylo, P., Davoli, G., Rzepka, M., Lason, A., & Cerroni, W. (2021). Unified and standalone monitoring module for NFV/SDN infrastructures. *Journal of Network and Computer Applications*, *175*, 102934.
15. Ahmed, J., Gharakheili, H. H., Russell, C., & Sivaraman, V. (2022). Automatic detection of DGA-enabled malware using SDN and traffic behavioral modeling. *IEEE Transactions on Network Science and Engineering*, *9*(4), 2922–2939.
16. Shirmarz, A., & Ghaffari, A. (2021). Automatic software defined network (SDN) performance management using TOPSIS decision-making algorithm. *Journal of Grid Computing*, *19*, 1–21.
17. Chowdhary, A., Sabur, A., Vadnere, N., & Huang, D. (2022). Intent-driven security policy management for software-defined systems. *IEEE Transactions on Network and Service Management*, *19*(4), 5208–5223.
18. Xu, L., Zhang, Y., Chinprutthiwong, P., & Gu, G. (2023, July). Automatic synthesis of network security services: A first step. In *2023 32nd International Conference on Computer Communications and Networks (ICCCN)* (pp. 1–10). Honolulu, HI: IEEE.
19. Karimi, M. (2022). *Contextual Decision Making and Action Enforcing Applications in Wireless Networks and IoT using SDN as a Platform* (Doctoral dissertation, University of Pittsburgh).
20. Shu, J., Zhou, L., Zhang, W., Du, X., & Guizani, M. (2020). Collaborative intrusion detection for VANETs: A deep learning-based distributed SDN approach. *IEEE Transactions on Intelligent Transportation Systems*, *22*(7), 4519–4530.

Part III

Human-centric risk mitigation approaches

Chapter 14

A human-centric approach to cyber risk mitigation

Ediomo Udofia

14.1 INTRODUCTION

Human factors play quite a significant role in staying formidable to cyberattacks and data breaches. Several human actions and inactions pose cyber risks, and it is essential to put these in perspective to achieve remarkably secure systems and data. While business executives, systems administrators, and cybersecurity professionals are obligated to set in place relevant mechanisms in the form of policies, processes, and technology to guard against potential breaches, end users are also responsible for following information security guidelines provided by the organization. Research has revealed that employees often underestimate the probability of cybersecurity breaches [1]. Another study sampling users from various countries shows that internet users with sufficient cyber threat awareness only apply minimal, relatively common, simple protective measures [2]. A global study of 50 countries conducted by the World Economic Forum indicates that 95% of cybersecurity challenges can be traced to human error and that insider threats (intentional or accidental) constitute 43% of all breaches [3]. There must be more advocacy on a more practical approach to improving human behaviors, especially across organizations, as this is instrumental in cyber defense.

The central challenge within organizations is inadequate cybersecurity awareness training programs or the complete lack of it. Secondary to this are poor business processes, policies, and technology.

This chapter attempts to identify common root causes for cyber breaches that involve the human element. With the above in perspective, this chapter also proposes proactive human-centric measures and best practices to guard against cyberattacks. These measures include understanding and building a healthy cybersecurity awareness culture, a healthy authentication culture, and ideal policies governing them. There are also detailed contributions on identity and access management (IAM) best practices, essentially attempting to point decision-makers in the right direction of handling authentication, identity, and access control within their businesses. Also discussed are a few emerging cyber threats plaguing businesses of various scales.

14.2 METHODOLOGY

There are several discourses around identifying and maximizing the human elements in cybersecurity to our advantage. This study elaborates on the perspectives of IT professionals of various roles across various sectors.

A qualitative research methodology with a thematic approach was utilized for this study. The themes that made up the sections of this work were carefully curated from the

DOI: 10.1201/9781032714806-17

discourses of the participants. To achieve this, a group of 20 IT professionals (consisting of IT managers, IT support engineers, systems administrators, cybersecurity engineers, cloud administrators, etc.) from various industries and countries (telecommunications, banking and finance, FMCG, power, hospitality, health and wellness, etc.) were engaged in a series of focus-group discussions. Participants were diverse in ethnicity, gender, and total length of professional experience.

The discussions first stemmed from individual points of view on the subject matter and then focus-group discussions. We first attempted to highlight human behaviors that undermine the cybersecurity posture of businesses and then actionable proactive approaches to prevent cyberattacks. The key points from these discussions formed the basis for this chapter. Providing a best practices template applicable to individuals and organizations.

14.3 HUMAN FACTORS AND IMPACT

14.3.1 Typical risky human factors

In cybersecurity, while the list of risky human factors and behaviors may appear seemingly inexhaustive, there are very common human practices that can put a user or an organization at cyber risk. These include: phishing susceptibility, inadequate password/management, usage of unauthorized software, clicking on suspicious links, ignoring software updates, poor Wi-Fi practices, insufficient physical security, social engineering susceptibility, sharing sensitive information, ignoring removable drive risks, insufficient employee awareness/training, using obsolete technology, shadow IT practices, ignoring two-factor authentication, poor-organizational cyberculture, bypassing organization's security policies, etc. [4]. Figure 14.1 classifies various employee practices and attributes based on risk level.

From the software developer's perspective, several cybersecurity risks are encountered during the development, release, and maintenance of software that are very critical. Very common threats include SQL injection, cross-site scripting (XSS), insecure APIs, unpatched vulnerabilities, etc. [5], others are runtime vulnerabilities, buffer overflows, data breaches, broken authentication, insecure data storages, input validation attacks, and other insecure coding practices. In addition to maintaining excellent secure coding practices among the development team, organizations that build software need other teams of professionals designated to ensure software security compliance and test for bugs or vulnerabilities. Such roles may include DevSecOps, QA Testers, Software Auditors, Penetration Testers, etc.

14.3.2 Impact of human factors on cybersecurity

Several human factors that have led to minor or severe cyberattacks or data breaches are avoidable. Still, of course, this is by being proactive with the correct and up-to-date processes, knowledge, and technology. In an exploratory analysis of past data breaches in healthcare organizations, it was discovered that the vast majority of health records were compromised due to poor human security. In addition, the findings also indicated that more patients' records were compromised from falling for a phishing scam than any other reason [6]. According to IBM's "Cost of a Data Breach Report 2023", globally, the average data breach cost in 2023 was valued at USD 4.45 million, signifying a 15% increase over 3 years [7]. The same report also indicates that 51% of organizations intend to increase their security budgets due to a breach [7]. Analyzing the cost of phishing attacks by the Ponemon Institute in their 2020 report "The Economic Value of Prevention in the Cybersecurity Lifecycle", it was indicated that the average total cost of a phishing attack is $832,500 [8].

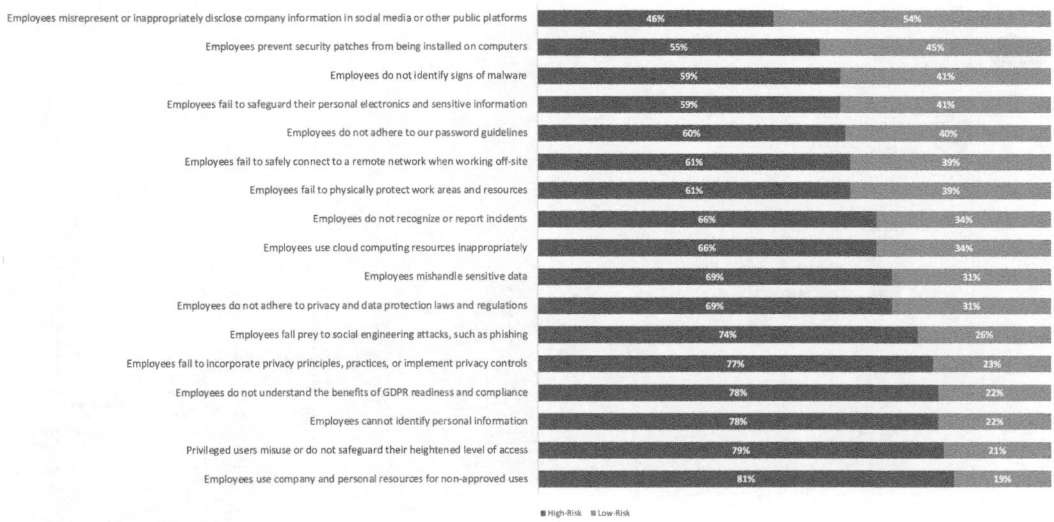

Figure 14.1 Classification of employee practices and attitudes based on risk level.

This provides a significant pointer to an approximate financial loss that could be averted by a human-centric cyberattack – a phishing attack. As indicated in the "Verizon 2023 Data Breach Investigations Report", from the 5,199 confirmed data breaches collected from various countries that were analyzed in the investigation, 74% of these breaches involved a human element, derived from factors such as human mistakes, privilege misuse, use of compromised credentials, or social engineering [9]. An upward trend has been observed in the total average cost triggered by insider elements from $11.54 million in 2019 [10] to $15.4 million in 2021 [11], according to Ponemon Institute's Cost of Insider Threats: Global Report 2020 and 2022. In the context of this research by the Ponemon Institute, "insider threats" can stem from negligence from an employee or contractor, a malicious insider, or a credential thief. This strongly indicates that human factors are the core causes of cyber breaches, ranging from gaps in knowledge, total negligence, policy lapses, or a disgruntled insider [7].

Figure 14.2 shows the cost of data breaches indicated by countries/regions valued in millions of USD. The figure also references the previous year's value, indicating an increase or a decrease.

14.3.3 Real-life cases trigged by human factors

Regardless of the year-on-year increase in investments in cybersecurity, organizations still suffer data breaches, not necessarily from technical flaws but from human events that create doorways for these incidents that may lead to breaches or exploitations. These human events may be intentional or accidental or may be caused by the absence of adequate organizational policies or strict compliance.

Human factors are associated with aspects of poor planning, lack of attention to detail, and ignorance are connected to the rise of "the unintentional insider" – in these cases, there are no malicious intents nor proper cybersecurity strategies, but their actions are equally as damaging and disruptive to the organization [12].

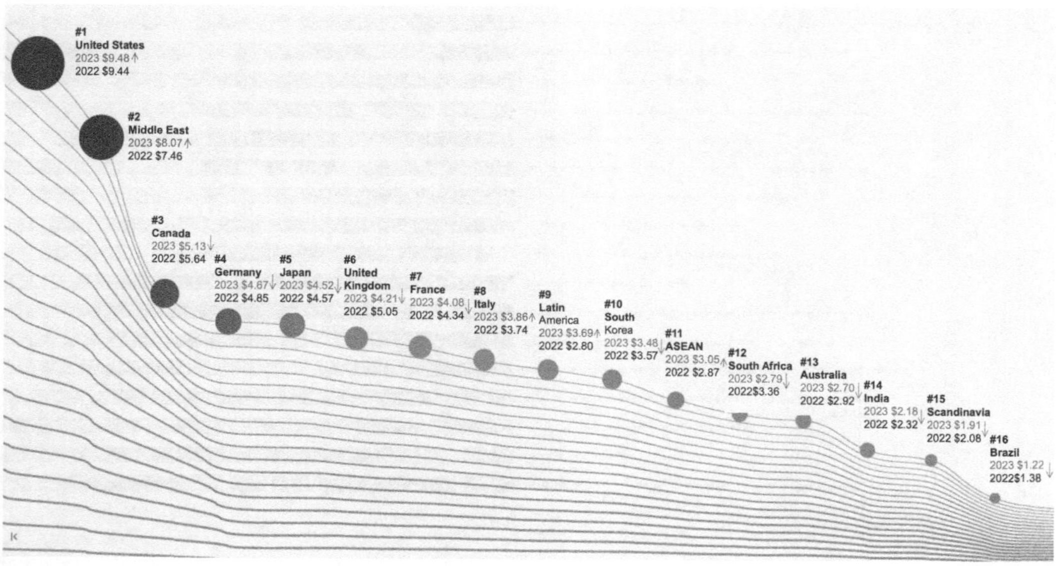

Figure 14.2 Value of data breaches by countries or regions in millions (USD).

For instance, a series of internal archive audits in 2021 revealed the massive data loss in the Dallas Police Department because of an employee's negligence. The employee had erroneously deleted 7.5 terabytes and 15 terabytes of police case files and secretary office materials on two separate occasions, as revealed by the audit [13]. Allegedly, the staff deleted data from DPD's archives without following procedures, and subsequent information gathered from the internal audit trail revealed a pattern of error caused by the same employee [13].

In another scenario, personal data including date of birth, salary information, and Social Insurance Number of more than 3,700 City of Calgary employees was shared with another municipality. Affected personal data included information such as Social Insurance Number, date of birth, Health Care Number, staff identification, address, salary information, etc. – the unencrypted transmission of such sensitive data places the data subject at high risk of identity theft and financial fraud [14].

A further examination of another real-life case would be the 2022 Uber data breach caused by a social engineering bait that a staff member fell for, assisting the bad actor disguised as the organization's IT personnel with an authentication request [15]. According to the same New York Times publication, the malicious actor accessed Uber's source code, Slack, and other internal systems, causing the organization to take several communication and engineering systems offline. In a separate but more revealing report of the same case study of Uber's 2022 data breach, it was revealed that the malicious actor purchased the stolen credentials from a dark web marketplace and went ahead to contact the employee in disguise as an IT staff to accept the overly persuasive authentication requests [16]. Further reports from the same UpGuard reveal that after successfully connecting to Uber's intranet, the hacker accessed the organization's VPN just to discover scripts of the organization's Privileged Access Management (PAM) solution containing credentials of a privileged user. A discovery that significantly increased the severity of the breach, thus, allowing further access to other relevant platforms such as Onelogin, Amazon Web Services (AWS), GSuite, DUO, etc. [16].

Mailchimp, an email and marketing automation platform alongside some of its partners, became targets to cybercriminals, suffering at least three attacks in the space of

12 months, the third being in January of 2023 and affecting 133 Mailchimp accounts [17]. Another source has it that this breach was a targeted social engineering attack focused on Mailchimp's employees and contractors [18]. This impacted several partnering organizations such as WooCommerce, Statista, FanDuel, and DigitalOcean. In fact, on the very next day, August 9, 2022, one of the impacted organizations DigitalOcean had to migrate their email services away from Mailchimp as seen in the statement released in the wake of the breach [19].

Human errors may not necessarily be malicious or have ill intentions. Some could come from the erroneous actions of employees responsible for handling and transmitting sensitive information to authorized entities. A typical case was the September 2015 incident that occurred in 56 Dean Street, a sexual health clinic run by Chelsea and Westminster Hospital NHS Foundation Trust. This email error caused 781 people who had attended the clinic to be able to view the names and email addresses of other patients who opted for the service [20]. In this case, the email sender who should have used the blind copy (BCC) feature of the email to prevent such failed to do so – and such resulted in a £180,000 fine against the NHS Trust, the same source says.

Most, if not all, of the above incidents could have been avoided with the right combination of tools, security awareness for users, security policies, and workplace culture, as will be discussed in the later parts of this chapter.

14.4 AWARENESS CULTURE: WHAT ORGANIZATIONS AND USERS SHOULD KNOW AND PRACTICE

14.4.1 Cybersecurity training programs for employees

The relevance of all-inclusive, regular, and intentional cybersecurity awareness and training programs for business employees cannot be over-emphasized. This is the bedrock for building a robust cybersecurity culture within an organization, regardless of its size. The primary goal should be that every user within the organization stays conscious of evolving loopholes that may be caused as the result of their actions or inactions and what the worst possible consequences may be.

A common mistake small- and medium-scale enterprises make is the belief that they are not prime targets of cyberattacks, hence, they treat cybersecurity with absolute levity. It is strongly recommended that cybersecurity training programs be factored into cybersecurity budgets. Structured training should be carried out, and unaware users should be tested against their knowledge with simulated practical scenarios. Organizations should also have training strategies, such as partnering with tested and trusted organizations that may have various models of delivering quality cybersecurity training programs. Another training strategy could be having a dedicated cybersecurity department that handles regular training and awareness programs for all staff members, including new hires, at their onboarding stages. A comprehensive study on a specific attack vector – phishing, recommends an awareness model delivered to employees in the form of interactive games, taking into consideration factors like the age and gender of the trainees [21].

According to EY Canada, "employees are a major access point for cyber attackers," organizations need to incorporate their staff into their security plan to have resiliency [22]. A study conducted on 5,000 businesses worldwide revealed that 52% of organizations agree that employees are the weakest link in their IT security strategies, putting these strategies at risk because of their actions [23]. Regular cybersecurity awareness can positively impact an organization's overall security posture. A study shows that organizations that implemented regular cybersecurity training experienced a 70% decline in security incidents [24]. In a separate study conducted by KnowBe4, it was ascertained that employees who are trained monthly are 34% more inclined to believe that clicking on a suspicious link or an email attachment from an unknown source is risky compared with the employees who receive less frequent training (not more than twice annually). Also, employees receiving monthly cybersecurity training are 26% more likely to believe that password reuse is a risky behavior than employees receiving training not more than twice annually [4].

14.4.2 Social engineering awareness and intervention

In straightforward terms, social engineering, according to the European Union Agency for Cybersecurity (ENISA), refers to every technique aimed at talking a target into revealing specific information or performing a specific action for illegitimate purposes [25]. Preventing social engineering attacks from being successful requires human cognitive intervention at the core – this means a good defense system requires appropriate knowledge, process, and technology. All employees, including nonusers, should be well enlightened on various social engineering concepts to build a strong defense mechanism against potential cyber criminals. This can only come through regular training, practical demo sessions, and policies culminating in a stronger cybersecurity culture within the workplace. It is recommended for all employees (end users or not regardless of their level of computer/device use or access) because there are various technical and nontechnical variants of social engineering, and the commonest of them will be discussed subsequently.

14.4.2.1 Phishing

Phishing merges social psychology, technical systems, security objects, and politics [21]. From the Cyber Security Breaches Survey 2022 carried out in the UK by the Department for Digital, Culture, Media, and Sports, of the 39% of UK businesses that identified an attack, 83% comprised phishing attempts [26].

A study on 3,500 adult working participants across seven countries showed that only 61% of these workers were most likely to recognize the term "phishing" [27]. For this same study by Proofpoint, 95% of survey respondents agreed that their organizations deliver phishing awareness training, while 30% perform awareness training only on targeted end users. Targeting phishing awareness training only for a selected user base may not be the best approach to delivering such awareness training, as this would put untrained users at risk of phishing susceptibility.

Cybercriminals build their phishing emails carefully to achieve the victims' susceptibility [28]. While mass and generic phishing emails may be common, more serious-minded cybercriminals now engage in dedicated and targeted phishing emails to improve the susceptibility of their targets [27].

Organizations could adopt several models to deliver phishing awareness training, including instructor-led sessions (physical or virtual), simulated phishing exercises, role-based tailored training, interactive workshops, videos, posters, newsletters, emails, etc.

As part of a phishing awareness campaign, organizations are encouraged to conduct phishing simulation tests to determine their employees' practical knowledge level and their typical responses. Proofpoint recommends that organizations regularly use a more personalized approach in their tests, as this can better prepare users more for targeted attacks [27].

While the common phishing attacks lure unsuspecting victims into clicking on malicious links in emails, malicious email attachments, and redirection to cloned login pages, there are evolving sophisticated phishing attacks that organizations need to be aware of and guard their users and systems against. An example of this kind of attack, as highlighted by the Microsoft Defender team, leverages a multistage Adversary-in-the-Middle (AiTM) phishing and Business Email Compromise (BEC) attacks to abuse trusted relationships among trusted partner organizations [29]. The AiTM attack aims to intercept multifactor authentication (MFA) requests from legitimate authentication services and capture the session cookie with the primary goal of successfully impersonating the victim's identity [29]. As acknowledged by the Esentire Threat Response Unit, AiTM, which is a variant of the traditional Man-in-the-Middle (MitM) attack, has trended lately for several reasons including a successful bypass of MFA requests, the capture of session cookies (and the ability to replay them), the evolution of Phishing-as-a-Service (PhaaS) tools – becoming more user friendly and Cloudflare's Turnstile tool redirection (CAPTCHA replacement) [30]. For this sort of sophisticated attack, the Microsoft Defender team recommends the following: enabling more granular authentication and access policies, e.g., conditional access (this a Microsoft-offered authentication policy that requires users to meet several set conditions before access is granted), organizations should implement continuous access evaluations, invest in antiphishing solutions, and continuously monitor suspicious or anomalous activities [29].

The National Cyber Security of the UK encourages a multilayered approach to phishing attack prevention by organizations, as this would drastically reduce the chances for a successful phishing attack to be executed [31]. This multilayered approach, in summary, addresses four key points: (a) making it stringent for attackers to reach users – by reducing digital footprinting, investing in email filtering service, etc., (b) helping users to identify and report suspected cases, (c) defending the organization against undetected phishing emails – by investing in anti-malware software, using updated software, etc., and (d) Swift incident response [31].

14.4.2.2 Whaling

This special phishing attack targets high-profile individuals, such as business seniors with the malicious aim of stealing the organization's money or sensitive data. This attack begins with a careful reconnaissance stage to gather vital information to help make the bad actor appear legitimate [32]. According to GreatHorn's survey in 2021, 59% of organizations acknowledge that an executive has been the target of a whaling attack [32]. A successful whaling attack that may culminate in financial or data loss can cause significant reputational damage to the organization and the individual. Several members of staff, including the CEO of FACC, an Austrian aerospace manufacturer, lost their jobs because of falling victim to a whaling attack and consequently losing € 50 million of the organization's funds [33].

A rudimental step to preventing a whaling attack within a corporation is sensitizing highly probable targets within the organization. These targets include but are not limited to C-level executives and other principal members who are custodians of sensitive information and those who can approve funds. When specific information about individuals within these categories is made publicly available on the internet, then the threshold for reconnaissance is lowered for cybercriminals. For instance, the business email and phone number of a CFO published in the About section of the company's website or LinkedIn profile makes it easier for the bad actor to have something to work with. Business executives of this caliber need to be specially and regularly sensitized on reducing their sensitive digital footprints on social media, such as joining risky forums and groups on platforms like WhatsApp and Telegram where their phone numbers can easily be harvested for social engineering purposes.

Investing in the right email defense tools is also a way to prevent whaling and other phishing attacks within the organization. Ironscales recommends the deployment of tools such as Sender Policy Framework (SPF), Domain Keys Identified Mail (DKIM), and Domain-based Message Authentication, Reporting, and Conformance (DMARC) and the implementation of "the 4-eye principle" among other best practices [34]. "The 4-eye principle" essentially decentralizes the authorizing power of a single employee for huge sums of money [34].

14.4.2.3 Tailgating

In simple terms, tailgating is gaining access to a restricted area by riding on the advantage of authorized personnel, often achieved through psychological gimmicks. This is one of the best no-tech means of gaining access to a building [35]. The bad actors take advantage of kindness or complacency to follow authorized users to these restricted areas [36]. To combat cybersecurity, businesses are implementing sophisticated solutions such as high-level encryption services, firewalls, advanced authentication systems, intrusion prevention systems, and intrusion detection systems. Nevertheless, they must also pay rapt attention to the potential cyber threats that physical loopholes may pose. Aside from data theft, the tailgater may have ulterior motives such as harming a staff, destroying properties, or committing other criminal acts.

The most effective measures to curb tailgating in a business environment would be

a. To create serious awareness among employees regarding tailgating within the premises: The concept of employee awareness can never be overemphasized. Every employee must have sufficient knowledge of various social engineering attacks, including nontechnical ones such as tailgating. Employees must always be conscious of their environment and report all suspicious activities to the appropriate authority. Employees need to be aware of typical scenarios, and locations where tailgating can most likely occur, so they do not unconsciously aid the tailgater in succeeding in the act.
b. **Improvement of physical access control [36]:** Organizations must deploy trained dedicated security personnel to man ingress and egress points. They must stay very vigilant, especially during peak rush hours. To prevent tailgating, organizations must also deploy access control systems at entrances and other sensitive areas that require staff members to use their access cards to gain entry. Access should require authentication using biometrics for very sensitive areas of the building. In shared offices with multiple businesses, owners should consider combining resources for a unified and robust security system [36].
c. **Visitor's identification:** There must be a defined process for individuals who are visitors. Their purpose of visit must be verified, and they must wear a visitor's badge for ease of identification. Their area of access must also be limited.
d. **Advanced video surveillance:** Investing in advanced video surveillance systems can remarkably complement the control efforts by human security [36]. These advanced video surveillance systems are AI-powered with video analytics capabilities such as motion detection, facial recognition, headcount, etc.

14.4.2.4 Smishing and vishing

SMS phishing (smishing) is a kind of attack in which fraudulent actors send phishing messages via text messages or other messaging apps to their victim's devices [37]. These messages often contain malicious links to infest one's device, cloned webpages done to steal one's credentials, or a psychological deception to steal the victim's funds or make them reveal sensitive information [37]. Some criminals use spoofed or compromised phone numbers and hacked accounts on popular social media platforms to fake their identities [38].

Data from the Federal Trade Commission (FTC) indicates that US customers lost $86 million to SMS fraud in 2022 only – with the number of reported cases for this contact method at 334,524 [39].

Voice phishing (vishing) is a kind of attack where the bad actor uses phone calls or voice messages to impersonate legitimate businesses to trick their victims into giving them money or divulging sensitive information [37]. To successfully achieve their vishing attacks, these criminals now use sophisticated techniques such as fake caller IDs, "war dialers" for mass calling, and artificial speech applications [40].

Data from the FTC also suggests that US customers lost $436 million to phone call fraud in 2022 only – with the number of reported cases for this contact method at 383,598 out of a total of 5.2 million reported cases [39].

During the COVID-19 pandemic, the Federal Bureau of Investigation (FBI) and Cybersecurity and Infrastructure Security Agency (CISA) issued a joint statement cautioning businesses about an ongoing vishing campaign targeted at remote workers in a bid to steal their VPN credentials [40].

Smishing and vishing attacks are presented in various baiting scenarios such as order/shipment confirmation, government institutions demanding payments, bank account verification, charities/prize claims, technical support from service providers, etc. [37,40].

Several prevention strategies could prevent an individual from falling victim to smishing and vishing attacks. These strategies are useful both for personal and business purposes. Organizations must incorporate them into their cybersecurity awareness training programs and have them communicated regularly to employees. While SMS and phone calls are common means of reaching out to existing and potential customers, attackers have become excellent at disguising themselves as legitimate businesses to exploit unsuspecting victims using SMS and voice calls as their attack vectors.

As a rule of thumb, do not respond to messages from unknown sources that bear in them a sense of urgency or require you to reveal sensitive information.

- Call back your bank or merchant directly using their official contact channels [41].
- Take precautions on other messaging apps like LinkedIn, WhatsApp, Facebook, etc. – smishing is not limited to SMS only [42].
- As a primary safety rule, just like emails, do not click on unsolicited links that arrive on your devices via SMS or other messaging apps.
- Never keep credit card information on your mobile phones [41].
- Never provide a password, account recovery, or authentication code via SMS [41].
- If in doubt or suspicious of an SMS or voice call, solicit the opinion of a trusted expert within your organization.
- Block the sender's number and delete suspicious messages to avoid accidentally clicking the link [43]. Do use third-party caller ID verification apps like Truecaller – they flag reported spam callers and can help you block them.
- Pay attention to unnatural synthetic voices. Scammers use AI-powered calling apps to impersonate others [43].

14.4.3 Misconfiguration

An inadequately configured system, application, or network setting is one human error that poses a security risk to an organization's system or data. This can be because of a configuration mistake, leaving configurations on default settings, enabling unnecessary features, broad permissions granted, insecure API configurations, etc. [44]. Typical examples are leaving a wireless router's login password as default, using a firewall appliance with its default configurations, enabling unnecessary ports, etc. IT professionals responsible for systems and network administration must reverify periodically to ensure that misconfiguration does not affect systems and networks. Manual or automated checks (such as vulnerability assessments) can reveal inadequately configured systems, networks, and applications. It is highly recommended that organizations use suitable Security Information and Event Management (SIEM) solutions to help gain security visibility in real time by monitoring and responding to potential threats [44].

14.4.4 Human email errors

In total, 361 billion business and consumer emails will be sent and received daily in 2024 [45]. With this number of emails sent globally every day, human errors with email handling are inevitable. While several social engineering attacks use email services as an attack vector, there are also human errors with email handling that can lead to data breaches. Some of these errors are as follows:

a. **Misdirected emails**

   According to Verizon's 2023 Data Breach Investigations Report, misdelivery accounts for 43% of breach-related errors [9]. Due to haste or nonchalance, users often send confidential emails to the wrong recipients. This is typically facilitated by the autocomplete feature in email clients such as Microsoft Outlook [46] or an accidental "Reply All" [47]. Often, unintended recipients share the same first name or surname as the intended recipients [46]. A misdelivered email containing sensitive information about a customer or client would be considered a data loss or breach [47].

   Some email client apps, like Microsoft Outlook, allow users to recall emails that the recipient has not read. This is a first step to help abate the event of email misdelivery. Therefore, this is important to consider when choosing an email service and client for an organization. As part of user training on data handling and cybersecurity, users need to be trained on how to utilize the email recall feature, this may come in handy at any time.

b. **Failure to use Bcc for bulk recipients**

   This is a common email error when an email is sent to bulk recipients, and the sender fails to use the Bcc field but uses the To/Cc field. This essentially exposes all the email addresses to all the recipients [46]. In the past, this has caused a severe data privacy breach for patients at a medical facility, resulting in huge fines [20].

c. **Using personal email for work**

   Unlike personal emails, corporate emails are often configured with security in mind. Therefore, sending sensitive information to personal emails puts the organization at risk of a data breach. Employees may send work emails for malicious or nonmalicious purposes – for instance, to continue their work at home [46]. Users do this often without realizing the exposure it poses. This act can be restricted or monitored using email security services within an organization [9].

Figure 14.3 shows a yearly progression of breaches associated with human errors (misdelivery, misconfiguration, and publishing errors).

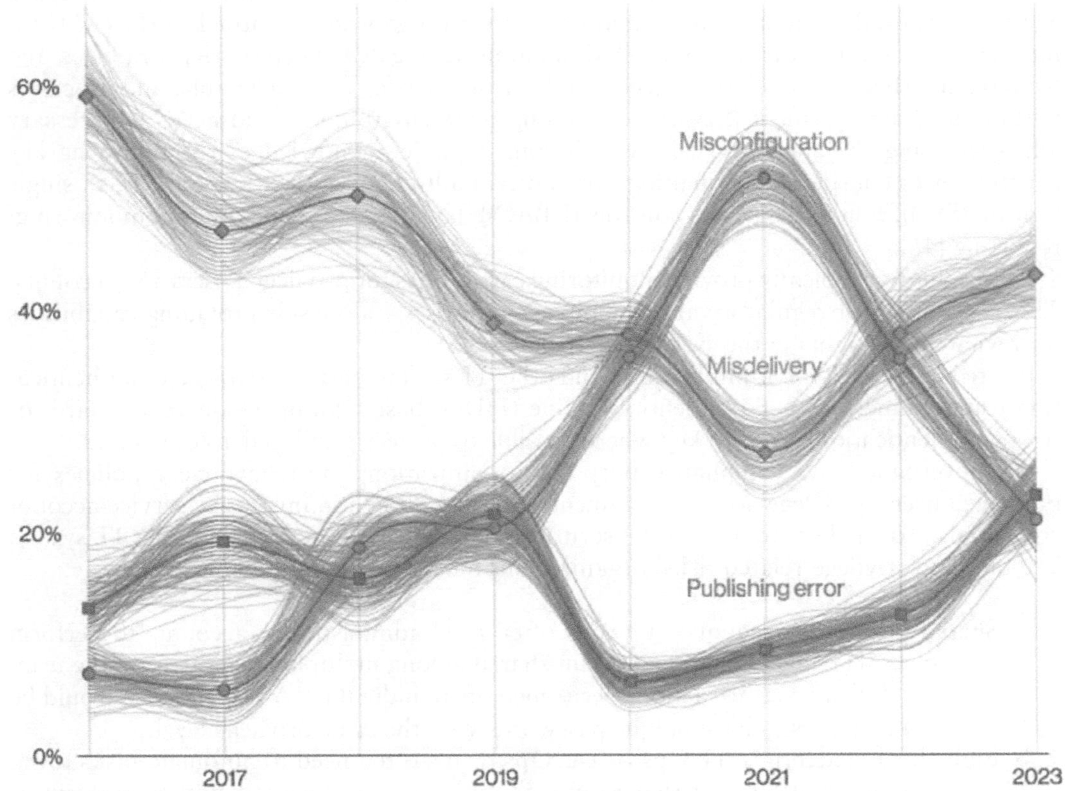

Figure 14.3 Miscellaneous error breaches associated with human errors.

14.4.5 Privileged users and accounts

According to Centrify, 74% of all breaches commence with compromised or stolen privileged credentials, which holds for SMEs and large-scale enterprises [48]. Local admin, domain admin, and cloud admin accounts are generally considered privileged accounts because they can usually install software, reset passwords, access sensitive data, and modify IT infrastructure [49]. These accounts are elevated above the standard accounts typically assigned to the rest of the users. The credentials of a privileged user getting into the wrong hands can cause varying organizational damages. These could be reputational, financial, or even cause IT infrastructure downtime. Cybercriminals can assign themselves new elevated privileges for continuous long-term reconnaissance or other malicious acts [50].

Providing specialized periodical training for privileged users within an organization is a fundamental human defense strategy to prevent cyber breaches. Elevated privileges mean elevated risks – and these users need to understand the risk level associated with their privileges. Privileged users must be trained adequately to avoid social practices or negligence that make them victims of social engineering attacks. The concept of least privilege should be strictly adhered to, especially for users with changing roles. Users must be assigned just enough access rights for their job functions.

Implementation of a PAM solution within an organization is another strategy to guard against privileged account misuse – and to reduce an organization's attack surface [51] for internal and external threats [51]. PAM solution aggregates cybersecurity strategies and technologies that control elevated access and permissions for users, accounts, and processes within an IT environment [51]. Understanding one's environment and needs is necessary when choosing the appropriate PAM solution. A guide acknowledges the following key capabilities to consider when making this choice: multifactor authentication (MFA), single sign-on (SSO), role-based access controls (RBACs), lateral access controls, and monitoring/reporting [48].

PAM solutions typically provide monitoring capabilities for privileged users and accounts. This is essential for regulatory and compliance purposes – alongside providing capabilities for reviewing accounts in the event of an incident [48].

It is recommended for administrators and other IT staff to perform two-factor authentication using biometrics or First Identity Online (FIDO)-based (an open-source standard for strong authentication) security key where possible to protect privileged accounts [52].

For governance and compliance purposes, organizations must have clear policies for privileged users, privilege access assignments, and handling administrative service account credentials. To further strengthen the security architecture of an organization's IT system and mitigate privilege-related risks, organizations must avoid the following:

a. **Shared privileged accounts:** IT teams often need administrative accounts to perform their tasks. Having a common account shared among multiple individuals on the team makes it difficult to trace actions performed on an individual. A best practice would be for team members to have unique passwords with the same privilege level.
b. **Embedded credentials:** IT Ops or DevOps teams often need to automate tasks using scripts or automation tools that require administrative service accounts to perform. It is a serious security risk to hardcode credentials in plain text embedded in scripts.

14.4.6 Emerging threats

As computer technology advances and the reliance on internet use grows, so do cyber threats exponentially. Over the years, cyber threats have evolved in model, sophistication, and attack vectors, making them increasingly difficult to contain in many cases. Hackers consistently devise new means to advance reconnaissance to discover vulnerabilities within a system, software, or platform – consequently birthing zero-day exploits. Recently, the Intrusion Detection System (IDS) and Intrusion Prevention System (IPS) have been built to machine learn and respond to zero-day exploits proactively.

One catastrophic emerging threat is the advancement of artificial intelligence (AI) and machine learning (ML). While ransomware and malware are utilized to execute a cyber exploit, AI and ML are technologies that aid in automating reconnaissance and actively exploring vulnerabilities in a system or network. The rapid advancement in AI-powered systems in various sectors has brought a new revolution in service efficiency and capabilities and AI-imposed cyber challenges [53]. This has significantly increased AI-powered attack vectors – like sophisticated phishing lures, deceptive chatbots, counterfeit social media profiles, etc. [53]. AI has, in very recent times, energized the explosive growth in concepts like deepfakes (digitally modified media using an algorithm to have a strikingly similar personality of someone, especially a public figure) and "quishing" (a specialized type of phishing that involves the alteration of QR codes to redirect victims to the malicious website for malicious reasons).

14.5 AUTHENTICATION CULTURE: SAFE AND BEST PRACTICES

14.5.1 Password practices

Password authentication is an age-long method for authentication and may never completely go out of use soon. Cybercriminals have become sophisticated in bypassing passwords by building on the existing password bypassing techniques such as brute-force attacks, dictionary attacks, keyloggers, credential stuffing, etc. Cybercriminals also use publicly available compromised credentials to build a foothold in their attack landscape. Because of these, it is imperative for users to strictly adapt to password best practices to avoid ease of credential compromise.

14.5.1.1 Strong passwords

As a rule of thumb, using strong passwords increases the degree of predictability even when using automated tools. The secret of strong passwords is embodied in complexity, length, and unpredictability [54]. Password complexity is achieved when passwords are composed using upper cases, lower cases, digits, and special characters. However, the National Institute of Standards and Technology (NIST), USA, no longer recommends complex passwords because it has been observed that users, out of frustration, tend to use very predictable methods to compose these passwords when imposed, and they use the same across multiple systems for ease of recollection. NIST recommends password managers instead [55]. Passwords should be 8–12 characters in length to help their strengths. Passwords should also be complicated to predict – avoid common names, organization names, usernames, dictionary words, etc. The combination of these three password characteristics will make a very strong password.

14.5.1.2 Password managers

A password manager will help you maintain unique passwords across various systems and platforms, minimizing the attack surface for a compromised credential. Good password managers also ensure you do not choose passwords from the myriad publicly available lists of compromised passwords. It further ensures that you meet password security requirements and sends reminders when passwords are close to expiration [54].

14.5.1.3 Never store in plain text

Users should never store passwords in plain text for any reason. It is also unsafe to reveal passwords to other users even when a temporary duration is intended. Sending them your password via SMS or email may be seen by an unintended party.

Avoid providing passwords in plain text to scripts or applications that need to interact with other systems. This is a major risk factor in cybersecurity.

14.5.1.4 Implement healthy password policies

As a measure of cyber defense, organizations must practice healthy password policies. These policies must be communicated appropriately and monitored for compliance and governance purposes. These policies include but are not limited to:

a. Enforce password complexity.
b. Enforce password age (90 days recommended) [56].

c. Enforce password backlist for guessable passwords.
d. Disallow password reuse.
e. Enforce password reset on the first login.
f. Avoid password sharing among team members.
g. Enforce lockout policy on several failed attempts.
h. Reset and Disable accounts immediately after an employee exits.
i. Enforce password encryption where possible.
j. Establish password audits.
k. Disable root logins for IT teams where possible.
l. Ensure a separate password policy for privileged users and privileged access assignments.

14.5.1.5 Other recommendations

Avoid dependence on knowledge-based authentication, also called "security questions". NIST recommends against using knowledge-based authentications as a means of authentication as hackers now easily discover answers to many knowledge-based questions [55].

Users must avoid password reuse as a healthy password security practice for personal and corporate accounts. This is a major risk for both individuals and companies. A password like "oreocookie," which may look difficult to guess, has been seen over 3,000 times [57]. The National Cyber Security Center (NCSC) in the UK acknowledges a breach analysis that found 23.2 million accounts globally using 123456 as passwords [58].

Organizations must have defined standard practices for authoritative accounts, such as the Global Administrator account in Microsoft Cloud services. Using the Global Administrator account as a case study, it should not be associated with any employee, MFA must be enforced, it should use an alternative recovery email, there should be at least two Global Administrators in the tenant and not more than five, passwords should not be saved in web browsers, and these accounts should be used only when very necessary.

14.5.2 Multifactor authentication (MFA)

Despite predictions of password use phasing out, over the years, the growth in password use has become exponential. There are increasingly complex password requirements and often unrealistic demands on users that have resulted in "password overload" [59]. Passwords alone cannot protect your data and systems, even when implemented correctly [59].

MFA essentially attempts to combine something that you know (password, PIN) with something that you have (for instance, a verification that is received via an SMS or an authentication app on your device), together with something that you are (biometrics, voice). This combination adds a remarkable layer of security to the authentication process and enhances security. Enforcing the use of MFA elevates an organization's security posture by requiring users to identify with additional methods other than usernames and passwords [60] – this is highly recommended as a baseline policy for organizations.

Many systems today make provision for MFA to be enabled for authentication purposes; individuals and organizations must take advantage of this to secure these systems better. Other than the traditional usernames and passwords, additional ways to authenticate include one time passwords (OTPs), smart cards (with security keys), USB devices (with encoded security keys), authentication apps (request-based, code generated), fingerprints, iris scanning, facial recognition, voice, etc.

Taking it a notch higher is adaptive authentication also referred to as risk-based authentication. This utilizes AI and requires more than the traditional MFA to authenticate.

In addition to the conventional username, password, and MFA, it uses additional information, such as location, registered devices, assigned role, etc., to grant the user access [61].

14.6 IDENTITY AND ACCESS MANAGEMENT (IAM) MEASURES

14.6.1 Least privilege principle

In an organizational structure, it is a risk to grant users access to resources or privileges they do not require or cease to require. The least privilege principle is essentially a cybersecurity architecture that grants users access to resources or privileges sufficient for them to perform their job functions. RBAC is one effective type of access control that organizations can use to implement the least privilege principle in practice. RBAC, as the name implies, is a security approach that allows or restricts a user's access based on their role in the organization [62]. For instance, in a M365/Azure Active Directory (Microsoft Entra ID) setting, you would think that the CISO of a large multinational firm using this service should be assigned the Global Administrator. This can be considered a security flaw because his job function in practice does not require that level of access regardless of his position in the firm – and of course, it is good practice not to assign a human entity the Global Administrator role. Instead, the CISO can be assigned the Global Reader role (a read-only counterpart to the Global Admin role), where they can view all settings and assigned privileges without modifying them.

Cloud providers like Amazon Web Services (AWS), Microsoft Azure, Google Cloud Platform (GCP), and Platform-As-A-Service (PaaS) providers have RBAC features, and organizations should strictly ensure the least privilege principle is implemented as a rudimentary access control model for accessing resources.

14.6.2 Advanced identity and access management (IAM) concepts

While Access Control List (ACL) and RBAC are security approaches to restrict access to resources and privileges, there are more advanced and adaptive approaches to IAM. While the following concepts are directly associated with Microsoft 365 and Microsoft Azure Active Directory (now Microsoft Entra ID), they are available on the other major cloud providers but have different names and configurations.

One such is Conditional Access, a core constituent of Microsoft's "Zero Trust" security model, built around a set of preconfigured policies and settings used to restrict or grant users access to services and resources on the platform [63]. Simply put, they give the ability to enforce access requirements when specific conditions occur. This provides a very granular approach to IAM layered on the existing RBAC and is based on various signals. Factors that make up these signals (assignments) include specific users or groups, specific cloud applications to be accessed, and specific conditions (e.g., sign-in risk level, device platform, location, client app in use, device state, IP information). When the specified conditions are met, then there are enforcements which could be an outright block, access granted but requiring further conditions (such as requiring MFA, requiring the device to be marked as compliant, requiring the device to be Hybrid Azure AD joined, requiring approved client app), and access granted but with limited session experience with the cloud application.

Another advanced IAM approach is Privilege Identity Management (PIM), a service that allows management, control, and monitoring of essential resources and privileges within an organization [64]. This service offering on Microsoft Entra ID and other Microsoft Cloud services provides just-in-time privileged access to Entra ID and Azure resources, provides time-bound access, requires approvals for access, enforces MFA, requires justification for access, and prevents removal of Global Administrator role assignment [64].

14.6.3 Access reviews and auditing

Monitoring users' access rights and privileges to data and applications is crucial to ensure that appropriate privileges are assigned to the right users [65]. Users' access can be reviewed for compliance purposes and to regularly ensure that only the right individuals have continued access [66].

Organizations should have predefined onboarding and offboarding access assignment templates for employees depending on their job functions. Periodic access reviews and audits are necessary because excess access rights can lead to compromise and indicate poor control over access [66].

Access Reviews in Microsoft Entra ID is done to sanitize employees' and guests' group memberships, organize and track reviews for compliance and risk management initiatives, and clean up guest users in groups. Access Reviews also ensures that groups and application owners get to review accesses on a timely or on-off basis so that penalties applied under Conditional Access can be decided upon.

14.7 CONCLUSION

Ultimately, humans do not have to be the weakest link to data breaches or cyberattacks. However, businesses can take absolute advantage of the right knowledge and tools, among other resources, to build a formidable human entity to combat breaches and cyberattacks. Different businesses would have unique cybersecurity models based on their modus operandi. Nevertheless, the human factor should be a significant consideration for building these models and should factor in all users regardless of their role in data handling within the organization. Organizations must be able to tell what the cumulative cybersecurity awareness posture of their end users is. This can be carried out through biannual in-house surveys, simulated tests, and verbal interrogations. This data is vital in planning or restrategizing training frequencies or training models to help improve the awareness level of employees. Organizations must ensure they have regularly reviewed IT security policies and procedures with practicable measures to enforce user compliance. While this chapter may not be very exhaustive in capturing all aspects of human-centric cyber defense approaches, there are, however other aspects worthy of mention. They include physical security, secure software development practices, policies for Company-Owned Devices (COD) and Bring Your Own Device (BYOD), incident response and reporting procedures, cultural/organizational factors, and legal/ethical considerations.

REFERENCES

1. T. Herath, R. Rao, "Protection Motivation and Deterrence: A Framework for Security Policy Compliance in Organisations," *European Journal of Information Systems*, v. 18, pp. 106–125, 2009.
2. M. Zwilling, D. Lesjak, Ł. Wiechetek, G. Klien, "Cyber Security Awareness, Knowledge and Behavior: A Comparative Study," *Journal of Computer Information Systems*, v. 62, pp. 82–97, 2022.

3. World Economic Forum, "The Global Risks Report 2022," Ed. 17, pp. 45–56 (January 2022). Available at: https://www.weforum.org/publications/global-risks-report-2022/
4. KnowBe4, "2021 State of Privacy and Security Awareness Report" (2021). Available at: https://www.knowbe4.com/hubfs/2021-State-of-Privacy-Security-Awareness-Report-Research_EN-US.pdf
5. O. Trad, "10 Cybersecurity Risks in Software Development and How to Mitigate Them" (October 2023). Available at: https://devtalents.com/cyber-security-during-software-development/
6. L. H. Yeo, J. Banfield, "Human Factors in Electronic Health Records Cybersecurity Breach: An Exploratory Analysis," *Perspectives in Health Information Management*, v. 19, p. 1i, March 2022.
7. IBM Security, "Cost of a Data Breach Report" (2023). Available at: https://www.ibm.com/reports/data-breach
8. Ponemon Institute Research Report, "The Economic Value of Prevention in the Cybersecurity Lifecycle" (April 2020). https://www.ponemon.org/research/ponemon-library/security/the-economic-value-of-prevention-in-the-cybersecurity-lifecycle.html
9. "Verizon 2023 Data Breach Investigations Report" Verizon (2023). Available at: https://www.verizon.com/business/resources/reports/2023-data-breach-investigations-report-dbir.pdf
10. Ponemon Institute, "Cost of Insider Threats: Global Report 2020" (2020). Available at: https://www.ibm.com/downloads/cas/LQZ4RONE
11. Ponemon Institute, "2022 Cost of Insider Threats Global Report" (2022). Available at: https://www.proofpoint.com/us/resources/threat-reports/cost-of-insider-threats#:~:text=As%20the%202022%20Cost%20of,a%20third%20to%20%2415.38%20million
12. J. McAlaney, L.A. Frumkin, V. Benson, *Psychological and Behavioural Examinations in cyber security*. USA: IGI Global, pp. 46–60 (2018).
13. "City Officials Fire IT Employee Involved in Massive Deletion of Dallas Police Evidence" emails show (2021, August 30). Available at: https://www.dallasnews.com/news/crime/2021/08/30/city-officials-fire-it-employee-involved-in-massive-deletion-of-dallas-police-evidence-emails-show/ [accessed January 22, 2024].
14. "Class-Action Lawsuit Claims City Shared Personal Information of 3,700 Employees" (2017, October 3). Available at: https://calgaryherald.com/news/local-news/class-action-lawsuit-claims-city-leaked-personal-information-of-3700-employees [accessed January 22, 2024].
15. "Uber Investigating Breach of Its Computer Systems" (2022, September 15). Available at: https://www.nytimes.com/2022/09/15/technology/uber-hacking-breach.html [accessed January 23, 2024].
16. "What Caused the Uber Data Breach in 2022?" (2023, March 2). Available at: https://www.upguard.com/blog/what-caused-the-uber-data-breach [accessed January 23, 2024].
17. "Information about a Recent Mailchimp Security Incident" (2023, January 13). Available at: https://mailchimp.com/newsroom/january-2023-security-incident/ [accessed January 23, 2024].
18. "Third MailChimp Data Breach Makes It Hard to Rebuild Trust" (2023, January 24). Available at: https://tech.co/news/mailchimp-breach-phishing-trust [accessed January 24, 2024].
19. "Impact to DigitalOcean Customers Resulting from Mailchimp Security Incident" (2022, August 15). Available at: https://www.digitalocean.com/blog/digitalocean-response-to-mailchimp-security-incident [accessed January 24, 2024].
20. "NHS Trust Fined for 56 Dean Streat HIV Status Leak" (2016, May 9). Available at: https://www.bbc.com/news/technology-36247186 [accessed January 25, 2024].
21. A. Zainab, H. Chaminda, N. Liqaa, K. Imtiaz, "Phishing Attacks: A Recent Comprehensive Study and a New Anatomy," *Frontiers of Computer Science*, v. 3, pp. 2–16, 2021. DOI: 10.3389/fcomp.2021.563060.
22. "Your Employees Are the Weakest Link in Your Cybersecurity Chain" (2022, March 11). Available at: https://www.ey.com/en_ca/cybersecurity/your-employees-are-the-weakest-link-in-your-cybersecurity-chain [accessed January 27, 2024].
23. "The Human Factor in IT Security: How Employees Are Making Businesses Vulnerable from Within" (n.d). Available at: https://www.kaspersky.com/blog/the-human-factor-in-it-security/ [accessed January 28, 2024].

24. D. Negussie, "Importance of Cybersecurity Awareness Training for Employees in Business," *Vidya - A Journal of Gujarat University*, v. 2, pp. 2–3, 2023. DOI: 10.47413/vidya.v2i2.206.
25. "What Is "Social Engineering"?" (n.d). Available at: https://www.enisa.europa.eu/topics/incident-response/glossary/what-is-social-engineering [accessed January 29, 2024].
26. "Cyber Security Breaches Survey 2022" (2022, July 11). Available at: https://www.gov.uk/government/statistics/cyber-security-breaches-survey-2022/cyber-security-breaches-survey-2022 [accessed February 3, 2024].
27. Proofpoint, "2020 State of the Phish" (2020). Available at: https://www.proofpoint.com/sites/default/files/gtd-pfpt-us-tr-state-of-the-phish-2020.pdf
28. G. Raywood-Burke, D. Jones, P. Morgan, "Human Factors in Cybersecurity," In *Proceedings of the 14th International Conference on Applied Human Factors and Ergonomics and the Affiliated Conferences,* Ed. A. Moallem, San Francisco, CA: AHFE International, pp. 40–82 (2023).
29. "Detecting and Mitigating a Multi-Stage AiTM Phishing and BEC Campaign" (2023, June 8). Available at: https://www.microsoft.com/en-us/security/blog/2023/06/08/detecting-and-mitigating-a-multi-stage-aitm-phishing-and-bec-campaign/ [accessed February 2, 2024].
30. "Exploiting QR Codes: AiTM Phishing with DadSec PhaaS" (2023, October 17). Available at: https://www.esentire.com/blog/exploiting-qr-codes-aitm-phishing-with-dadsec-phaas [accessed February 2, 2024].
31. "Phishing Attacks: Defending Your Organization" (2018, February 5). Available at: https://www.ncsc.gov.uk/guidance/phishing [accessed February 1, 2024].
32. GreatHorn, "Whaling Attacks" (2021). Available at: https://info.greathorn.com/hubfs/eBooks/greathorn-whaling-attacks-ebook.pdf
33. "Whaling: How It Works, and What Your Organization Can Do about It" (2016, October 6). Available at: https://www.ncsc.gov.uk/guidance/whaling-how-it-works-and-what-your-organisation-can-do-about-it [accessed February 2, 2024].
34. "Phishing Prevention Best Practices: Whaling Attack" (2023, May 15). Available at: https://info.greathorn.com/hubfs/eBooks/greathorn-whaling-attacks-ebook.pdf [accessed February 3, 2024].
35. L. Long, S. Pinzon, J. Wiles, K. Mitnick, *No Tech Hacking*, pp. 13–26 (2008), DOI: 10.1016/B978-1-59749-215-7.00002-0.
36. "A Step-By-Step Guide to Preventing Tailgating Attacks" (2022, February 3). Available at: https://cyberready.com/a-step-by-step-guide-to-preventing-tailgating-attacks [accessed February 4, 2024].
37. "Vishing and Smishing: What You Need to Know" (2023, February). Available at: https://www.cisecurity.org/insights/newsletter/vishing-and-smishing-what-you-need-to-know [accessed February 6, 2024].
38. "Five Tips to Help Avoid Smishing Scams" (n.d). Available at: https://business.bofa.com/en-us/content/what-is-smishing-how-to-prevent-it.html [accessed February 7, 2024].
39. Federal Trade Commission, "Consumer Sentinel Network—Data Book 2022" (2023). Available at: https://www.ftc.gov/system/files/ftc_gov/pdf/CSN-Data-Book-2022.pdf
40. "Smishing and Vishing: What You Need to Know about These Phishing Attacks" (2021, November 11). Available at: https://www.tessian.com/blog/what-is-smishing-and-vishing/ [accessed February 6, 2024].
41. "What Is Smishing and How to Defend Against It" (n.d). Available at: https://www.kaspersky.com/resource-center/threats/what-is-smishing-and-how-to-defend-against-it [accessed February 7, 2024].
42. "SMS Phishing: How to Avoid Smishing Attacks" (2021, September 27). Available at: https://blog.icorps.com/5-ways-to-avoid-sms-phishing [accessed February 7, 2024]
43. "Phishing vs. Smishing vs, Vishing: Key Differences and Prevention" (2023, December 2). Available at: https://nordvpn.com/blog/phishing-vs-smishing-vs-vishing/ [accessed February 7, 2024].
44. "Top 5 Security Misconfigurations Causing Data Breaches" (2024, January 22). Available at: https://www.upguard.com/blog/security-misconfigurations-causing-data-breaches#:~:text=This%20could%20be%20due%20to,attacks%2C%20among%20other%20malicious%20activities [accessed March 3, 2024].

45. The Radicati Group Inc., "Email Statistics Report, 2021–2025" (2021). Available at: https://www.radicati.com/wp/wp-content/uploads/2021/Email_Statistics_Report,_2021-2025_Executive_Summary.pdf?trk=article-ssr-frontend-pulse_little-text-block
46. "10 Email Mistakes that Lead to Security Incidents" (2023, April 12). Available at: https://www.egress.com/blog/data-loss-prevention/the-10-most-common-email-mistakes [accessed February 8, 2024].
47. "You Sent an Email to the Wrong Person. Now What?" (2021, October 4). Available at: https://www.tessian.com/blog/consequences-of-sending-email-to-the-wrong-person/ [accessed February 8, 2024].
48. Solutions Review, "2021 Privileged Access Management Buyer's Guide" (2021). Available at: https://solutionsreview.com/dl/bg/2021_6539765637_PAM_BuyersGuide.pdf
49. "Privileged User Awareness: Defend Your Most Valuable Targets" (2022, July 29). Available at: https://frsecure.com/blog/privileged-user-awareness/ [accessed February 8, 2024].
50. "How Privileged Users Become Privileged Targets for Hackers" (2021, January 13). Available at: https://solutionsreview.com/identity-management/how-privileged-users-become-privileged-targets-for-hackers/ [accessed February 9, 2024].
51. "Privileged Access Management" (n.d). Available at: https://www.beyondtrust.com/resources/glossary/privileged-access-management-pam [accessed February 9, 2024].
52. Duo Cisco Secure, "Multi-Factor Authentication Evaluation Guide 2022" (2022). Available at: https://duo.com/assets/ebooks/Duo-Cisco-Secure-Multi-Factor-Authentication-Guide-2022.pdf
53. "The Emerging Landscape of AI-Driven Cybersecurity Threats: A Look Ahead" (2023, December 28). Available at: https://www.securityweek.com/the-emerging-landscape-of-ai-driven-cybersecurity-threats-a-look-ahead/ [accessed February 10, 2024].
54. "The Ultimate Guide to Password Best Practices: Guarding Your Digital Identity" (2023, November 24). Available at: https://blog.netwrix.com/2023/11/15/password-best-practices/ [accessed February 10, 2024].
55. "NIST Special Publication 800-63: Digital Identity Guidelines—Frequently Asked Questions" (2022, March 3). Available at: https://pages.nist.gov/800-63-FAQ/#q-b07 [accessed February 10, 2024].
56. "20 Password Management Best Practices 2023" (2023, March 10). Available at: https://www.cybertalk.org/2023/03/10/20-password-management-best-practices-2022/ [accessed February 10, 2023].
57. "Passwords, Passwords Everywhere" (2019, April 21). Available at: https://www.ncsc.gov.uk/blog-post/passwords-passwords-everywhere [accessed February 10, 2023].
58. "Most Hacked Passwords Revealed as UK Cyber Survey Exposes Gaps in Online Security" (2019, April 21). Available at: https://www.ncsc.gov.uk/news/most-hacked-passwords-revealed-as-uk-cyber-survey-exposes-gaps-in-online-security [accessed February 10, 2023].
59. "Password Administration for System Owners" (2018, November 19). Available at: https://www.ncsc.gov.uk/collection/passwords/updating-your-approach [accessed February 10, 2023].
60. "What Is Multi-Factor Authentication (MFA) and How Does It Work?" (n.d). Available at: https://www.onelogin.com/learn/what-is-mfa [accessed February 10, 2024].
61. "What Is Adaptive Authentication?" (n.d). Available at: https://www.onelogin.com/learn/what-why-adaptive-authentication [accessed February 10, 2024].
62. "The Definitive Guide to Role-Based Access Control (RBAC)" (2023, December 22). Available at: https://www.strongdm.com/rbac [accessed February 11, 2024].
63. "What Is Conditional Access?" (2024, February 1). Available at: https://oxfordcomputertraining.com/glossary/conditional-access/ [accessed February 11, 2024]
64. "What Is Microsoft Entra Privileged Identity Management?" (2023, October 27). Available at: https://learn.microsoft.com/en-us/entra/id-governance/privileged-identity-management/pim-configure [accessed February 11, 2024]
65. "What Is an Access Review?" (2023, January 4). Available at: https://www.vanta.com/resources/what-is-an-access-review [accessed February 11, 2024].
66. "What Are Access Reviews?" (2023, October 23). Available at: https://learn.microsoft.com/en-us/entra/id-governance/access-reviews-overview [accessed February 11, 2024].

Chapter 15

Human factors in cyber defense

Qasem Abu Al-Haija

15.1 INTRODUCTION

Cybersecurity (CSec) has become a key field incorporating the traditional fields of technology and the intricate relationship between HFs and protective systems. To improve digital landscapes in this stage of amplified cyber dangers, it is essential to possess a comprehensive understanding of human performance, cognition, and corporate culture. This introductory part stipulates the framework for a further cutting-edge debate on "HFs in Cyber Defense." We will shed light on the critical role that individuals, their behaviors, and the greater organizational environment play in determining the success of CSec approaches, providing practical insights and recommendations for the field.

As cyber threats become more sophisticated, it is cardinal to change away from tactics solely focused on technology and toward a more holistic approach encompassing the human component. Dhamija et al. [1] demonstrate the dominance and efficiency of SEng attacks and emphasize the need to address human vulnerabilities through grown user knowledge and education. This shift is seen in these works, demonstrating the ubiquity of SEng attacks. Anderson's [2] study of cognitive biases in decision-making procedures indicates the profound impact of human psychology on CSec consequences, requiring a nuanced assessment of these factors in protection policies.

The importance of UCD concepts in security systems must be considered, particularly since corporations strive to overcome the challenges posed by human-centric vulnerabilities. This technique balances strong security measures and user-friendly, intuitive interfaces. Moreover, examining all of the psychological aspects that lead to internal risks is crucial. This is highlighted by Ackerman et al.'s [3] evaluation of factors and indicators. They specified a guide for identifying and preventing hostile activities that launch within an organization.

A comprehensive understanding of HMC is essential in developing AI and automation in CSec [4]. Furnel et al. [5] contend that AI's trust-building techniques and apparent decision-making procedures are necessary for ensuring harmony between human intuition and machine efficiency.

CSec is beyond technology resolutions; the obscure correlation between technological resolutions and HFs is highlighted. To develop security measures that are both powerful and effective, it is necessary to comprehend and manage the HFs involved in cyber defense. This chapter examines the human actions, cognition, and psychological factors that influence CSec. It also examines how defenses could be adjusted against electronic threats.

This chapter navigates this intricate area by inspecting these foundational studies and others to arrange perceptions into the interaction between HFs and CSec. By combining multiple viewpoints, we can contribute to the ongoing debate and guide the design of flexible and robust CSec approaches. Figure 15.1 illustrates the major human factors involved in CSec.

 DOI: 10.1201/9781032714806-18

15.2 RELATED RESEARCH

In its most basic form, CSec is defined by the dynamic interplay of technological protections and the HFs that generate and interact with these defenses. This literature review aims to assess and synthesize the most relevant findings from various studies to shed light on the complicated relationship between HFs and cyber protection.

Dhamija et al. [1] initiated a comprehensive investigation of why phishing attacks are successful. They emphasized the necessity of comprehending human flaws to strengthen user education and awareness. Anderson [2] established the notion of security engineering as an underlying work that stipulated an inclusive approach to developing trustworthy distributed systems. The conclusion of this study emphasizes the multidimensional nature of CSec, which is beyond the specific technical elements.

Adams and Sasse [6] also stimulated the UCD approach in security systems, accentuating the crucial role of users. They asserted that users are not the adversaries, demonstrating the significance of users. Their views are essential for balancing user-friendly interfaces and security defenses. In addition, Ackerman et al. [3] disseminated internal threats, categorizing stimuli and identifying behavioral indicators. Their study establishes a firm basis for evaluating internal organizational risks and implementing mitigation strategies. Moreover, Liu et al. [4] investigated several XAI models for CTI and requested that AI decision-making activities be considered. This task is essential to ensure that the decisions made by automated systems are accurate.

Furnell and his colleagues conducted extensive research in AI and ML in CSec. Their findings provided a vital insight into HMC's development. The study examines how human intuition and technical advancements could be utilized to combat cyberattacks effectively. Furthermore, Herley and Florencio [7] reviewed the economics of CSec, highlighting the difficulties of seeking encouragement for users, developers, and hackers. This economic approach could provide a more comprehensive overview of the consequences of cyber threats.

Von Solms and Van Niekerk [8] were instrumental in developing a human-oriented approach to information security. They focus on human involvement when developing appropriate CSec awareness programs. Moreover, Sasse et al. [9] studied the authentication methods and the challenges of establishing protected and intelligible authentication systems.

According to Blythe et al. [10], they analyzed the impact of security cautions on user decision-making. Their findings contribute to the ongoing debate about the importance of security cautions and the need to develop effective solutions that meet human cognition. Whitty and Doodson [11] analyzed the psychological influences of individuals vulnerable to online romance scams. Their research outcomes disclosed precious perceptions of the SEng aspects of cyber risks and the impact of targeted awareness.

Campbell and Gordon analyzed the impact of OCC processes on organizational culture. Their proposed model emphasizes the need for a security-conscious philosophy and emphasizes the responsibility of leadership in establishing a resilient CSec environment. Furnell and Clarke have investigated the challenges of using user-centric privacy and security solutions. This study examines the ongoing debate about allowing individuals to engage in CSec.

Bonneau et al. [12] analyzed password security and perceived significant differentiations between endorsed and actual user behaviors. This study sheds light on the challenges of forming strict password requirements without triggering any inconveniences for users. Furthermore, Mayer-Schonberger and Cukier [13] assayed "big data" and its consequences for CSec. Their research focuses on the potential and challenges of employing significant amounts of data to identify and prevent cyberattacks.

This review of the relevant state of the art brings together a variety of perspectives, establishing the framework for more thorough research of the HFs that radically influence

Figure 15.1 The main HFs around CSec.

CSec. By examining previous studies, this chapter intends to provide a more comprehensive understanding of the multifaceted nature of human involvement in cybersecurity. We present Table 15.1 to provide readers with information on the evaluated studies discussed in this chapter.

15.3 HUMAN-CENTRIC APPROACH TO CSEC

When striving to develop a robust CSec framework, it is critical to prioritize a human-centered approach. This section focuses on cognitive biases in decision-making, SEng exploits, and the usefulness of user awareness and education; it also studies fundamental aspects of understanding human behavior in CSec. The section focuses on the significance of this insight.

15.3.1 Understanding human behavior

- Cognitive biases in decision-making might unintentionally cause weaknesses in CSec. Heuristics and biases may lead people to deviate from rational decision-making, and Tversky and Kahneman's research [14] provided a framework for analyzing these biases. Egelman et al. [15] stated that it is essential to be knowledgeable of these intellectual errors to design security techniques that realize and mitigate specific biases.
- **Experiential skills in software engineering:** Using human psychology to convince people that they are presenting sensitive information or contributing to security-related

activities is an instance of an SEng exploit. Mitnick and Simon [16], in their classical book, delve into the psychology of SEng. It underlines the need for CSec strategies that address the human component. When one thoroughly knows the techniques attackers use, developing effective countermeasures and educating users about potential threats becomes easier.
- **Acquiring knowledge and user awareness:** To establish a human-centric CSec approach, user awareness and education are critical pillars. Hadnagy's "SEng: The Art of Human Hacking" [17] highlights the importance of proactive education to equip users to detect and protect themselves against SEng attacks. Implementing good training programs and awareness campaigns, as advocated by Camp [18], greatly aids in developing a user base that is aware of security and can make decisions based on that understanding.

This section underlines the need to understand human behavior while building CSec procedures. Organizations may significantly increase their defenses against expanding cyber threats by addressing cognitive biases, anticipating SEng vulnerabilities, and increasing user awareness, primarily via education. This allows businesses to increase their defenses drastically.

15.3.2 UCD in security systems

The UCD approach is one of the most significant components in building effective security systems that provide reliable protection while ensuring a positive user experience. This section discusses the core components of UCD. These components include design considerations for user-friendly security interfaces and the delicate balance between usability and security.

15.3.2.1 Principles of design for security interfaces (friendly to end users)

User-friendly security interfaces are one of the most fundamental considerations in prompting individuals to engage in safe activities. Using notorious design concepts to enhance the effectiveness and efficiency of security components is advantageous. Nielsen's usability criteria [19] provide a crucial foundation, incorporating system state visibility, system compatibility with the real world, and user control and autonomy. Integrating these concepts into the design of security interfaces confirms that interfaces are user-friendly and reduces the likelihood of user errors.

The study "Why Johnny Can't Encrypt" [20] by Whitten and Tygar shows the usability issues with encryption software. This underlines the need for security systems to meet users' mental abilities and capabilities, imposing the design of interfaces that can accommodate users with different levels of experience.

15.3.2.2 Striking a happy medium between usability and safety

One of the most significant concerns in UCD for security systems is realizing optimal usability and security. Adams and Sasse [6] accentuate that those aggressive practices damage consumers' productivity. Their research encourages systems to integrate security features efficiently without requiring unwarranted cognitive requirements on users or asking them to do additional tasks. Continuous repetition, usability assessment, and a full understanding of user workflows can only accomplish this objective. This is accomplished to ensure that security measures meet user expectations.

Table 15.1 The summary of surveyed papers

Ref.	*Key contribution*	*Advantages*	*Limitations*
	Examined why phishing attempts are successful, highlighting the need to know human weaknesses.	• Offers insights on SEng exploits.	• Limited emphasis on the technical components of phishing attempts.
	Provided a detailed approach to developing reliable distributed systems, emphasizing the multifaceted nature of CSec.	• Offers a comprehensive view of security engineering.	• Generalized advice; needs more specifics in some situations.
	Emphasized the importance of users and pushed for a UCD approach in security systems.	• Advocates for a balanced approach to security and usability.	• Balancing security measures and user-friendly interfaces presents challenges.
	Investigated insider risks by classifying reasons and finding behavioral markers.	• Builds a framework for recognizing and minimizing insider risks.	• Limited investigation of technological elements of insider risks.
	Investigated XAI for CTI and advocated openness in AI decision-making processes.	• Addresses the need for openness in AI-powered cybersecurity (CSec).	• There needs to be more discussion of the practical issues of building explainable AI.
	CSec performed a detailed poll on AI and ML.	• Provides insights about the changing landscape of HMC.	• It gives a general overview but needs to go into detail on individual AI applications.
	Critically studied the economics of CSec, identifying issues in matching incentives for consumers, developers, and attackers.	• Offers an economic viewpoint on CSec.	• CSec has limited practical recommendations for resolving economic inequality.
	Proposed a human-centric approach for information security awareness, highlighting the significance of addressing the human factor in CSec.	• Emphasizes the importance of human elements in building effective awareness initiatives.	• The suggested human-centric paradigm is not empirically validated.
	Investigated the usability and security trade-offs in authentication systems, offering insights into creating secure, user-friendly authentication processes.	• Provides insight into the issues of combining security and usability.	• It focuses mostly on authentication, with little treatment of larger CSec concerns.
	A study was conducted on the influence of security alerts on user decision-making processes, adding to the debate over their usefulness.	• Gives insight into the psychological factors of responding to security warnings.	• Limited research addresses the long-term usefulness of security alerts.
	Investigated the psychological features of individuals vulnerable to online romance frauds, offering insights into the SEng components of cyber dangers.	• Provides insight into the SEng elements of cyber dangers.	• Focusing on a single sort of cyber danger limits generalizability.

(*Continued*)

Table 15.1 (Continued) The summary of surveyed papers

Ref.	*Key contribution*	*Advantages*	*Limitations*
	Investigated the impact of company culture on CSec practices, highlighting the need for a security-conscious culture.	• Emphasizes the role of organizational culture in CSec.	• There are limited tangible techniques for developing a security-conscious society.
	Examined the problems and possibilities given by user-centric privacy and security controls, contributing to the conversation around empowering users in CSec.	• Contributes to the discussion on user empowerment in CSec.	• There needs to be more coverage of particular privacy and security control systems.
	Conducted an empirical investigation of password security, revealing discrepancies between suggested methods and user behavior.	• Provides insight into the difficulty of implementing strong password regulations.	• It focuses exclusively on password security, with little treatment of broader CSec topics.
	Explored the notion of "big data" and its implications for CSec, examining the benefits and drawbacks of harnessing massive volumes of data.	• Provides insights into using big data for CSec.	• There must be more debate on the ethical and privacy issues of employing big data in CSec.

Cranor and Garfinkel's [21] "Security Usability Principles for Vulnerability Analysis and Policy Enforcement" investigate the conditions for implementing efficient security into the design framework. These conditions provide ample information by providing security measures that enhance the user's enjoyment.

In security systems, it is essential to utilize the UCD methodology [18] to create user interfaces that are straightforward to use while providing excellent protection. Organizations may accomplish a secure user infrastructure by following established design principles and achieving stability and safety.

15.4 HUMAN ERROR AND CSEC

Human errors in CSec can cause different security problems and undermine organizational defense (discussed in this section). Such errors could differ from unreliable behavior to ineffective knowledge and awareness. Mistake-tolerant skills, education programs, and simulations are engaged to lower the likelihood of human errors.

15.4.1 Types of human errors

- Unintentional activities can result in various security troubles. Many CSec concerns are caused by users' actions, such as browsing malicious websites or revealing confidential information. Sheng et al. [22] postulate comprehension of how unintentional incidents caused by user attitudes lead to security concerns. To execute preventative procedures, it is vital to recognize the nature of unintentional errors.
- The lack of knowledge and expertise. In CSec, the likelihood of human error decreases significantly when personnel lack knowledge and training. Williams and Camp [23] demonstrate the need to address security awareness's psychological issues and highlight inadequacies in users' mental models. They emphasize the prominence of thorough training programs that transfer knowledge and initiate a secure user environment.

15.4.2 Mitigating human error

- **Simulations and employee training programs:** Effective training programs alleviate human-induced errors by increasing user awareness and understanding. Kim and Solomon [24] propose simulations that provide an effective and effective learning environment. Users may learn to identify and respond to security issues by incorporating real-world scenarios into their training program.
- **Implementing error-tolerant systems in organizations:** Creating error-tolerant systems is essential to tackle human errors. Leveson [25] introduces the ideas of system safety engineering, advocating for establishing systems that can survive and recover from mistakes. Redundancy, fail-safe features, and user-friendly interfaces help build systems resistant to human error [18].

Finally, addressing human errors in CSec necessitates a multifaceted approach that implies recognizing the diverse types of errors, inducing awareness, and implementing approaches to mitigate the impact of these errors. Corporations may be able to safeguard their users' inadvertent actions by using training, simulations, and error-tolerant design.

15.5 INSIDER THREATS

Insider threats arise within an organization and involve intentional and unintentional actions that can lead to security breaches. Insider attacks are a significant risk to all aspects of CSec. This section characterizes insider risks, distinguishes between adverse insiders and accidental risks, and identifies behavioral factors. In addition, the psychological aspects of insider threats are discovered, such as the sources of such attacks and systems to identify and avoid them.

15.5.1 Identifying insider threats

- The ability to distinguish between malevolent insiders and unintentional hazards is essential for effective mitigation. Malicious insiders are not devised to cause harm. The research by Ackerman and colleagues [3] provides significant insights into many insider threats. When businesses fully realize these characteristics, they may tailor their detection and prevention approaches to various factors and behaviors.
- The use of behavioral markers can help identify insider attacks. Bishop and Engle [26] investigated various behavioral indicators, including changes in work habits or sudden demands for access. Recognizing such indicators enables businesses to address potential concerns and take proactive preventative measures.

15.5.2 Psychological aspects of insider threats

- A successful preventative approach requires understanding the reasons behind insider attacks. Stottelaar and Shishika [27] highlighted the industry's specific incentives to understand the underlying reasons for developing targeted countermeasures. Those reasons may be philosophy, financial gain, or political hatred.
- To create strategies for identifying and preventing internal risks, a multifaceted strategy must be necessary. Almukaynizi and his colleagues [28] provide a comprehensive overview of detection mechanisms and emphasize the importance of ongoing monitoring. There is a correlation between combining advanced technology and user behavior analytics and an organization's ability to effectively identify and respond to hazards within the organization [18].

The conclusion states that a comprehensive understanding of the various components is essential to combat insider threats effectively. Corporations can recognize adverse insiders and accidental risks, evaluate behavioral indications, and assess the psychological aspects of insider incidents to improve their defenses against this complex CSec threat.

15.6 HMC IN CYBER DEFENSE

The development of collaboration between humans and computers has been a central factor of contemporary CSec, with AI and automation being the elementary tasks. This section focuses on using AI and automation to boost human capacities and the challenges correlated with their consolidation. It also considers approaches for establishing confidence in security systems via openness and human-based surveillance.

15.6.1 Role of AI and automation

- It is challenging to utilize AI and automation in CSec due to its ability to enhance human threat detection, analysis, and response capabilities. Liu et al. examine how AI can improve CTI by employing ML algorithms that enable the lengthy analysis of large datasets. This provides significant insights while allowing individuals to prioritize strategic decisions.
- In contrast, the theoretical integration of AI and automation in CSec poses various risks and obstacles. Anderson's [2] work shows the need to consider automated systems' vulnerabilities and unintended consequences. Concerns regarding adversarial attacks on AI models and the risk of bias in decision-making highlight the significance of utilizing AI thoughtfully and constantly monitoring its performance.

15.6.2 Building trust in security systems

- In the Automated Decision-making Process, transparency in automated systems' decision-making processes is essential to enhance trust in AI-powered security solutions. Adams et al. [6] emphasize the importance of explainability in AI systems (i.e., XAI systems), expressing that they enable users and stakeholders to recognize the motives behind decisions. Encompassing transparency approaches expand accountability and user acceptability of AI-powered security solutions.
- **Human oversight and intervention:** While AI and automation contribute to CSec, preserving human management is essential. Camp et al. [18] contend for a balanced approach in which individuals are not disqualified from decision-making. To provide a safety net against unplanned occurrences and escalating threats, humans must be able to intervene and reject automated system decisions.

The synergy between humans and robots in CSec is essential to sustaining a competitive advantage over sophisticated adversaries. When establishing confidence in automated security systems, addressing potential threats, providing transparency, and having human involvement are crucial. AI and automation both strengthen security efficiency.

15.7 ORGANIZATIONAL CULTURE AND CSEC

The organizational culture contributes considerably to the effectiveness of CSec measures. In this section, we examine the key elements that must be present to create a secure environment within an organization, focusing on leadership commitment, employee participation, and accountability. Furthermore, the absorption of security into the development process, the integration of CSec objectives with corporate objectives, and the integration of security into business operations are examined.

15.7.1 Fostering a security-conscious culture

- The Commitment of Leadership to CSec Establishing a security-conscious environment requires leadership. The study conducted by Campbell and Gordon [29] emphasizes the importance of leadership in establishing the fundamental principles that govern a company's culture. Leadership commitment to CSec guides on the importance of security measures, which promotes a collective mentality that emphasizes security and incorporates it into daily operations.
- **Responsibility and employee engagement:** Employee involvement and responsibility are essential to creating an effective CSec culture. Furnell and Clarke's [30] essay discusses the pros and cons of using user-centric privacy and security measures. A more resilient corporation may be created by promoting employee participation in awareness campaigns and enhancing a sense of ownership in the organization's security procedures. Organizations may benefit from employee insights and attentiveness by having them actively participate in CSec operations, thus increasing overall security.

15.7.2 Integrating security into business processes

- **Strengthening the coherence between CSec objectives and organizational objectives:** To maximize the success of CSec initiatives, security objectives must be aligned with larger business objectives. Von Solms and Van Niekerk [31] present a human-centered approach to information security, highlighting the importance of a comprehensive approach. To foster a culture where security is crucial to the overall performance, businesses must ensure that CSec's objectives align with the company's objectives.
- Incorporating security throughout development is essential for preventing software and system vulnerabilities. Adams and Sasse assert that a UCD approach in security systems is essential. Using security issues from the beginning of the development process facilitates companies' development of robust and secure systems. This reduces the likelihood of vulnerabilities occurring after installing the system and ensures an effective CSec strategy is implemented.

Eventually, forming a security-aware philosophy inside a company demands a widespread strategy incorporating leadership dedication, employee commitment, and the continuous assimilation of security into business processes. Organizations may adopt robust and preemptive CSec principles by linking CSec objectives with more corporate objectives and integrating security into the development process.

15.8 CONCLUSIONS

The investigation into "HFs in Cyber Defense" has highlighted crucial elements to enhance CSec resilience. Understanding human behaviors, reducing cognitive biases, and establishing a secure business environment are essential to developing effective defensive strategies. The combination of AI and automation has the potential to be transformative; however, it necessitates a rigorous assessment of the risks involved, highlighting the importance of transparency and human oversight. Future initiatives must prioritize multidisciplinary research, user-centric approaches, and ethical considerations to prevent the development of threats and secure the digital future. This is because the CSec environment is constantly evolving and changing.

Politics, industries, and academics must collaborate to achieve this dynamic area. We can collaborate with our defenses and create a CSec environment that matches the current challenges of the digital world by examining human aspects, technological innovations, and organizational dynamics.

REFERENCES

1. S. Dhamija, J. D. Tygar, and M. Hearst, "Why phishing works," In *Proceedings of the SIGCHI Conference on Human Factors in Computing Systems,* 2006, pp. 581–590. doi: 10.1145/1124772.1124861.
2. R. Anderson, *Security Engineering: A Guide to Building Dependable Distributed Systems*, John Wiley & Sons, 2008.
3. L. Ackerman, M. Carbone, C. S. Furnell, and A. R. Leeson, "A survey of insider threat detection in cloud computing," *Computers & Security*, vol. 68, pp. 157–176, 2017.
4. Y. Liu, X. Tang, X. Chen, and Y. Huang, "Explainable artificial intelligence for cyber threat intelligence: A survey," *IEEE Access*, vol. 7, pp. 127204–127222, 2019.
5. C. S. Furnell, L. J. Camp, M. M. A. Rahman, M. E. M. Law, and A. Y. A. Alsudais, "Artificial intelligence and machine learning in cybersecurity: A review," In *Proceedings of the 2nd International Conference on Computing and Artificial Intelligence (ICCAI 2020)*, 2020, pp. 63–70. doi: 10.1007/978-3-030-31703-4_16.
6. A. Adams and M. A. Sasse, "Users are not the enemy," *Communications of the ACM*, vol. 42, no. 12, pp. 41–46, 1999.
7. N. Herley and P. Van Oorschot, "So long, and no thanks for the externalities: The rational rejection of security advice by users," In *Proceedings of the 2009 Workshop on New Security Paradigms Workshop,* 2009, pp. 133–144. doi: 10.1145/1719030.1719050.
8. J. Von Solms and J. Van Niekerk, "From information security to cybersecurity," *Computers & Security*, vol. 38, pp. 97–102, 2013.
9. M. A. Sasse, S. Brostoff, and D. Weirich, "Transforming the 'weakest link'-a human/computer interaction approach to usable and effective security," *BT Technology Journal*, vol. 19, no. 3, pp. 122–131, 2001.
10. C. Blythe, S. Coventry, and P. Adams, "A hindsight analysis of the psychological determinants of clicking on phishing emails," *Computers in Human Behavior*, vol. 91, pp. 311–318, 2019.
11. M. Whitty and R. Doodson, "Behavioural characteristics of individuals who fall victim to phishing attacks," *Cyberpsychology, Behavior, and Social Networking*, vol. 23, no. 3, pp. 179–185, 2020.
12. J. Bonneau, S. Preibusch, and R. Anderson, "A birthday present every eleven wallets? The security of customer-chosen banking PINs," In *Proceedings of the 2012 ACM Conference on Computer and Communications Security,* 2012, pp. 657–666. doi: 10.1007/978-3-642-32946-3_3.
13. V. Mayer-Schönberger and K. Cukier, *Big Data: A Revolution that Will Transform How We Live, Work, and Think*, Houghton Mifflin Harcourt, 2013.
14. A. Tversky and D. Kahneman, "Judgment under uncertainty: Heuristics and biases." *Science*, vol. 185, no.4157, pp. 1124–1131, 1974.
15. S. Egelman, E. Peer, and A. P. Felt, "When permission models collide: End users' mental models of permission & authority." In *Proceedings of the SIGCHI Conference on Human Factors in Computing Systems,* 2013, pp. 1391–1400.
16. K. D. Mitnick and W. L. Simon, *The Art of Deception: Controlling the Human Element of Security*, Wiley, 2002.
17. C. Hadnagy, *Social Engineering: The Art of Human Hacking*, John Wiley & Sons, 2011.
18. L. J. Camp, *CSec Awareness: Building a Culture of Defenders*, CRC Press, 2019.

19. J. Nielsen, "Heuristic evaluation," In *Usability Inspection Methods*, John Wiley & Son, New York, 1994, pp. 25–62.
20. A. Whitten and J. D. Tygar, *Why Johnny Can't Encrypt: A Usability Evaluation of PGP 5.0*, USENIX Security Symposium, 1999, pp. 169–184.
21. L. F. Cranor and S. Garfinkel, "Security usability principles for vulnerability analysis and policy enforcement," *IEEE Security & Privacy*, vol. 3, no. 5, pp. 24–33, 2005.
22. S. Sheng, M. Holbrook, P. Kumaraguru, and L. F. Cranor, "The weakest link: An empirical analysis of the security perceptions and habits of smartphone users," In *Proceedings of the SIGCHI Conference on Human Factors in Computing Systems,* ACM, Montréal Québec, Canada, 2007, pp. 1389–1398.
23. L. J. Williams and L. J. Camp, "The myth of the security mind," *Information Systems Security*, vol. 20, no. 1, pp. 33–41, 2011.
24. J. Kim and M. Solomon, "Using simulated security games to analyze and enhance human decision-making in security environments," *IEEE Transactions on Systems, Man, and Cybernetics: Systems*, vol. 45, no. 4, pp. 545–556, 2015.
25. N. Leveson, *Safeware: System Safety and Computers*, Addison-Wesley, 1995.
26. M. Bishop and S. Engle, "Insider threat detection: A review and integration of theory and practice," *Journal of Organizational and End User Computing*, vol. 30, no. 2, pp. 18–37, 2018.
27. B. Stottelaar and D. Shishika, "Insider threats in the healthcare sector: A literature review," *Journal of Medical Systems*, vol. 43, no. 6, p. 148, 2019.
28. Choudhary, A., & Bhadada, R. (2022). Insider threat detection and cloud computing. In *Advances in Data and Information Sciences: Proceedings of ICDIS 2021*, Editors: Shailesh Tiwari, Munesh C. Trivedi, Mohan Lal Kolhe, K.K. Mishra, Brajesh Kumar Singh. (Singapore: Springer Singapore. pp. 81-90, Agra, India.
29. D. Campbell and S. Gordon, "Culture as culprit: Four steps to effective change," *Cutter IT Journal*, vol. 31, no. 11, pp. 6–12, 2018.
30. C. S. Furnell and N. L. Clarke, "Evaluating the usability and security of a graphical authentication mechanism," *International Journal of Human-Computer Studies*, vol. 70, no. 1, pp. 65–77, 2012.
31. R. Von Solms and J. Van Niekerk, "From information security to cybersecurity," *Computers & Security*, vol. 38, pp. 97–102, 2013.

Chapter 16

Security operation center

Toward a maturity model

Achraf Samir Chamkar, Yassine Maleh, and Noreddine Gherabi

16.1 INTRODUCTION

A security operations center (SOC) represents a sophisticated infrastructure pivotal in providing comprehensive oversight and situational awareness for an enterprise, bolstering security by swiftly detecting anomalies, threats, and potential intrusion attempts through continuous monitoring. It is a centralized hub where all IT-generated events converge for analysis by a team of skilled security analysts. When searching for "SOC" on Google, one is likely to encounter images depicting rooms adorned with large screens, actively displaying data and graphs—a visual representation of a typical SOC setup. However, the intricacies of an SOC extend beyond mere visualizations, encompassing a diverse array of components and processes. The concept of SOC has evolved over the past fifteen years as a strategic defense mechanism against increasingly sophisticated cyberattacks. SOCs vary in scale and scope, ranging from smaller, internally managed setups to expansive operations staffed by numerous analysts operating round the clock. Most SOCs operate as managed services, as establishing and maintaining an in-house SOC entails considerable expenses. While imperative for organizational security in today's landscape, financial barriers often deter small enterprises from establishing their SOC (Taqafi et al., 2023).

A Security Operations Center (SOC) can be a physical or virtual (cloud-based) facility that is established to protect an organization's information security posture. The main objective of an SOC is to detect, analyze and respond to potential threats to an organization's assets. SOCs are staffed by a combination of cybersecurity professionals and are made up of specialized tools and technologies, processes and people, all focused on the same objective: mitigating security threats. It is an integral part of any organization's cybersecurity strategy, and ensures that security teams have the information and tools they need to detect, respond to and remediate potential threats (Muniz, 2021).

The purpose of this study is to argue that an SOC is the best managerial response to the issue of cyber protection. To accomplish the overarching goal, it is essential to attend to the first particular purpose, which is to present security operations. This includes explaining why an SOC was created and what services it offers. The second specific goal is to describe the process–people–technology triangle that helps with SOC management via their interaction and synergy. At last, a third particular aim will be created to detail the steps to take when developing an SOC.

This chapter is structured into six main sections to provide a comprehensive understanding of SOCs and their maturity assessment. Beginning with exploring the background, Section 16.2 delves into the various functions, challenges, benefits, and types of SOCs, setting the stage for a deeper examination. Section 16.3 outlines the methodology employed in developing the proposed maturity framework, elucidating the research approach, and data collection methods. Section 16.4 presents the theoretical framework underpinning the

DOI: 10.1201/9781032714806-19

SOC maturity model, offering insights into its conceptual foundation. The focal point of the chapter, Section 16.5, introduces the proposed SOC maturity framework, elucidating its components and domains in detail. Finally, Section 16.6 encapsulates the findings and implications of the study, providing a conclusive overview of the research outcomes and potential avenues for future exploration.

16.2 BACKGROUND

16.2.1 SOC functions

To provide insights into the essential functions required in designing your operating model, we have outlined the most common core functions typically found in SOCs (Forte, 2003; Tsoutsos & Maniatakos, 2014). It's worth noting that these functions may go by different names, and some functions might be combined for efficiency. Figure 16.1 presents the functions of a typical SOC.

- **Threat intelligence (TI):** This function furnishes the SOC with up-to-date information concerning cyber threats targeting the organization and its IT infrastructure. This encompasses indicators of compromise (IOCs) and qualitative data regarding threat actors.
- **Threat hunting:** This function focuses on proactively identifying and investigating cyber threats that may have bypassed existing security controls through iterative and human-centric approaches.
- **Content development or analytics:** This team transforms actionable TI into programmatic detection rules within SOC tools. These rules are instrumental in identifying suspicious behavior across systems and services.
- **Engineering:** Responsible for maintaining SOC tooling, technical onboarding of logs (commonly referred to as plumbing), and ensuring smooth operation of all systems.
- **Incident response or handling:** This team investigates the root cause of security events and determines whether they should be escalated to security incidents, aiming to mitigate their impact.
- **Incident management (IM):** Once a security event is confirmed as an incident, the SOC forwards it to the IM team. This team manages various responsibilities, from communication to technical responses, to minimize the impact and control the damage caused by the incident.

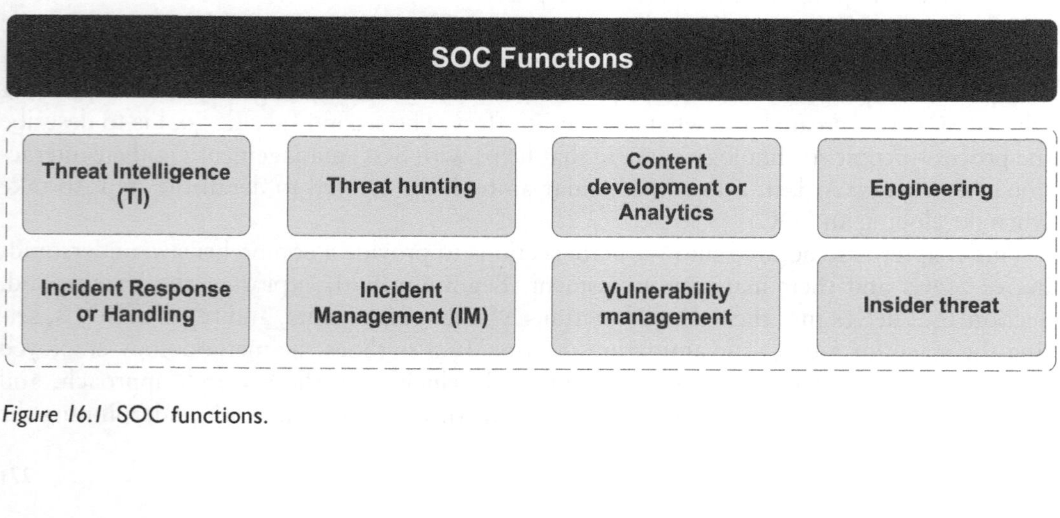

Figure 16.1 SOC functions.

- **Vulnerability management:** An SOC tasked with identifying and managing vulnerabilities within the organization's infrastructure requires distinct resources, expertise, and tools for testing, patching, and evaluating system builds. This function demands more significant interaction with systems and heightened responsibility, necessitating careful consideration in the SOC design.
- **Insider threat:** An SOC tasked with detecting and responding to insider threats will have a different configuration compared with one solely focused on external threats. While not mutually exclusive, monitoring employees requires sensitive handling, necessitating segregation of any capability, specifically toolsets, involved in such tasks. This segregation ensures that the SOC, its staff, and other business systems and staff subject to security monitoring and investigations maintain appropriate boundaries.

16.2.2 People, process, technology, and governance

It is necessary for an SOC to have people, procedures, technology, and governance to accomplish its goals, given the functional areas. There is a parallel between this paradigm and the three-layer model proposed by van den Berg (2018). In this approach, the layers of governance, social–technical, and technical have been identified. More information on these points of view is provided in the phrases that are to come.

16.2.2.1 People

Numerous assertions have been made on the significance of individuals in an SOC (Mansfield-Devine, 2016). It is possible for people to offer context for judgments while they are responding to alarms or events. The responsibilities that are performed at an SOC determine the number of employees that are employed there. With that being said, the majority of SOCs include both a manager and security analysts (Figure 16.2). The level of expertise required for a security analyst varies according to the tasks and activities that they perform (Miloslavskaya, 2016).

It is possible to categorize security analysts according to the following five levels:

- **Security analyst level 1:** Furthermore, the primary duty of the initial security analyst is to maintain continuous surveillance over a company's infrastructure and systems. This entails managing alarms, assessing the seriousness of alerts, and furnishing the level 2 security analyst with pertinent information for deeper investigation.
- **Security analyst level 2:** When analyzing scenarios, level 2 security analysts take into consideration a large number of security inputs relating to the data and take the right actions to mitigate or maybe avoid any more damage to the firm might be considered.
- **Security analyst level 3:** There are a few different names for the level 3 security analyst, one of which is the subject matter expert. The individual in question is an authority in a particular field of expertise, such as TI or malware development. The primary goal of this initiative is to adopt preventative actions to avoid incidents. Furthermore, a security analyst with a level 2 capability can at any moment make a request for professional help from a security analyst with a level 3 capability.

16.2.2.2 Process

The examination of processes within an SOC is approached from diverse viewpoints in the literature. Numerous perspectives on the operational functioning of an SOC are evident in the published scientific literature. "Computer Security Incident Handling Guide" is the

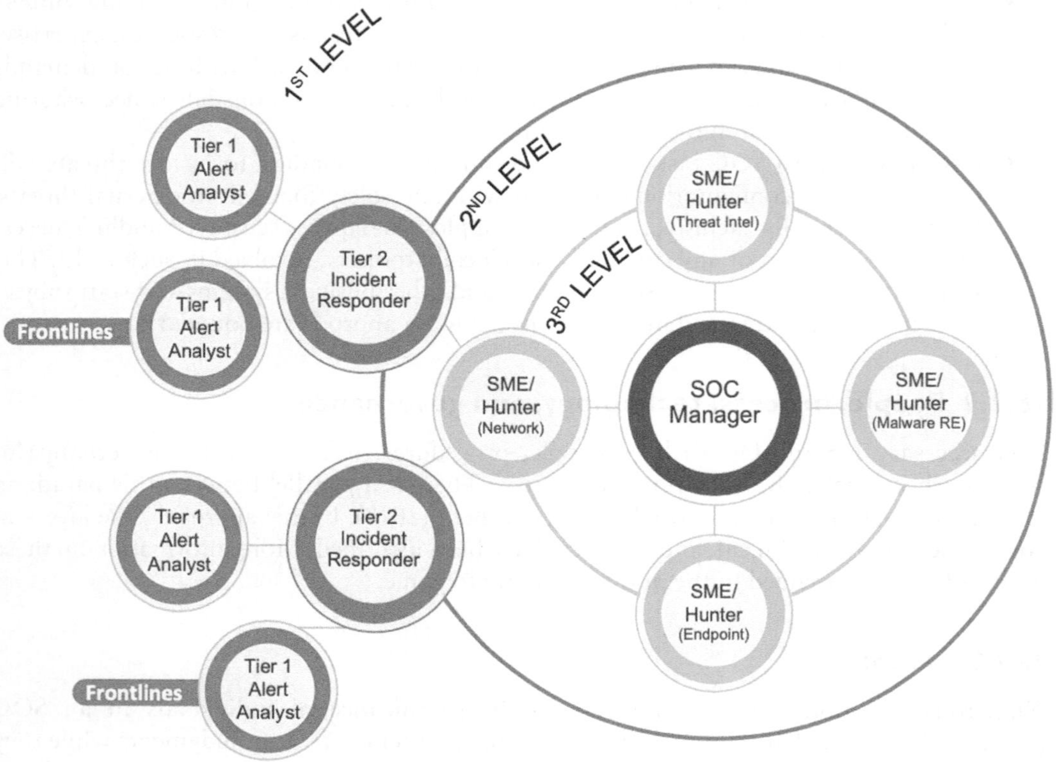

Figure 16.2 SOC levels.

title of a white paper released by the Escal Institute of Advanced Technologies (Scarfone et al., 2008). Within the body of published work, one may discover a wide variety of viewpoints concerning the procedures of an SOC. A white paper issued by the Escal Institute of Advanced Technologies provides detailed insights on the topic (Sundaramurthy et al., 2014). The organization's discretion guides the establishment of policies and procedures beyond the security incident response protocols, tailored according to its specific responsibilities and functions.

16.2.2.3 Technology

When it comes to efficient SOCs, a monitoring system is the most important component. In addition, this monitoring solution can gather, identify, aggregate, and analyze log data from a variety of different systems. Some examples of these include network devices, firewalls, mail filters, antivirus software, intrusion detection systems, intrusion prevention systems, proxy services, and business applications. The majority of SOCs make use of a monitoring system that is known as a Security Information and Event Management system (Chamkar et al., 2022).

16.2.2.4 Governance and compliance

An SOC plays a vital role in ensuring compliance with regulations, standards, and guidelines, serving as a valuable resource for auditors to assess and evaluate a company's IT infrastructure. Utilizing metrics, the SOC can effectively monitor various aspects of its performance, including the breadth of coverage (e.g., number of devices monitored), efficiency metrics (such as average analysis time and remediation enforcement time), incident response metrics (number of incidents handled), and compliance-related metrics (number of policy violations detected). These metrics provide valuable insights into the SOC's effectiveness and its ability to safeguard the organization's assets and uphold regulatory requirements.

16.2.3 SOC challenges

An SOC is typically an add-on service for enterprises, not originally part of their infrastructure. Integrating an SOC into the company's environment can be complex but is essential for maximizing its effectiveness. Mutemwa et al. (2018) propose integrating IM and other processes by establishing a council to oversee integration and assess IT environment changes. SOC playbooks serve as valuable resources for guiding workers on incident response procedures. Analysts working in an SOC often face a high volume of alerts daily, primarily manually validated due to SOC operations still largely reliant on manual processes. This influx of alerts, many of which are false positives, poses challenges, potentially leading to alarm fatigue, and desensitization (Wall & Rodrick, 2021). To address this issue, modern SOCs integrate TI sources and employ artificial intelligence algorithms to enhance event classification and security detection processes (Agyepong et al., 2020).

Effective alert management is crucial, requiring customized detection rules and alarms tailored to each organization's unique networks and systems (Vielberth et al., 2020). Setting up an SOC involves an initial onboarding phase where all technologies and business services are integrated into the SOC platform for real-time monitoring. However, current onboarding approaches, particularly for on-premises SOC services, can be time-consuming, expensive, and challenging (Taqafi et al., 2023). Leveraging cloud infrastructure can streamline this process, offering a faster, more efficient, and cost-effective onboarding procedure.

Looking ahead, SOC processes and technologies are expected to become increasingly automated, with cloud solutions enhancing speed and service delivery. Despite technological advancements, human error remains a significant concern, with SOC analysts being susceptible targets for cyberattacks. Employee training plays a crucial role in bolstering security awareness and reducing the risk of attacks. However, statistics show that a significant percentage of employees, especially in government sectors, lack sufficient training on ransomware prevention (Yassine et al., 2017).

At last, the operational costs of maintaining an SOC, especially for 24/7 monitoring solutions, can be substantial due to the need for a sizable analyst workforce.

16.2.4 SOC benefits

SCO Advantages: While it is true that this technology has its fair share of problems and room for development, SOCs have more benefits than drawbacks and that running one is beneficial and efficient—even though achieving a flawless system would be too costly and unachievable. Here are a few advantages that an SOC may provide:

- **Faster incident response times:** The SOC can keep tabs on all the devices from one central location and step in quickly if necessary. The Security Information and Event Management (SIEM) displays alerts together with all pertinent incident data.
- **Reduced cost:** Although there is an upfront cost associated with establishing an SOC, it is far lower than the expense of recovering from a ransomware assault.

- **Operational efficiencies:** The ratio of identified occurrences might approach 100% with a 6 inside the SOC, a wide range of expertise can be brought to bear on the analysis.
- **Enhanced visibility:** As more and more workers do their jobs from the comfort of their homes, it has grown increasingly difficult to monitor every device that has access to the company network. SOC makes monitoring easier by collecting and processing all the data.

Without SIEM, an SOC would be unable to operate. It offers a "big picture" view of the activities occurring inside the monitored IT infrastructure and stores data for use in audit operations, among its numerous other functions. Intrusion attempts, botnet operations, web application assaults, distributed denial of service attacks, data exfiltration, system bottlenecks, and vulnerability exploitation may all be detected, and unwanted access across several systems can be monitored. Another feature is that it monitors connections and notifies users when it detects any suspicious ones.

16.2.5 Types of SOC

Companies have varying infrastructures and budgets, therefore, it is important to choose an SOC architecture that works for them.

- **Internal or dedicated SOC:** Cybersecurity is the only emphasis of this department inside the firm. It offers centralized insight into all network activities and continuous monitoring around the clock every day of the year. Because staff members are familiar with the systems, they are keeping tabs on, threat reaction time is incredibly rapid. Cost is the biggest drawback; big businesses and governments can afford it, but tiny groups may not be able to do so.
- **Virtual SOC:** Because it is accessible, scalable, and hosted on the cloud through a web interface, it is also relatively inexpensive. Due to the lack of physical hardware, it is less expensive than a conventional SOC; nonetheless, its reliability is compromised. It is primarily reactive and often does not provide 24/7 service.
- **Hybrid SOC:** The goal of a hybrid SOC model is to combine the two approaches to their greatest potential. Compared with a completely specialized SOC, the expenses are lower, and alerts are evaluated by both internal and external analysts. Particularly helpful when a business depends on outside help due to a lack of in-house knowledge.
- **SOC as a service (SOCaaS):** Right now, this one is everyone's favorite. Countless benefits accrue when the SOC is outsourced to a business that handles this for several customers. By offering a solution for round-the-clock monitoring and top-notch protection, it lowers the hefty expenses of a specialized SOC.
- Many SOCaaS include proactive threat mitigation capabilities, numerous layers of analysts, and the ability to react in the event of an issue. In addition, they provide reports on a regular basis to inform you of the current state of your environment's security.

16.3 RELATED WORKS

In this section, we reviewed the landscape of SOC-specific maturity models, focusing on the SOC–CMM (Security Operation Center–Capability Maturity Model) introduced by Van Os (Rob van Os, 2018) and further elaborated by Erdur (2019). Van Os's model, developed through the application of the Design Science Research (DSR) principle and the Systematic

Literature Review methodology, encompasses five crucial domains: business, people, process, technology, and services, with an emphasis on technology and services evaluation. The model's foundation lies in 901 elements across different domains, with a notable focus on process-related aspects. Van Os highlights the challenge of creating a generalized model applicable to all SOCs, acknowledging the uniqueness of each SOC and the difficulty in covering all SOC activities comprehensively. In addition, suggestions for future research include the creation of a baseline for SOC comparison and balancing the model's focus to ensure effectiveness across all areas. Erdur builds upon existing continuous improvement methodologies, mainly focusing on comparing different approaches applicable to SOCs. Through a comparative analysis of frameworks such as PDCA, DMAIC, and DMADV, Erdur concludes that DMAIC is best suited for SOCs due to its reliance on statistical measurements, aligning well with maturity and capability assessment models offered by SOC-specific frameworks.

Furthermore, Safarzadeh et al. (2019) proposed an evaluation methodology for the SIEM element within SOCs. Their methodology aims to provide a more objective comparison of SIEM systems by evaluating specific criteria across dimensions such as capability, architectural components, and common features. This approach offers a structured method for assessing and comparing SIEM systems beyond manufacturer claims. Revaclier (2021) proposed SOC-AM, a tailored maturity model specifically designed for SOCs. While maturity models are commonly utilized in information technology (IT), such models are relatively scarce in the context of SOCs, and academic research in this area is limited. SOC-AM aims to fill this gap by providing a comprehensive evaluation framework that balances detailed assessment criteria with ease of use. By leveraging SOC-AM, organizations can gain valuable insights into the maturity of their SOC operations, enabling them to identify areas for improvement and make informed decisions to enhance their security posture. Through its systematic approach and adaptability, SOC-AM has the potential to become a valuable tool for SOC practitioners seeking to optimize their security operations effectively.

16.4 METHODOLOGY

The research design approaches will inform the pragmatic paradigm that will be utilized in this investigation (Hasteer et al., 2013). Inductive reasoning will be utilized throughout the study process, and the majority of the data collected will be qualitative. During the data collection process, academic literature, reference books, and a wide variety of references and best practices, including NIST, ISACA, OWASP, and ISO 27000, are frequently utilized.

To reach a conclusion, this study will base its findings mostly on qualitative data and employ inductive reasoning. Many best practices and reference standards from organizations such as NIST, ISACA, OWASP, and ISO 27000, in addition to scholarly journals and publications, served as the foundation for the knowledge that was gathered into this collection (Maleh et al., 2021).

This research will be carried out in the form of a qualitative descriptive study, with data collected from a representative sample of the community, but this information will be confirmed by cybersecurity subject matter experts. Design science will be utilized throughout the solution phase of this inquiry, which is the second point. In the subject of design, research must ultimately result in a practical conclusion, which may take the form of a construct, model, approach, or instantiation, respectively (Creswell & Creswell, 2017). It is conceivable that a solution to the identified root causes already exists based on the best practices that are currently in place; if this is not the case, then creating a new "artifact" will be necessary.

For gathering accessible data, a variety of observational approaches are utilized. The following techniques are utilized in the process of data collection.

- **Review of the relevant literature:** A comprehensive review of academic literature, reference books, and reputable sources such as NIST, ISACA, OWASP, and ISO 27000 will be conducted. This approach ensures that the research is grounded in existing knowledge and best practices within the cybersecurity domain.
- **Interviews with a semistructured mode:** To gain nuanced insights into the current state of SOC components and operations, semistructured interviews will be conducted with cybersecurity professionals. These interviews will provide qualitative data regarding SOC practices, challenges, and areas for improvement.

The chosen methodology is tailored to address the research goals of evaluating SOC effectiveness and maturity. By focusing on qualitative data collection methods and drawing insights from diverse sources, this research aims to provide a comprehensive understanding of SOC operations and their alignment with best practices.

16.5 THEORETICAL FRAMEWORK

As a result of standards being developed in response to unfavorable occurrences rather than in response to proactive measures, the path that led to the current existence of laws and standards was difficult to navigate. As a similar point of reference, the day-to-day tasks of an SOC are the product of years of growth and restructuring in the roles of the areas that comprise a department that is responsible for systems and telecommunications.

There have been particular areas established throughout the course of time to manage information technology security. Initially, the duties of these areas may be found within those of a data network administrator or a server administrator. When looking back, it is feasible to tell the difference between previous generations of SOCs or different levels of maturity.

The first generation of what we now refer to as an SOC consisted of a collection of responsibilities that were dispersed across different areas of the systems department. These responsibilities were carried out by individuals whose primary task was something else, and as a result, they lacked formal training or awareness of security concerns. The monitoring of network devices and servers to assure the availability of services, the management of antivirus systems on a fundamental level, and the collecting of audit logs for network devices were the responsibilities that they were responsible for.

To meet the requirements that have been defined, it is necessary to articulate the mix of human resources, technological resources, and procedures via cooperation and communication. It will be a continuing challenge for middle and senior management to find a way to shape a SOC in accordance with global principles while simultaneously providing the services that are instantly required by the firm.

It is necessary for the plan for the SOC to be in accordance with the strategic plan for information systems and technology, which must, in turn, be in accordance with the strategic plan for the organization to provide support for the operating model that has been determined by senior management. The organization will be able to safeguard the assets that are of importance to it thanks to this alignment, which will guarantee that both human and financial resources are deployed appropriately. In the absence of this direction, it is possible to develop an erroneous sense of safety. The scopes, missions, operational, and development models of the SOCs will then be created from the strategic plans at each step. These models will be used to determine the activities of the centers.

In the process of outlining these standards, they will act as a guide for each stage of the planning process as well as the analysts' day-to-day operations. When budgetary limits and limited resources are taken into consideration, risk management appears to be an effective method for prioritizing and distributing resources. Understanding the cybersecurity challenges that are currently being faced by the organization, the roles and responsibilities of an SOC, the processes that need to be implemented, and the technologies that will support management are all necessary components of the process of outlining the strategy for an SOC.

A list of transitory IT, which can be widespread, serves as the foundation for the SOC infrastructure. Let's concentrate on the most essential collection of SOC software and hardware, which will serve to automate its operations. Automating the process of analyzing observed events, incidents, and information security vulnerabilities is the most labor-intensive and critical aspect of information security governance. It is essential to do this to provide a central location for the collection and administration of any extra information with the purpose of ensuring the entirety of the lifetime of an event, incident, or vulnerability, beginning with the assignment of an analyst to be accountable for managing it and ending with the action that is required to resolve it. For the purpose of adhering to the best practices for the automation of business processes, the system of GRC class (Governance, risk management, and compliance) that is integrated with other software and hardware means that SOC and external systems have the potential to serve as the primary means of automating SOC activity. Because of their capabilities, GRC systems are able to:

- Identify and keep a centralized accounting of the assets that need to be protected, as well as determine the worth of those assets;
- Archive and maintain the most recent version of the SOC administrative document collection;
- Maintain an automated database that has a consistent database of incidents and a history of how those events were processed;
- It is important to provide staff with training and knowledge testing;
- Implement automated processes for the classification and recording of incidents, as well as the notification of responsible parties and the escalation of incidents;
- It is important to keep track of and assess how well SOC initiatives are working.

Almost every single SOC procedure may be automated by GRC solutions (Maleh, Sahid, et al., 2018; Maleh, Zaydi, et al., 2018), in addition to the aforementioned benefits. These systems amass information on the assets of the firm, the risks that have been predetermined, and the workers. Through the utilization of GRC systems, SOC personnel can swiftly access critical asset information pertinent to incident impact and asset ownership. Furthermore, these systems facilitate prompt report generation, enabling top management, and asset owners to oversee SOC effectiveness. An alternative, cost-efficient approach to GRC involves leveraging existing Service Desk or Help Desk systems within the organization, albeit with necessary modifications to meet new requirements such as automated processing of events from external analytical systems. While such systems may offer a more budget-friendly option, they typically provide minimal functionality essential for IM. Central to the automated collection, storage, and analysis of information security events is the SIEM, which segregates these events from those generated by the broader spectrum of the company's IT systems. Serving as a cornerstone for both detecting information security incidents and conducting operational and retrospective investigations, SIEM plays a pivotal role in maintaining IS integrity.

The SOC provides a comprehensive toolkit capable of identifying a wide range of illicit activities or causes behind technical malfunctions within an organization's IT network. Specialized scanners and configuration analysis technologies are instrumental in detecting vulnerabilities within the SOC framework. For instance, these tools enable automated compliance checks of network equipment configurations against specific regulatory requirements, asset inventory management, and identification of susceptible software versions. Timely accessibility of gathered data to SOC personnel is imperative for ensuring compliance with policies. Full engagement of all employees within the protected organization is essential for the SOC to operate effectively. It's worth emphasizing that SOC personnel play a pivotal role in ensuring the SOC's proper functioning, as the company's security hinges on their actions and adherence to protocols. Through their vigilance, SOC personnel can identify and alert others to potential information security incidents and vulnerabilities, underscoring their pivotal role in initiating protective measures and enabling SOC staff to fulfill their duties effectively.

The systems that have been detailed are both a "must-have" collection of systems that may be utilized to evaluate the potential outcomes and scenarios that may result from assaults and security breaches, as well as to detect and address the factors that contribute to these occurrences. A "must-have" set is what we refer to as the systems that have been specified. On the other hand, it is also possible to modify it according to the level of growth of the response center or the maturity of the firm. The following types of systems are included in other toolkits: systems that prohibit the disclosure of private information; systems that investigate events and collect evidence; and systems that prevent denial of service attacks. A diagram depicting the interaction scheme of the required set of technological means and information flows may be seen in Figure 16.3.

In the contemporary digital landscape, the evolution of technological demands has necessitated a reevaluation of core infrastructure, notably in the realm of Big Data. The exponential growth in data volume, often exceeding terabytes, has engendered a landscape characterized

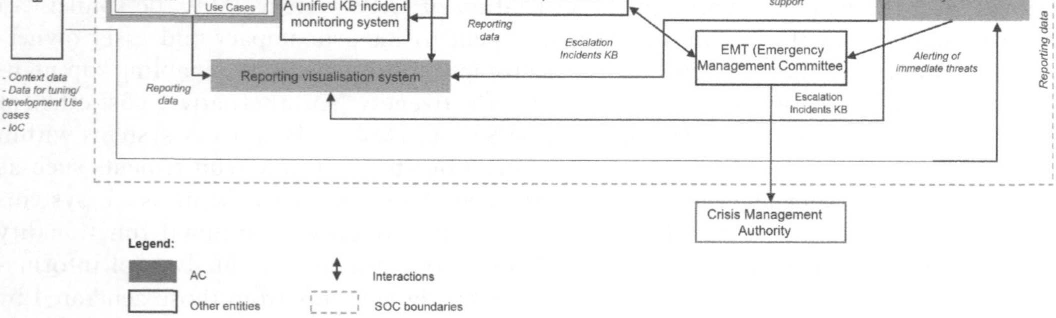

Figure 16.3 SOC architecture.

by disorganized and fragmented information, commonly referred to as "big data." This phenomenon transcends information security, permeating diverse domains reliant on processing vast amounts of structured and unstructured data to yield human-readable outcomes. Traditional approaches, such as event collection from select devices for firewalls and intrusion prevention systems, are no longer sufficient. The imperative now extends to defending against and identifying malicious or anomalous activities across both perimeter and internal networks, encompassing assets directly or indirectly. Evolving detection paradigms emphasize the transition from known threats to unknown ones, necessitating the aggregation, analysis, and retention of diverse, seemingly disparate data sets. Predicting the precise data required for incident investigation ex ante is untenable. Potential data needs may span reconstructing network sessions, accessing video surveillance recordings, cross-referencing storage and physical security systems data, among others. Consequently, the technical arsenal of security centers must be adept at processing vast data volumes, enabling the consolidation of disparate technologies into a unified monitoring and investigative tool. Such integration facilitates the identification of attacks or malicious activities based on correlational evidence gleaned from varied equipment and information systems.

16.6 DEVELOPING THE SOC MATURITY MODEL

16.6.1 Framework overview

It is challenging to model the growth of skills in the SOC category. This is as a result of the fact that secure operations centers make use of a wide range of technology and offer a wide array of services. On the other hand, modeling is necessary for measurement. For the purpose of developing the recommended SOC model for maturity, which is an artifact that bridges the gap between theory and practice, Design Science was utilized (Rob van Os, 2018). Both the model itself and a self-assessment tool for quantitative evaluation are included in the artifacts that are associated with the SOC maturity model. Following an extensive amount of study into the existing body of literature, the SOC–CMM was successfully built. Following this, a survey was administered to 16 different participant organizations in order to determine whether or not any of the components discovered in the literature were present in actual SOCs. With the help of the information obtained from the study, a capacity maturity model for SOC was established. Table 16.1 presents a comprehensive framework for an SOC, outlining five domains and their corresponding aspects or functions, totaling 25 individual features or elements.

The proposed SOC model encapsulates the multifaceted nature of modern security operations, acknowledging the diverse range of functions and responsibilities within an SOC environment. By categorizing these functions into distinct domains and aspects, organizations can better understand the holistic approach required to establish and maintain an effective SOC. For example, under the governance domain, strategic alignment emphasizes the importance of aligning SOC objectives with broader business goals, ensuring that security initiatives support organizational priorities. Meanwhile, the people domain highlights the significance of talent optimization and organizational dynamics in maximizing the potential of SOC personnel and fostering a culture of collaboration and knowledge sharing. Process-related aspects such as operational excellence and scenario planning underscore the importance of standardized procedures and proactive threat mitigation strategies in mitigating security risks and maintaining SOC resilience. In parallel, the technology domain emphasizes the critical role of advanced analytics and automation in enhancing threat detection and response capabilities, reflecting the growing reliance on artificial intelligence and machine learning technologies in cybersecurity operations.

Table 16.1 The proposed SOC model components

Domain	*Aspects*	*Description*	*Examples/Case studies*	*Dependencies*	*Metrics/KPIs*
Governance	Strategic alignment	Aligning SOC objectives with business goals	Implementing SOC initiatives to support company objectives	Business strategy, risk management	Alignment score, strategic initiative success rate
	Customer experience	Enhancing customer satisfaction through SOC services	Providing timely incident response and effective communication	Customer feedback, service-level agreements	Customer satisfaction score, response time
	Data governance	Ensuring compliance and protection of sensitive data	Implementing data protection policies and access controls	Regulatory requirements, data classification	Compliance audit results, data breach incidents
People	Talent optimization	Maximizing workforce potential and skills	Investing in training and development programs	Skills assessment, talent acquisition	Employee satisfaction, skill proficiency
	Organizational dynamics	Analyzing and optimizing team structures	Creating multidisciplinary teams for incident response	Team performance, communication channels	Team effectiveness, collaboration metrics
	Human capital development	Investing in employee growth and expertise	Providing opportunities for career advancement	Performance evaluations, mentorship programs	Employee retention, professional certifications
	Knowledge cultivation	Fostering a culture of knowledge sharing	Establishing knowledge sharing platforms	Knowledge repositories, collaboration tools	Knowledge contributions, content relevance
	Learning and development	Providing continuous learning opportunities	Offering training sessions and workshops	Training attendance, skill acquisition	Training completion rates, skill assessment
Process	Operational excellence	Driving efficiency and effectiveness in SOC operations	Implementing standardized processes and procedures	Process documentation, automation tools	Incident response time, process adherence
	Infrastructure management	Managing SOC facilities and resources	Maintaining SOC tooling and equipment	Resource utilization, asset management	Equipment uptime, resource allocation
	Performance measurement	Evaluating and improving SOC performance	Monitoring key performance indicators (KPIs)	Performance benchmarks, maturity assessments	SOC maturity level, performance improvement rate
	Scenario planning	Anticipating and preparing for potential threats	Conducting tabletop exercises and simulations	Threat intelligence (TI), risk assessments	Incident response time, scenario effectiveness

(Continued)

Table 16.1 (Continued) The proposed SOC model components

Domain	*Aspects*	*Description*	*Examples/Case studies*	*Dependencies*	*Metrics/KPIs*
Technology	Threat management	Centralized threat detection and response	Utilizing SIEM and IDS/IPS technologies	TI feeds, incident response tools	Threat detection rate, mean time to detect (MTTD)
	Intrusion prevention	Proactive defense against malicious activities	Deploying firewalls and endpoint protection solutions	Network traffic analysis, endpoint detection	Intrusion attempts blocked, malware detection rate
	External integration	Integrating external services for enhanced security	Incorporating TI feeds and incident response services	API integration, security standards compliance	Integration success rate, data accuracy
	Advanced analytics	Leveraging AI and machine learning for threat analysis	Applying machine learning algorithms for anomaly detection	Data analytics platforms, AI models training data	Accuracy of anomaly detection, false-positive rate
	Automation and response	Automated response to security incidents	Implementing SOAR platforms for automated incident response	Workflow automation, integration with security tools	Incident response time, automation rate
Services	Continuous surveillance	24/7 monitoring of security posture	Utilizing SOC analysts for real-time monitoring	Network traffic analysis, log management	Alert volume, mean time to respond (MTTR)
	Incident resolution	Swift and effective response to security incidents	Implementing incident response procedures	Incident severity, resolution time	Incident closure rate, customer satisfaction
	TI	Gathering and analyzing actionable TI	Utilizing TI feeds and platforms	Threat feeds accuracy, TI sources	Actionable intelligence rate, threat identification rate
	Proactive threat hunting	Actively seeking out and neutralizing hidden threats	Conducting threat-hunting exercises and investigations	Hunting success rate, threat detection rate	Number of identified threats, threat-hunting coverage
	Vulnerability remediation	Promptly addressing and mitigating system vulnerabilities	Implementing vulnerability scanning and patching processes	Vulnerability severity, patching effectiveness	Mean time to patch (MTTP), vulnerability closure rate
	Log aggregation and analysis	Aggregating and analyzing system logs for security insights	Utilizing log management tools for centralized log storage	Log volume, log retention period	Log analysis accuracy, log correlation rate

16.6.2 Framework maturity profile

To put it another way, the nature of the dangers that need to be dealt with in 2022 is constantly evolving, which makes it impossible to maintain cybersecurity in that year. If you want to have a fighting chance of surviving, the only way to do it is to upgrade to a more sophisticated SOC model that has several layers of protective technology.

By utilizing the maturity stages of the CMMI version 2.0 (Ramírez-Mora et al., 2020), we propose an approach for SOC that is both well-developed and systematic. A comprehensive suite of integrated and fully automatable protection, administration, and defense capabilities is required to attain cybersecurity maturity. This is the only way to achieve this level of maturity (Figure 16.4).

Table 16.2 outlines a maturity model for assessing the capabilities and characteristics of an SOC across five maturity levels, ranging from 0-Adhoc to 4-Resilient. Each maturity level is associated with specific scores, security operations capabilities, organizational characteristics, and risk characteristics, providing organizations with a structured framework to evaluate and improve their SOC capabilities. At the 0-Adhoc level, organizations exhibit an absence of defined security operations capabilities, relying primarily on basic security measures such as firewalls and antivirus software. Incident response is reactive and lacks systematic procedures, with limited awareness of security threats and compliance requirements. As organizations progress to higher maturity levels, they demonstrate a gradual improvement in their security operations capabilities and organizational characteristics. This includes mandated log data centralization, establishment of formal incident response procedures, investment in threat detection and response improvement, and the establishment of a formal SOC with trained staff and proactive threat-hunting capabilities. At the highest maturity level of 4-Resilient, organizations exhibit advanced capabilities, including extensive automation of investigation and mitigation, 24/7 SOC functionality, proactive threat management stance, and investment in top-tier resources and processes. These organizations demonstrate effective detection and response across all threat categories, resilience against cybercriminals and nation-state adversaries, and high resistance to breaches and service disruptions.

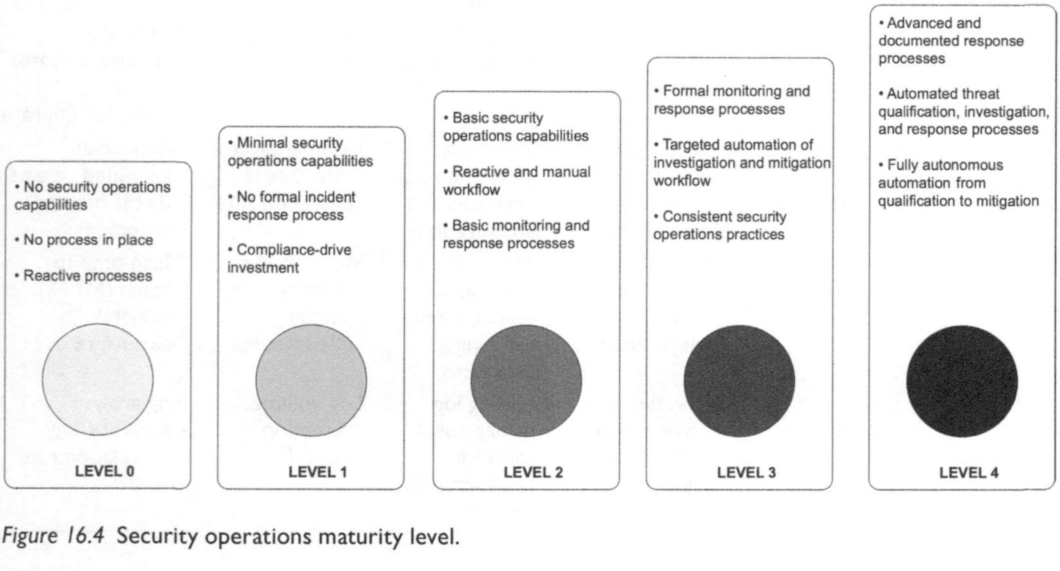

Figure 16.4 Security operations maturity level.

Table 16.2 Security operations maturity level

Maturity level	*Score*	*Security operations capabilities*	*Organizational characteristics*	*Risk characteristics*
0-Adhoc	0–25	• Absence of any defined security operations capabilities.	• Reliance on basic security measures like firewalls and antivirus software. • Presence of threat indicators and compromises, but lack of centralized logging and insight due to technological and functional silos. • Response to incidents based on individual bravery rather than systematic procedures.	• Noncompliance with security standards. • Lack of awareness regarding insider threats, external threats, advanced persistent threats (APTs), and potential IP theft.
1-Minimally Compliant	26–50	• Centralization of security event data and required logs. • Server forensics with an emphasis on compliance, including endpoint detection response and file integrity monitoring (EDR). The ability to monitor and respond to basic events.	• Investment driven by compliance requirements or identified areas needing protection. • Risks identified through compliance audits with varying degrees of management. • Improved visibility into threats, albeit with limitations in evaluation and prioritization. • Absence of formal incident response procedures, relying instead on individual efforts.	• Reduced compliance risks, depending on audit depth. • Limited awareness of insider threats, external threats, APTs, and potential IP theft.
2-Securely Compliant	51–75	• Targeted log data and security event centralization. • Targeted server and endpoint forensics. • Basic machine analytics for correlation and alarm prioritization. • Establishment of basic monitoring and response processes.	• Transition from a minimal compliance focus to seeking efficiencies and enhanced assurance. • Recognition of organizational vulnerability to threats, with efforts aimed at detection and response improvement. • Establishment of formal processes and responsibilities for monitoring and responding to high-risk alarms. • Basic incident response framework in place.	• Strong compliance posture. • Enhanced visibility into insider and external threats, with some blind spots. • Limited awareness of APTs, albeit with increased likelihood of detection. • Resilience against cybercriminals, except those utilizing APT-type attacks or targeting blind spots. • Vulnerability to nation-states, albeit with improved detection and response capabilities.

(Continued)

Table 16.2 (Continued) Security operations maturity level

Maturity level	*Score*	*Security operations capabilities*	*Organizational characteristics*	*Risk characteristics*
3-Vigilant	76–90	• Endpoint and server forensics; centralized holistic log data and security events. • Advanced machine analytics for known threat detection. • Formal and mature monitoring and response processes with standard playbooks. • Functional physical or virtual SOC.	• Acknowledgment of organizational vulnerability to high-impact threats. • Investment in processes and personnel to enhance threat detection and response capabilities. • Establishment of a formal SOC with trained staff and proactive threat hunting. • Utilization of automation for improved incident response efficiency.	• Robust compliance posture. • Comprehensive visibility into insider and external threats, with improved detection and response capabilities. • Adequate awareness of APTs, though some blind spots remain. • Resilience against cybercriminals, except for those leveraging APT-type attacks targeting blind spots. • Enhanced ability to detect and respond to nation-state threats.
4-Resilient	91–100	• Holistic log data and security event centralization. • Advanced machine analytics for known threat detection. • Established response processes with standard playbooks. • Functional 24/7 SOC. • Extensive automation of investigation and mitigation. • Advanced MTTD/MTTR operational metrics.	• High-value target for nation-states, cyber terrorists, and organized crime. • Continuous attacks across all potential vectors. • Zero tolerance for service disruption or breaches. • Proactive threat management stance. • Investment in top-tier resources and processes. • 24/7 Alarm monitoring with redundancies. • Extensive proactive capabilities for threat prediction and hunting.	• Effective detection and response across all threat categories. • Early identification and strategic management of APT activities. • Resilience against all classes of cybercriminals. • Ability to defend against the most extreme nation-state adversaries. • High resistance to breaches and service disruptions.

16.6.3 Framework maturity profile

The validation of the proposed SOC maturity framework was conducted via a comprehensive pilot study aimed at assessing its effectiveness, applicability, and alignment with industry standards. The pilot study involved deploying the framework within select organizations and evaluating its performance across various dimensions of SOC capabilities. During the pilot study, participating organizations implemented the SOC maturity framework within their existing security infrastructure. This involved conducting assessments across different domains, including governance, people, process, technology, and services. Evaluation criteria were derived from the framework's components, focusing on the maturity levels of specific capabilities relevant to SOC operations.

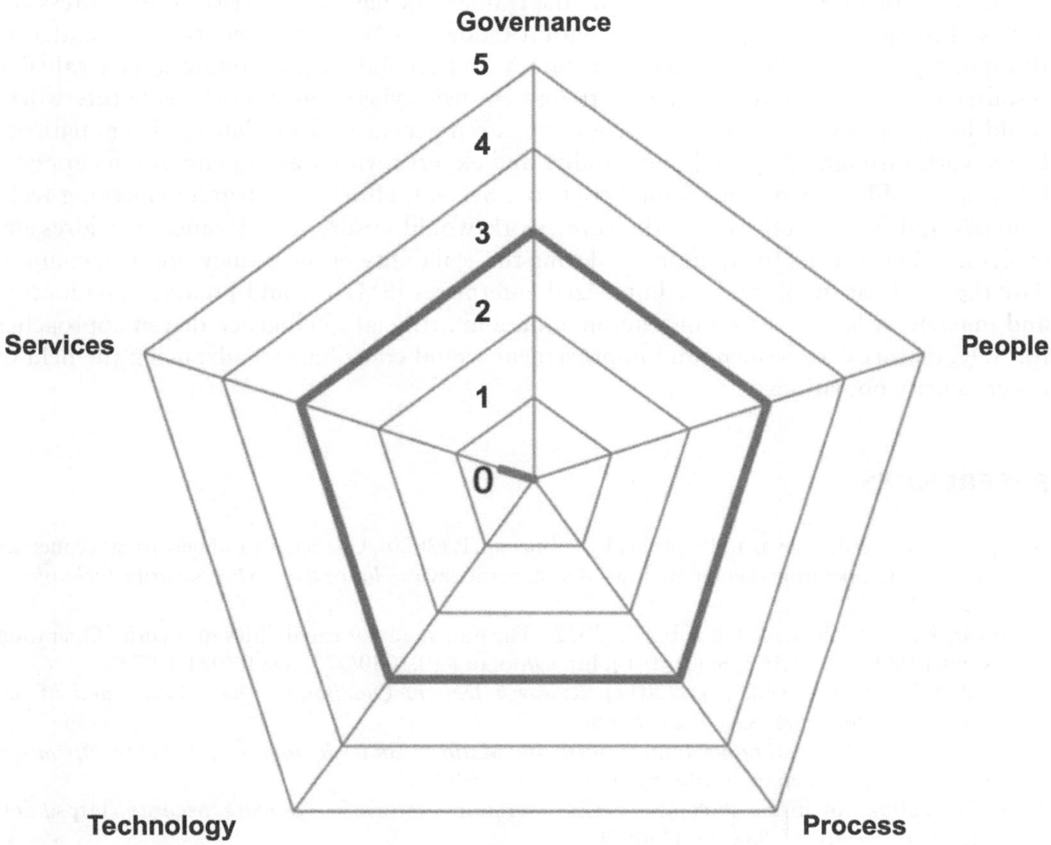

Figure 16.5 Capability maturity scoring analysis.

Upon completion of the assessments, scores were compiled and analyzed to identify gaps between actual and ideal maturity levels. The findings were presented in a comprehensive report, detailing the strengths and weaknesses of each domain and highlighting areas for improvement. A graphical representation, such as a radar chart, was utilized to visualize the overall maturity rating and disparities across domains.

Figure 16.5 showcases the specifics of each domain, facilitating a nuanced assessment of performance. Weak facets within domains were identified, emphasizing their impact on overall maturity. The plan for improvement was crafted with a risk-based strategy, prioritizing areas with the greatest impact on SOC effectiveness and resilience.

The SOC maturity model was found to be in sync with well-known frameworks, such the Cybersecurity Framework developed by the National Institute of Standards and Technology (NIST CSF). To guarantee uniformity and conformity with industry standards, every evaluation question was double-checked for relevance to the NIST CSF and assigned to the appropriate components.

16.7 CONCLUSION

This chapter has presented a robust maturity framework for SOCs, offering a structured approach to assess and enhance cybersecurity capabilities. Through rigorous methodology, including literature review, survey administration, and alignment with established frameworks like CMMI and NIST CSF, the framework has been developed to address the diverse functions and responsibilities within modern SOC environments. The validation through a pilot study underscores its accuracy and reliability, positioning it as a valuable resource for organizations seeking to strengthen their cybersecurity posture. Future works could focus on several areas. First, continuous refinement and validation of the maturity framework through additional pilot studies and expert reviews would enhance its applicability across different organizational contexts. Second, efforts to integrate emerging technologies and best practices into the framework would ensure its relevance in addressing evolving cyber threats. In addition, exploring the scalability of the framework to accommodate the needs of small- and medium-sized enterprises (SMEs) could broaden its adoption and impact. At last, research into automation and artificial intelligence-driven approaches for SOC maturity assessment and improvement would contribute to advancing the field of cybersecurity operations.

REFERENCES

Agyepong, E., Cherdantseva, Y., Reinecke, P., & Burnap, P. (2020). Challenges and performance metrics for security operations center analysts: A systematic review. *Journal of Cyber Security Technology*, *4*(3), 125–152.

Chamkar, S. A., Maleh, Y., & Gherabi, N. (2022). The human factor capabilities in Security Operation Center (SOC). *EDPACS*, *66*(1), 1–14. https://doi.org/10.1080/07366981.2021.1977026

Creswell, J. W., & Creswell, J. D. (2017). *Research Design: Qualitative, Quantitative, and Mixed Methods Approaches*. Sage Publications.

Erdur, E. S. (2019). *Continuous Improvement on Maturity and Capability of Security Operation Centers*. https://open.metu.edu.tr/handle/11511/45040

Forte, D. (2003). An inside look at security operation centres. In *Network Security*. https://doi.org/10.1016/S1353-4858(03)00509-9

Hasteer, N., Bansal, A., & Murthy, B. K. (2013). Pragmatic assessment of research intensive areas in cloud: A systematic review. *SIGSOFT Software Engineering Notes*, *38*(3), 1–6. https://doi.org/10.1145/2464526.2464533

Maleh, Y., Sahid, A., & Belaissaoui, M. (2021). A maturity framework for cybersecurity governance in organizations. *EDPACS*, *64*(2), 1–22. https://doi.org/10.1080/07366981.2020.1815354

Maleh, Y., Sahid, A., Ezzati, A., & Belaissaoui, M. (2018). A capability maturity framework for IT security governance in organizations. In A. Abraham, A. Haqiq, A. K. Muda, & N. Gandhi (Eds.), In *Innovations in Bio-Inspired Computing and Applications* (pp. 221–233). Springer International Publishing.

Maleh, Y., Zaydi, M., Sahid, A., & Ezzati, A. (2018). Building a maturity framework for information security governance through an empirical study in organizations. In *Security and Privacy Management, Techniques, and Protocols*. https://doi.org/10.4018/978-1-5225-5583-4.ch004

Mansfield-Devine, S. (2016). Creating security operations centres that work. *Network Security*, 2016(5), 15–18. https://doi.org/10.1016/S1353-4858(16)30049-6

Miloslavskaya, N. (2016). Security operations centers for information security incident management. *2016 IEEE 4th International Conference on Future Internet of Things and Cloud (FiCloud)*, pp. 131–136. https://doi.org/10.1109/FiCloud.2016.26

Muniz, J. (2021). *The Modern Security Operations Center*. Addison-Wesley Professional.

Mutemwa, M., Mtsweni, J., & Zimba, L. (2018). Integrating a security operations centre with an organization's existing procedures, policies and information technology systems. *2018 International Conference on Intelligent and Innovative Computing Applications (ICONIC)*, Mon Tresor, Mauritius: IEEE, pp. 1–6.

Ramírez-Mora, S. L., Oktaba, H., & Patlán Pérez, J. (2020). Group maturity, team efficiency, and team effectiveness in software development: A case study in a CMMI-DEV Level 5 organization. *Journal of Software: Evolution and Process*, *32*(4), e2232. https://doi.org/10.1002/smr.2232

Revaclier, A. R. (2021). *SOC-AM: An Accessible Maturity Model for Security Operation Centers.* Master thesis, Eindhoven university of technology, Netherlands. https://pure.tue.nl/ws/portalfiles/portal/198120514/Revaclier_A.R..pdf

Rob van Os. (2018). *SOC-CMM: Measuring Capability Maturity in Security Operations Centers.* Master Thesis, Luleå University of Technology, Sweden. URN: urn:nbn:se:ltu:diva-59591

Safarzadeh, M., Gharaee, H., & Panahi, A. H. (2019). A novel and comprehensive evaluation methodology for SIEM. In *Information Security Practice and Experience: 15th International Conference, ISPEC 2019*, Kuala Lumpur, Malaysia, November 26–28, Proceedings 15, pp. 476–488.

Scarfone, K., Grance, T., & Masone, K. (2008). Computer security incident handling guide. *NIST Special Publication*, *800*(61), 38.

Sundaramurthy, S. C., Case, J., Truong, T., Zomlot, L., & Hoffmann, M. (2014). A tale of three security operation centers. In *Proceedings of the 2014 ACM Workshop on Security Information Workers*, pp. 43–50. https://doi.org/10.1145/2663887.2663904

Taqafi, I., Maleh, Y., & Ouazzane, K. (2023). A maturity capability framework for security operation center. *EDPACS*, *67*(3), 21–38.

Tsoutsos, N. G., & Maniatakos, M. (2014). HEROIC: Homomorphically encrypted one instruction computer. In *Proceedings of the Conference on Design, Automation & Test in Europe*, pp. 246:1–246:6. https://dl.acm.org/citation.cfm?id=2616606.2616907

van den Berg, J. (2018). Cybersecurity for everyone. In Cybersecurity Best Practices*: Lösungen Zur Erhöhung Der Cyberresilienz Für Unternehmen Und Behörden In: Bartsch, M., Frey, S. (eds) Cybersecurity Best Practices. Springer Vieweg, Wiesbaden.* https://doi.org/10.1007/978-3-658-21655-9_40, (pp. 571–583).

Vielberth, M., Böhm, F., Fichtinger, I., & Pernul, G. (2020). Security operations center: A systematic study and open challenges. *IEEE Access*, *8*, 227756–227779.

Wall, T., & Rodrick, J. (2021). Jump-start your SOC analyst career. In *Jump-Start Your SOC Analyst Career*. https://doi.org/10.1007/978-1-4842-6904-6

Yassine, M., Abdelkebir, S., Abdellah, E. (2017). A capability maturity framework for IT security governance in organizations. *13th International Symposium on Information Assurance and Security (IAS 17)*. https://doi.org/10.1007/978-3-319-76354-5_20

Index